SENSOR-BASED QUALITY ASSESSMENT SYSTEMS FOR FRUITS AND VEGETABLES

Postharvest Biology and Technology

SENSOR-BASED QUALITY ASSESSMENT SYSTEMS FOR FRUITS AND VEGETABLES

Edited by
Bambang Kuswandi, PhD
Mohammed Wasim Siddiqui, PhD

First edition published 2021

Apple Academic Press Inc.
1265 Goldenrod Circle, NE,
Palm Bay, FL 32905 USA

4164 Lakeshore Road, Burlington,
ON, L7L 1A4 Canada

CRC Press
6000 Broken Sound Parkway NW,
Suite 300, Boca Raton, FL 33487-2742 USA

2 Park Square, Milton Park,
Abingdon, Oxon, OX14 4RN UK

First issued in paperback 2021

Library and Archives Canada Cataloguing in Publication

Title: Sensor-based quality assessment systems for fruits and vegetables / edited by Bambang Kuswandi, PhD, Mohammed Wasim Siddiqui, PhD.

Names: Kuswandi, Bambang, editor. | Siddiqui, Mohammed Wasim, editor.

Series: Postharvest biology and technology book series.

Description: Series statement: Postharvest biology and technology series | Includes bibliographical references and index.

Identifiers: Canadiana (print) 20200307932 | Canadiana (ebook) 20200308017 | ISBN 9781771889353 (hardcover) | ISBN 9781003084174 (ebook)

Subjects: LCSH: Fruit—Quality. | LCSH: Vegetables—Quality. | LCSH: Detectors.

Classification: LCC SB360.6 .S46 2021 | DDC 664/.8—dc23

Library of Congress Cataloging-in-Publication Data

Names: Kuswandi, Bambang, editor. | Siddiqui, Mohammed Wasim, editor.

Title: Sensor-based quality assessment systems for fruits and vegetables / edited by Bambang Kuswandi, Mohammed Wasim Siddiqui.

Other titles: Postharvest biology and technology book series.

Description: First edition. | Palm Bay, FL, USA : Appae Academic Press, 2021. | Series: Postharvest biology and technology | Includes bibliographical references and index. | Summary: "Sensor-Based Quality Assessment Systems for Fruits and Vegetables provides an abundance of valuable information on different sensing techniques for fruits and vegetables. It covers emerging technologies, such as NMR, MRI, wireless sensor networks (WSN), and radio-frequency identification (RFID) and their potential for industrial applications. Currently, a number of new sensor technologies based on electromagnetic properties have emerged, and great progress has also been made on other existing sensor technologies. These include X-ray, nuclear magnetic resonance (NMR) or magnetic resonant imaging (MRI), fluorescence, and other related developments, such as electrical impedance and permittivity (mainly microwave), thermal sensing, and selective gas/volatile sensing. These developments have opened new areas of research as well as new applications for sensing quality of fruits and vegetables. Key features of the volume: Provides an inclusive review of the developments of sensors for quality analysis and inspection of fresh fruits and vegetables. Fosters an understanding of the basic sensing techniques for quality assessment of fresh fruits and vegetables, including further technological development. Covers advanced sensing technologies, including computer vision, spectroscopy, X-rays, magnetic resonance, mechanical contact, wireless sensor networks, and radio-frequency identification sensors. Reviews the significant progress in the sensor development of noninvasive techniques for quality assessment of fruits and vegetables. Presents the current status of different sensing systems in the context of commercial application and potentials of these sensing technologies for fruits and vegetables. This volume is dedicated solely to the application of sensor technology for evaluation of quality of fruits and vegetables and will be a rich resource for professors/lecturers teaching sensor technology for food technology courses; researchers who work on sensor development, agriculture and food technology; students in related areas; and industry professionals"-- Provided by publisher.

Identifiers: LCCN 2020035805 (print) | LCCN 2020035806 (ebook) | ISBN 9781771889353 (hardcover) | ISBN 9781003084174 (ebook)

Subjects: LCSH: Fruit--Quality. | Vegetables--Quality. | Detectors.

Classification: LCC SB360.6 .S46 2021 (print) | LCC SB360.6 (ebook) | DDC 634--dc23

LC record available at https://lccn.loc.gov/2020035805

LC ebook record available at https://lccn.loc.gov/2020035806

ISBN: 978-1-77188-935-3 (hbk)
ISBN: 978-1-77463-886-6 (pbk)
ISBN: 978-1-00308-417-4 (ebk)

BOOKS IN THE POSTHARVEST BIOLOGY AND TECHNOLOGY SERIES

Postharvest Biology and Technology of Horticultural Crops: Principles and Practices for Quality Maintenance
Editor: Mohammed Wasim Siddiqui, PhD

Postharvest Management of Horticultural Crops: Practices for Quality Preservation
Editor: Mohammed Wasim Siddiqui, PhD, Asgar Ali, PhD

Insect Pests of Stored Grain: Biology, Behavior, and Management Strategies
Editor: Ranjeet Kumar, PhD

Innovative Packaging of Fruits and Vegetables: Strategies for Safety and Quality Maintenance
Editors: Mohammed Wasim Siddiqui, PhD, Mohammad Shafiur Rahman, PhD, and Ali Abas Wani, PhD

Advances in Postharvest Technologies of Vegetable Crops
Editors: Bijendra Singh, PhD, Sudhir Singh, PhD, and Tanmay K. Koley, PhD

Plant Food By-Products: Industrial Relevance for Food Additives and Nutraceuticals
Editors: J. Fernando Ayala-Zavala, PhD, Gustavo González-Aguilar, PhD, and Mohammed Wasim Siddiqui, PhD

Emerging Postharvest Treatment of Fruits and Vegetables
Editors: Kalyan Barman, PhD, Swati Sharma, PhD, and Mohammed Wasim Siddiqui, PhD

Emerging Technologies for Shelf-Life Enhancement of Fruits
Editors: Basharat Nabi Dar, PhD and Shabir Ahmad Mir, PhD

Sensor-Based Quality Assessment Systems for Fruits and Vegetables
Editors: Bambang Kuswandi, PhD, and Mohammed Wasim Siddiqui, PhD

ABOUT THE EDITORS

Bambang Kuswandi, PhD
Head, Chemo and Biosensors Group, Faculty of Pharmacy,
University of Jember, Indonesia

Bambang Kuswandi, PhD, is currently Head of the Chemo and Biosensors Group, Faculty of Pharmacy, University of Jember, Indonesia, and Visiting Professor at IChemTech, Faculty of Science and Technology, and Fellow at IHRAM (Institute of Halal Research & Management), Universiti Sains Islam Malaysia. He was formerly Head of School of Pharmacy and Dean of Faculty of Pharmacy at the University of Jember, Indonesia. Prof. Kuswandi was a Visiting Research Scientist at the Department of Chemistry, Hacettepe, Ankara, Turkey; Department of Instrumentation and Analytical Science, UMIST, UK; Department of Chemistry, University of Indonesia; Department of Chemistry, University of Florence, Italy; Laboratory of Supramolecular Chemistry and Technology, MESA+ Research Institute for Nanotechnology, University of Twente, the Netherlands; and Department of Chemistry and Biochemistry, Georgia Institute of Technology, Atlanta, Georgia, USA. He was also a Visiting Professor at the School of Chemical Sciences & Food Technology, Faculty of Science & Technology, UKM Malaysia. He has an H index of 18 and has been cited 1219 times (Google Scholar) for his over 60 papers published in refereed journals. Prof. Kuswandi received his PhD in Instrumentation and Analytical Science from the University of Manchester Institute of Science and Technology (UMIST), Manchester, UK in 1999 under supervisor Prof. Dr. R. Narayanaswamy. He also did his MSc degree (1996) in the same department, at UMIST, UK, and his undergraduate program in Chemical Education (1993) at IKIP Malang (now UM, State University of Malang), Indonesia.

Mohammed Wasim Siddiqui, PhD

Assistant Professor and Scientist, Department of Food Science and
Post-Harvest Technology, Bihar Agricultural University, Sabour, India

Dr. Mohammed Wasim Siddiqui is an Assistant Professor and Scientist in the Department of Food Science and Post-Harvest Technology, Bihar Agricultural University, Sabour, India and author or co-author of more than >50 journal articles, >50 book chapters, and several conference papers. He has >25 books to his credit published by Elsevier, USA, CRC Press, USA, Springer, USA, & Apple Academic Press, USA.

He is the Founder Editor-in-Chief of two book series entitled "Postharvest Biology and Technology" and "Innovations in Horticultural Science" being published from Apple Academic Press, New Jersey, USA, where he is a Senior Acquisitions Editor for Horticultural Science as well. He also established an international peer reviewed journal "Journal of Postharvest Technology." Dr Siddiqui has been serving as an editorial board member and active reviewer of several international journals including Horticulture Research (Nature Publishing Group), Postharvest Biology and Technology (Elsevier), PLoS ONE, (PLOS), LWT- Food Science and Technology (Elsevier), Food Science and Nutrition (Wiley), Journal of Plant Growth Regulation (Springer), Acta Physiologiae Plantarum (Springer), Journal of Food Science and Technology (Springer), Indian Journal of Agricultural Science (ICAR) as so on.

Dr Siddiqui is one of key members in establishing the WORLD FOOD PRESERVATION CENTER (WFPC), LLC, USA. Presently, he is an AMBASSADOR of WFPC, LLC, USA. He also served the Postharvest Education Foundation, USA as Member of Board of Directors for the year 2019-2020. Dr Siddiqui has received numerous awards and fellowships in recognition of research and teaching achievements. Recently, he is conferred with the Glory of India Award-2017, Best Researcher Award-2016, Best Citizens of India Award-2016, Bharat Jyoti Award-2016, Outstanding Researcher Award-2016, Best Young Researcher Award – 2015, Young Scientist Award-2015, and the Young Achiever Award- 2014 for the outstanding contribution in research and teaching from several organizations of national and international repute. He was also awarded Maulana Azad National Fellowship Award from the University Grants Commission, New-Delhi, India. He is an Honorary Board Member and Life Time Author Society for Advancement of Human and Nature (SADHNA), Nauni, Himachal Pradesh, India. He has been an active member of organising committee of several national and international seminars/conferences/summits.Considering his outstanding contribution in science and technology, his biography has been published in "Asia Pacific Who's Who", Famous Nation: India's Who's Who, "The Honored Best Citizens of India" and Emrald Who's Who in Asia.

Dr. Siddiqui acquired B.Sc. (Agriculture) degree from Jawaharlal Nehru Krishi Vishwa Vidyalaya, Jabalpur, India. He received the M. Sc. (Horticulture) and Ph. D. (Horticulture) degrees from Bidhan Chandra Krishi Viswavidyalaya, Mohanpur, Nadia, India with specialization in the Postharvest Biotechnology. He has received several grants from various funding agencies to carry out his research projects. He is dynamically indulged in teaching (graduate and doctorate students) and research, and he has proved himself as an active scientist in the area of Postharvest Biotechnology.

CONTENTS

CONTRIBUTORS

Giovanni Agati
Institute of Applied Physics "Nello Carrara" (IFAC) – National Research Council (CNR),
50019 Sesto Fiorentino, Florence, Italy

Maryam Amini
Department of Chemistry, Faculty of Science, Vali-e-Asr University of Rafsanjan,
P.O. Box 518, Rafsanjan, Iran

Bindvi Arora
Division of Food Science and Postharvest Technology, ICAR-Indian Agricultural Research Institute,
New Delhi, India

Vasudha Bansal
Department of Food Engineering and Nutrition, Center of Innovative and Applied Bioprocessing
(CIAB), Knowledge City, Sector-81, Mohali – 140306, Punjab, India,
E-mail: vasu22bansal@gmail.com

S. Bashir
Department of Food Technology, IUST Kashmir – 192122, India

Wolfgang Bilger
Botanical Institute, Christian-Albrechts-University Kiel, Olshausenstraße 40, 24098 Kiel, Germany

Zoran G. Cerovic
Paris-Saclay University, CNRS, AgroParisTech, Ecology Systematics and Evolution,
91405, Orsay, France

Kartikey Chaturvedi
Basic and Applied Sciences, National Institute of Food Technology and Entrepreneurship Management,
Sonepat, Haryana, India

Aamir Hussain Dar
Department of Food Technology, Islamic University of Science and Technology, Awantipora,
Pulwama, Jammu and Kashmir, India, E-mail: daraamirft@gmail.com

Ali A. Ensafi
Department of Chemistry, Isfahan University of Technology, Isfahan – 84156–83111, Iran,
E-mail: ensafi@cc.iut.ac.ir

G. Gajanan
ICAR-Indian Agricultural Research Institute, New Delhi – 110 012, India

Esmaeil Heydari-Bafrooei
Department of Chemistry, Faculty of Science, Vali-e-Asr University of Rafsanjan,
P.O. Box 518, Rafsanjan, Iran

A. Jabeen
Department of Food Technology, IUST Kashmir – 192122, India

Shafat Khan
Department of Food Technology, Islamic University of Science and Technology, Awantipora,
Pulwama, Jammu and Kashmir, India

Bambang Kuswandi
Chemo and Biosensors Group, Faculty of Pharmacy, University of Jember, Jl. Kalimantan 37,
Jember – 68121, Indonesia, E-mail: b_kuswandi@farmasi.unej.ac.id

D. Majid
Department of Food Technology, IUST Kashmir – 192122, India

Hilal Ahmad Makroo
Department of Food Technology, Islamic University of Science and Technology, Awantipora,
Pulwama, Jammu and Kashmir – 192122, India

F. Mehraj
Department of Food Technology, IUST Kashmir – 192122, India

Neelma Munir
Department of Biotechnology, Lahore College for Women University, Lahore, Pakistan,
E-mail: neelma.munir@yahoo.com

Swarajya Laxmi Nayak
Division of Food Science and Postharvest Technology, ICAR-Indian Agricultural Research Institute,
New Delhi, India

Shagufta Naz
Department of Biotechnology, Lahore College for Women University, Lahore, Pakistan,
E-mail: drsnaz31@hotmail.com

S. Vijay Rakesh Reddy
ICAR-Central Institute for Arid Horticulture, Bikaner, Rajasthan, India

Burera Sajid
Queen Marry College, Lahore, Pakistan, E-mail: burerasajid123@gmail.com

Shruti Sethi
Division of Food Science and Postharvest Technology, ICAR-Indian Agricultural Research Institute,
New Delhi, India, E-mail: docsethi@gmail.com

Shafaq Shah
Department of Food Technology Jamia Hamdard University, New Delhi, India

Nadia Sharif
Department of Biotechnology, Lahore College for Women University, Lahore, Pakistan,
Tel.: +92 3237501948, E-mail: nadiasharif.s7@gmail.com

Nitya Sharma
Department of Food Engineering and Nutrition, Center of Innovative and Applied Bioprocessing (CIAB),
Knowledge City, Sector-81, Mohali – 140306, Punjab, India

R. R. Sharma
Division of Food Science and Postharvest Technology, ICAR-Indian Agricultural Research Institute,
New Delhi – 110 012, India

Mohammed Wasim Siddiqui
Department of Food Science and Post-Harvest Technology, Bihar Agricultural University, Sabour, India
Email: wasim_serene@yahoo.com

Sucheta
Center of Innovative and Applied Bioprocessing, Mohali, Punjab, India

ABBREVIATIONS

3D	three-dimensional
A/D	analog-to-digital
Ab	antibody
ABP1	auxin-binding protein receptor
AChE	acetylcholinesterase
AFM	analog firmness meter
AFM1	aflatoxin M1
AgNP	silver nanoparticles
ALA	α-linolenic acid
AMC	acetyl methyl carbinol
ANN	artificial neural network
AOTF	acousto-optic tunable filter
ATCl	acetylthiocholine chloride
ATR	attenuated total reflection
AuNP	gold nanoparticles
BPB	bromophenol blue
CCD	charge-coupled device
CCP	critical control points
CHIT	chitosan
ChlF	chlorophyll fluorescence
ChlFES	chlorophyll fluorescence excitation screening
CI	chemical image
CIE	Commission Internationale de' Eclairage
CLA	conjugated linoleic acid
CP	carbon paste
CPR	chlorophenol red
CSII	carbosieve SII
CT	computed tomography
CTIFL	Center Techniques Inter Professionnel des Fruits et Légumes
CV	cyclic voltammetry
DAG	diacylglycerols
DEA	di-electric analysis
DFM	digital firmness meter

DLE	delayed-light emission
EAS	electronic article surveillance
EIS	electrochemical impedance spectroscopy
EN	electronic nose
FA	fatty acids
FANOVA	functional analysis of variance
Fc	ferrocene
FDA	factorial discriminant analysis
FF	far field
FFA	free fatty acids
FR	far-red emission
FSCM	fruit supply chain management
FTIR	Fourier transform IR spectrometers
FTM	flexible tag microlab
GA	glutaraldehyde
GAPDH	glyceraldehyde 3-phosphate dehydrogenase
GC	gas chromatography
GCE	glassy carbon electrode
GFSI	Global Food Safety Initiative
GMP	good manufacturing practices
GO	graphene oxide
Gs	graphene sheets
HACCP	hazard analysis critical control points
HR/MAS	high resolutions angle spinning
HSI	hue saturation intensity
ICT	information and communication technology
IoT	internet of thing
IPM	integrated postharvest management
IR	infrared spectroscopy
kVp	kilovoltage peak
LCTF	liquid tunable channel
L-Cys	L-cysteine
LDV	laser Doppler vibrometer
mA	milliamperes
MAP	modified atmosphere packaging
MCM	multi-part chemical media
MEMS	micro-electro-mechanical system
MG	mature green
MIPs	molecularly imprinted polymers

MIR	mid-infrared
MLR	multivariate linear regression
MOS	metal oxide semiconductor
MR	methyl red
MRI	magnetic resonance imaging
MWCNTs	multi-walled carbon nanotubes
NADH	nicotinamide adenine dinucleotide
ND	non-destructive
NF	near-field
NG	nitrogen-doped graphene
NIR	near-infrared
NIRS	near-infrared spectroscopy
NMR	nuclear magnetic resonance
OMC	ordered mesoporous carbon
PAM	pulse-amplitude-modulation
PANI	polyaniline
PC	personal computer
PCA	principal component analysis
PEA	plant efficiency analyzer
PECH	poly epychloro hydrine
PLS	partial least squares
PLSR	partial least squares regression
RF	radio frequency
RFID	radio frequency identification
rGO	reduced graphene oxide
SAM	self-assembled monolayer
SAW	surface acoustic wave
SCM	supply chain management
SELEX	systematic evolution of ligands by exponential enrichment
SERS	surface-enhanced Raman spectroscopy
SFR	simple fluorescence ratio
SIM	selected ion monitoring
SIMCA	soft independent modeling of class analogy
SPCE	screen-printed carbon electrode
sRGB	standard Red Green Blue
SRS	space-resolved spectroscopy
SVM	support vector machine
SWV	square wave voltammetry
TAG	triacylglycerols

TCA	trichloroanisole
THI	thionine
TI	temperature indicator
TIC	total ion current
TMA	trimethylamine
ToF	time-of-flight
TRS	time domain reflectance spectroscopy
TRS	time-resolved spectroscopy
TTIs	time-temperature indicators
UHF	ultra-high frequency
ULPH	ultra-low power consumption hotplates
UV	ultra-violet
VIR	visible infrared
VOC's	volatile organic compounds
WSN	wireless sensor networks

PREFACE

Currently, a number of new sensor technologies based on electromagnetic properties have emerged, whereas great progress has also been made on other existing sensor technologies. They include X-ray, nuclear magnetic resonance (NMR) or magnetic resonant imaging (MRI), fluorescence, and other lesser sensor developments, such as electrical impedance and permittivity (mainly microwave), thermal sensing, and selective gas/volatile sensing. These development of these sensors has opened new areas of research as well as new applications for sensing the quality of fruits and vegetables.

A number of sensors have been developed for noninvasive and prompt assessment of the quality of fruits and vegetables. Many of these sensors include a broad range of sensing techniques, with various applications. Moreover, several sensors are focused on physical sensors, such as mechanical methods for firmness measurement, size characterization, computer vision, and others with chemical and biosensors, such as chemical sensing, biosensing, near-infrared (NIR) spectroscopy, NMR, as well as wireless sensing.

The needs for sensor developments are mainly driven to meet increasing consumer demand for better quality and safer fresh products of fruits and vegetables. A large number of sensors for nondestructive (ND) detection of food quality are related to the utilization of electromagnetic radiation in a wide range of frequencies. Electromagnetic radiation-based technologies have shown great potential; some of them have been successfully used for monitoring the quality of fruits and vegetables.

Successful application of sensor technologies requires the combination of effective sensors with sophisticated mathematical models and computer algorithms to establish relationships between selected physical/chemical properties and quality attributes of the product. As a result, as a number of sensors developed recently are focused on utilizing different ND optical techniques for quality detection of fruits and vegetables. Great advances have been made in spectroscopy and computer vision, and these techniques are being widely used for quality inspection and control of products in many food industries. As sensor technologies based on VIS, NIR, mid-infrared (MIR), and ultra-violet (UV) are becoming more affordable and equipped with more user-friendly data treatment and calibration capabilities, they have fostered further development of detection procedures for different quality- and composition-related properties of fruits and vegetables.

   This book, *Sensor-based quality assessment systems for Fruits and Vegetables*, is intended to provide information on different sensing techniques, including on those emerging technologies such as NMR, MRI, wireless sensor networks (WSN) and radio-frequency identification (RFID) for fruits and vegetables and their potential for industrial applications.

CHAPTER 1

REAL-TIME QUALITY ASSESSMENT OF FRUITS AND VEGETABLES: SENSING APPROACHES

BAMBANG KUSWANDI

Chemo and Biosensors Group, Faculty of Pharmacy, University of Jember, Jl. Kalimantan 37, Jember – 68121, Indonesia, E-mail: b_kuswandi@farmasi.unej.ac.id

ABSTRACT

Ripeness sensors that can be used for ripeness monitoring of fruits and vegetables could be as a sensor or an indicator where this device has the ability to detect the ripeness level of fruits and vegetables related to freshness and firmness conditions of the fruits and vegetables. This device is introduced into the packaging, as it has intelligent functions to monitor ripeness or freshness of fruits and vegetables, commonly it is called intelligent or smart packaging. The sensors can be created based on various sensor types, designs, and constructions for ripeness monitoring of fruits and vegetables that reflects food quality in term of ripeness level, freshness, and consumer preferences, e.g., crunchiness, firmness, juiciness, aroma, etc., which are discussed with the addition of RFID and nanosensors as the emerging technology in this field. This chapter describes the different concepts of the ripeness sensor, including freshness for fruits and vegetables. This technology has the potential to improve food freshness, shelf-life, and quality control throughout the supply chain and distribution system. Since it is dedicated to meet the need for desired ripeness, freshness, longer shelf-life, and quality as well as the safety of fruits and vegetables. Moreover, the device will be able to directly communicate to consumers using sensors or indicators that provide visual information on ripeness, freshness, and quality of fruits and vegetables. Thus, this device acts as active labeling for real-time shelf-life monitoring combining with the traditional inspection, which in turn, lowering fruits and vegetables lost.

1.1   INTRODUCTION

A recent trend in the increasing need of consumer preferences for good food in terms of fresh and safe have brought innovative technology in food packaging, such as intelligent packaging (Smolander, 2003; Yam, Takhistov, and Miltz, 2005; Kuswandi et al., 2011). Intelligent packaging incorporated sensing device that gives status regarding the freshness, the time-temperature history and the integrity of the food inside packaging, where it will be useful quality and safety status of the food (Kerry, O'Grady, and Hogan, 2006; Ahvenainen, 2003), specifically for perishable foods, including fruits and vegetables (Kuswandi et al., 2011; Rokka et al., 2004; Taoukis and Labuza, 2003; Welt, Sage, and Berger, 2003). Therefore, while protecting, containing, and preserving of food as the main rules of packaging (Robertson, 2006), it provides quality status on ripeness, freshness, tamper indication, safety, or traceability, using its sensing devices. In this regard, this package sensor technology offers to deliver desired ripeness, freshness, quality, and safer fruits and vegetables from the farm to the consumers.

The sensor used in smart packaging, it is described as a sensing device or tool employed to sense, determine marker (matter or energy) as a response or signal for the determination of chemical or physical properties that it gives response (Kerry, O'Grady, and Hogan, 2006; Kress-Rogers, 1998). Sensor response as signals is an output toward food conditions, where the devices contain consist of a receptor and a transducer as two major components. Another term that commonly found is an indicator that also used in intelligent or smart packaging. The indicator is described as a compound or material where indicates the absence, presence or level of a substance called an analyte/marker or even the degree of reaction between two or more substances based on a change in its property, e.g., mass or energy/color (Hogan and Kerry, 2008). Thus, between sensor and indicator commonly changeable when it is used, however, indicator mostly refers to colorimetric sensor either the chemical sensor or biosensors as a color indicator.

The chemical sensor is a chemically sensitive membrane or film that enables the detection of the presence, concentration, or activity of analyte (target chemical or marker) via adsorption and diffusion through it. Then the analyte converted into measurable signals or response by the transducer, either passive signal or active signal depends on the power supply applied in the measurement (Vanderroost et al., 2014). The current trend in the optical sensors usually does not need the power supply, since it has been detected remotely via optical fiber using UV/Vis, NIR, IR, or even nude eye. Here, the transducers based on silicon devices consist of circuits that are integrated optically into semiconductor-based silicon materials (Yebo et al., 2012). Similarly, the biosensor is consisting of bioreceptor and transducer as two major components (Alocilja and Radke, 2003; Kuswandi, Andres, and Narayanaswamy, 2001). A biosensor is a device or tool that employed to detect, determine, and transduce signals based on biochemical reactions (Kuswandi, Andres, and Narayanaswamy, 2001; Yam, Takhistov, and Miltz, 2005). The bioreceptor reacts to the marker or analyte and converts signals via the transducer to produce the measurable signal (Kuswandi et al., 2001; Yam, Takhistov, and Miltz, 2005). The bioreceptor could be an organic or biological material, e.g., nucleic acid, enzyme, antigen, hormone, microbes, etc. The transducer could be electrochemical, optical, or mechanical.

For example, the ripeness sensors for fruit applications contain sensor or indicator that detects the ripeness level of fruits in the packaging to show food status associated with the ripeness level, including the fruits quality during distribution, transportation, and storage as well as a display (Kuswandi et al., 2011; Yam, Takhistov, and Miltz, 2005). This sensor technology produced various kinds of ripeness sensor or indicator designs, which are appropriate for ripeness, freshness, and quality of fruits and vegetables monitoring in real-time. These ripeness sensors or indicators are targeted to fulfill the increasing demand for ripeness preferences, freshness, and quality of fruits and vegetables. The market demands on the ripeness sensor for fruits and vegetables are estimated to increase in the near future, and it can be achieved by its integration into food packaging systems. Due to the increasing demands for food information on packaging as a result of consumers need to know what components or ingredients are in the food product and how they should be stored and used, could also mean that there has to be an innovation in serving this information, and this could boost the need for ripeness and freshness sensor in future for fruits and vegetables. Furthermore, ripeness sensors will be able to communicate with the consumer via sensor or indicator giving visual status on freshness, ripeness, and quality of fruits and vegetables. Hence, the ripeness indicator or sensor acts as active shelf-life labeling devices in combination with the traditional inspection on fruits and vegetables or could be employed to control distribution, the stock rotation, and lowering fruit and vegetable loss.

1.2 RIPENESS SENSORS

Generally, a ripeness indicator or sensor is described as an indicator or a sensor that detects or determines the ripeness level of fruits and vegetables (showing the level of ripeness or firmness), it also informs the freshness and quality of fruits and vegetables that can be consumed. The ripeness sensor could also be described as an on-package indicator or sensor, including an on-food sensor that able to detect or determine the ripeness or freshness level associated to the condition inside or outside the food packaging and give information on the ripeness and freshness level of fruits and vegetables (Kuswandi et al., 2011), as shown in Figure 1.1.

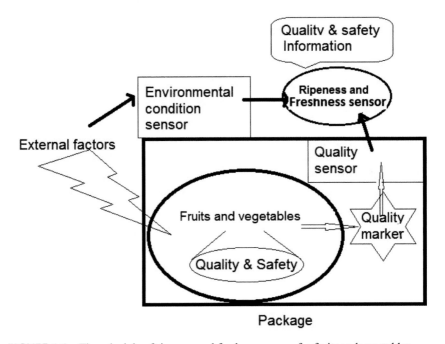

FIGURE 1.1   The principle of ripeness and freshness sensor for fruits and vegetables.

Basically, there are two types of developed ripeness sensors or indica-tors, i.e., direct ripeness sensor; and indirect ripeness sensor. The first type of ripeness sensor is the direct ripeness sensor works based on the detec-tion of a certain target or marker directly for the ripeness level of fruits and vegetables. In this context, various concepts and designs for ripeness markers have been described for the ripeness of fruits and vegetables in the literature. The second type, the indirect sensor works based on detec-tion indirectly of marker or analyte, such as based on a system that reacts or changes similar to the fruit or vegetable ripeness or freshness. In the second type, the device is estimated to follow or mimic the change in ripeness or freshness parameter of fruits and vegetables under a similar condition to temperature or time that has been exposed. Some of them are constructed to follow the gas evolution and change due to temperature and time during the distribution, while others are constructed to be employed in the packages. Furthermore, since ripeness or freshness devices (sensors and indicators), the change rate in the devices should be related correctly with the rate of ripeness stage of fruits or vegetables. The sensors change should also be related to the variation in temperature along with time from transportation, distribution, and storage (Seldman, 1995).

The ripeness level could be stated in various parameters depend on the fruit or vegetable types. For instance, the first type of ripeness sensor is a sensor or an indicator for monitoring of ripeness stage of fruits. Here, various concepts, design, and construction regarding ripeness indicators or sensors have been reported in the literature. Mostly, the ripeness sensors have been constructed for monitoring of ethylene (Lang and Hübert, 2012), acids (Kuswandi et al., 2013a), pH (Kuswandi et al., 2013a, 2014), ethanol (Cameron and Talasila, 1995), and CO_2 (Mattila, Tawast, and Ahvenainen, 1990). By incorporating the device into the food package or directly on fruits or vegetable, the ripeness sensor can be produced as a color indicator or label or tag that change the color in the presence of marker for the ripeness of fruits and vegetables.

The time-temperature indicator (TTI) is the example of the non-direct freshness sensor or indicator. Commercially TTIs available are working based on various mechanisms, such as physical, chemical, and biological mechanisms. In a physical or chemical mechanism, its response is based on a physical change or chemical reaction over time and temperature, such as melting point, polymerization reaction, acid-base reaction, etc. In the biological mechanism, its response is based on the rate of biological activity change, e.g., enzymes. Microorganism, or spores overs time and temperature. When the TTIs are exposed to a higher temperature than recommended storage temperature, it will change color as the product reach to the end of the shelf-life. Hence, TTIs that work based on physical, chemical, or enzymatic activity employed for processed food, e.g., puree, juice, jump, etc., it shows a correct, accurate, and reliable freshness indication of the product for the indication of quality and safety of the food condition.

The ripeness sensors, based on its design can be categorized into three types, i.e., (1) single sensor; (2) dual sensors; and (3) multiple sensors. The first design, a single sensor is a ripeness sensor that employed a single sensor or indicator to detect and monitor the ripeness level of fruits and vegetables. The second design, dual sensors are the ripeness sensors that are used two sensors or indicators that reference each other in detecting or sensing and informing the ripeness or freshness level of fruits and vegetables. The third design, multiple sensors are the sensors that employed three or more sensors or indicators as an array to build a pattern in detecting or monitoring and informing the level of ripeness or freshness of the fruits and vegetables, for example, e-nose. Currently, a commercially available ripeness sensor is mostly a single sensor. The dual, triple, and multiple sensors are still undergoing testing in the lab scale. Figure 1.2 depicted the ripeness sensors based on type, design, and application for fruits and vegetables.

The ripeness indicators or sensors can play a major role in detecting ripeness or freshness levels of fruits and vegetables to inform its desired ripeness, firmness, freshness, aroma, and quality. Thus, they sense, monitor, and communicate in which fruits and vegetables are in the desired level of ripeness, firmness, and freshness to be consumed. This becomes crucially important when fruits or vegetables are distributed and stored in less than preferred conditions such as extreme temperature, both too hot and too cold. Moreover, in the case of processed foods, where originated from fruit and vegetable that should not be frozen, instead of a ripeness sensor, a freshness indicator or sensor would be important to show and indicate whether they had been improperly exposed to freezing conditions. In the reverse case, a freshness indicator or sensor could also specify where the processed food products heat sensitive that had been exposed to high temperatures over their duration time.

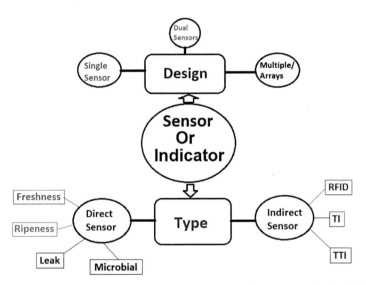

FIGURE 1.2 Classification of sensors or indicators for fruit and vegetable including their processed food packaging based on type and design.

Various ripeness sensors or indicators in terms of their types, designs, and applications, including freshness sensors for the processed food products that available commercially are described in Table 1.1. In addition, the description of the ripeness and freshness sensors are discussed in the subsections.

TABLE 1.1   The Commercially On-Package Sensor or Indicator for Monitoring of Fruits and Vegetables

Trade Name	Company	Type
RipeSense®	RipSense™ and ort Research	Ripeness indicator
Freshtag®	COX Technologies	Freshness indicator
Sensorq®	DSM NV and Food Quality Sensor International	Freshness indicator
Novas®	Insignia Technologies Ltd.	Integrity indicator
Ageless Eye®	Mitsubishi Gas Chemical Inc.	Integrity indicator
Timestrip®	Timestrip Ltd.	Integrity indicator
Novas®	Insignia Technologies Ltd.	Integrity indicator
Best-by™	FreshPoint Lab.	Integrity indicator
Smartlid	Smart Lid Systems	Temperature indicator
Timestrip® PLUS Duo	Timestrip UK Ltd.	Temperature indicator
Thermochromic ink	Tetrapak Ltd	Temperature indicator
Timestrip Complete®	Timestrip UK Ltd.	Time-temperature indicators
MonitorMark™	3M™, Minnesota	Time-temperature indicators
Fresh-Check®	Temptime Corp	Time-temperature indicators
Onvu™	Ciba Specialty Chemicals and Freshpoint	Time-temperature indicators
Checkpoint®	Vitsab	Time-temperature indicators
Cook-Chex	Pymah Corp	Time-temperature indicators
Color-Therm	Color-Therm	Time-temperature indicators
Thermax	Thermographic Measurements Ltd.	Time-temperature indicators
Easy2log®	CAEN RFID Srl	RFID
Intelligent Box	Mondi Plc	RFID
CS8304	Convergence Systems Ltd.	RFID
Temptrip	Temptrip LLC	RFID

1.3 DIRECT SENSORS

The direct ripeness sensors or indicators detect directly for a particular marker or compounds related to an indication of the food ripeness stage. In this regard, various different ripeness indicators or sensors have been reported in the literature along with freshness sensors that could be expressed for good quality of fruits and vegetables, e.g., ripeness indicator, freshness indicator, spoilage indicator, and even leak indicator. Mostly the indicators or sensors were constructed as a color indicator or colorimetric sensor for simple visual detection by the nude eye, where development of the color change in the sensor or indicator correctly related with the development of ripeness process of fruits and vegetables or the rate of deterioration of the food products in the case of freshness sensors resulted by variation in temperature overtime during distribution, transportation, and storage as well as display.

1.3.1 ON-FOOD RIPENESS INDICATORS

Generally, the ripeness stage in fruits and vegetables reflects their crunchiness, firmness, juiciness, and freshness. For preferred ripeness levels in fruit, sometimes for the consumer, it is hard to determine when it has achieved their preferred ripeness level. Therefore, it often becomes a barrier when the fruits should be bough, stored, or consumed. In this situation, the ripeness indicators or sensors could help consumers to solve this problem, where the indicator detects gas or volatile compound or aromas released by the ripened fruit at a certain state. The related studies on-food sensors or indicators for ripeness of fruit are not many. Even though, there are some works to construct ripeness sensors or indicators for fruits based on ethylene detection released by the ripening process, where it can be employed as an analyte or marker for fruit ripeness, e.g., Apple (Lang and Hübert, 2012; Kim and Shiratori, 2006). The colorimetric sensor was developed based on molybdenum (Mo) as chromophores and develop color change when it reacted with ethylene released by ripened apple, changing to blue from white, due to the Mo(VI) to Mo(V) reduction. The sensitivity of Mo color change reactions can be tuned based on the pH values (pH 1.4–1.5) and composition of ammonium molybdate used. The sensor attached directly to fruit and it can also be coupled with color recognition for quantitative measurements (Figure 1.3). For stone fruits that are bigger like apple, kiwi fruits, mangoes, melon, etc., the on-food sensor could be visible, but for smaller fruits like grape, strawberry, cherry, etc., the on-package ripeness sensors could be more suitable and reliable.

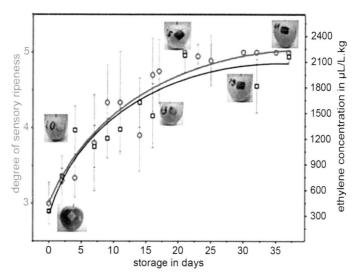

FIGURE 1.3 (A). Design of on-food label with a color indication for ripeness level related to ethylene concentration released by apple during its ripening process (Reprinted with permission from Lang and Hübert, 2012. © Springer Nature.)

1.3.2 ON-PACKAGE RIPENESS INDICATOR

The *ripeSense*™ (www.ripesense.com), a New Zealand company has produced the ripeness sensor that helps consumers to detect ripeness level of fruits, where the sensor detects the aromas or gas produced by the ripened fruit. Originally, the sensor color is red and develops to orange, then finally yellow. By detecting the sensor color change, consumers could select fruit at their preferred ripeness level. Significantly, this sensor decreases shrinkage and even damage-in due to consumers, particularly when they inspect fruit before purchase. In addition, the sensor pack that can be recycled provides an enhancement in hygiene security. The sensor has been used for pear, and also can be used to other stone fruits, e.g., mango, kiwifruit, melon, avocado, etc. (www.ripesense.com). The gas or aroma releases from fruits or vegetables can show the freshness and products quality. Moreover, the aromas or gases released during ripeness processes of fruit have been investigated using e-nose and used for ripeness detection in tomato (Gómez et al., 2006).

Another approach that developed for the on-package sensor was using bromophenol blue (BPB) as a color indicator and applied for the ripeness sensor of guava (*Psidium guajava* L.) (Kuswandi et al., 2013b). The simple absorption method was used to immobilized BPB onto the nata decoco (bacterial cellulose membrane). The colorimetric sensor as a ripeness indicator works based on decrease of pH due to the volatile organic acids (e.g., ascorbic acid) released gradually in the headspace of packaging during guava ripening stages. As a result, the color of the indicator would develop to green from blue as the indication overripe. The study shows that the indicator could be applied to monitor guava ripeness, at both 4 and 28°C. At these conditions, the indicator color develops related to the several parameters change, such as pH, texture softness, soluble solids contents, and sensory evaluation that commonly applied to study the guava ripeness degree.

In the non-climacteric fruits, e.g., strawberries, low-cost, and simple on-package indicator or sensor has been developed based on methyl red (MR) for monitoring strawberries (*Fragaria vesca* L.) ripeness (Kuswandi, 2013). The MR was also immobilized onto nata-decoco membrane and it works based on pH increase. Here, gradually the gas or volatile acids decreased in the headspace of packaging as the enzymatic process of esters produces in the ripening process. Consequently, the indicator will develop color to red-purple from yellow for an indication of overripe. The study shows that the indicator can be applied to detect the strawberries ripeness due to the correlation of the color develops toward strawberries ripeness was a similar trend. The sensor was successfully applied as an on-package sensor for monitoring of strawberries ripeness in real-time at both room and chiller temperatures.

Just recently, an on-package indicator has been developed for monitoring of grapes (*Vitis vinifera* L.) ripeness using chlorophenol red (CPR) (Kuswandi and Murdyaningsih, 2017). The CPR was impregnated onto the filter paper using the adsorption method to produce the CPR membrane as a ripeness indicator. The indicator acts as a colorimetric sensor for the detection of grape ripeness degree that works based on pH change when the volatile organic acids reduced inside the headspace of packaging as a result of the sugars formation during the grape ripening process. Consequently, the indicator develops a color originally to beige from white and lastly to yellow for an overripe indication that can easily be viewed by nude eye. Here, the study shows that the sensor could be employed for ripeness monitoring, as the indicator color developments are in a similar trend. Thus, the sensor or indicator can be employed as a label device for monitoring of grapes ripeness at ambient and chilled conditions as given in Figure 1.4.

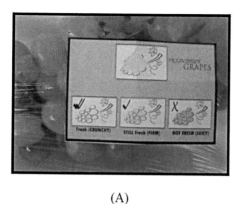

(A)

(B)   (C)

FIGURE 1.4 Application of on-package indicator for grape ripeness monitoring, (a) white for fresh (crunchy), (b) beige for maturity (firm), and (c) yellow for overripe (juicy) (Reprinted with permission from Kuswandi and Murdianingsih, 2017. © Springer Nature.)

1.3.3 FRESHNESS SENSOR

The process in which fruits, vegetables, or its processed food products deteriorate to the state of not fresh or its edibility quality decreases and it cannot be consumed by humans anymore, it is called spoilage or not fresh. Fruits and vegetables as well as its processed products, such as juice, puree, jump, etc, are easily getting spoilage. Many external factors could cause spoilage of these products, e.g., temperature, humidity, shelf life, pH, etc. For example, to detect food, freshness, or spoilage is based on its pH change. Many sensor or indicator designs which able non-invasive and real-time freshness monitoring of these products based on a pH change have been reported in the literature.

This color indicator or colorimetric sensor could be developed as a single sensor design based on mixed-dye for spoilage indicator (Nopwinyuwong, Trevanich, and Suppakul, 2010) which capable decreasing margins of error, by having an effective shelf-life of food product via allowing dynamic freshness to be detected visually alongside the best-before date. The application of a single sensor design concept based on mixed-dye for food spoilage indicator to other food products is a possible field for future freshness sensor developments, e.g., for easily prepared foods, perishable food products, bakery products, desserts, and fresh-cut fruits and vegetables.

Multiple sensors or sensor array, like e-nose (electronic nose), are the concept used to mimic or imitate the system in the mammalian olfactory. Here, an instrument is constructed to obtain the reproducible measurements that able to identify and classify the aroma mixtures contained in the odor. The instrument produces a pattern as a specific signal response to each odor, flavor, or savor. The system in E-nose consists of an array of either biosensors or chemical sensors with specificity and the chemometric methods capable to recognize the simple or complex odor, flavor, or savor (Gardner and Bartlett, 1994; Vanderroost et al., 2014). The e-nose was successfully applied for ripeness and freshness detection of tomato (Gómez et al., 2006). Moreover, the volatile gases released during various freshness states of fruits and vegetables could be monitored using e-nose.

Other approaches have been developed for other kinds of freshness sensors, various innovative and novel platforms have been constructed for the food spoilage in term of detection of microbial pathogens. Most of them are integrated within sensors and need the sample extraction to determine the target molecule presence in it. When applying such sensors for food packaging, mostly they are focused on the microbial contaminant growth detection. The challenges of this sensor are that it must be able to be integrated within the packaging, provide a simply distinctive response (like color development), and low-cost for mass production. In this case, the presence of microbial contamination could be indirectly detected by measuring of gas composition changes within the package headspace due to microbial growth, employing the gas sensor. However, the reported concepts of package sensors for pathogens or contaminants are still limited. Even if CO_2 used as the indication of microbial growth in MAPs is still difficult, due to it is already often, consist of CO_2 at high concentration. In addition, it is only possible to use the CO_2 concentration increase in microbial contamination or pathogen detection, if in packages not also contain protective gas using CO_2 as well (Mattila, Tawast, and Ahvenainen, 1990).

The indicators or colorimetric sensors based on reactions of microbial metabolites and other principles for the detection of contamination have been proposed in the literature (DeCicco and Keeven, 1995; Kress-Rogers, 1993). Generally, the colorimetric sensors are based on a color development of chromogenic substrates during the enzymatic reaction that resulted by contaminating microbes (DeCicco and Keeven, 1995), the certain nutrients consumption of the product or on the microorganism detection directly (Kress-Rogers, 1993). In this context, instead of the electrochemical biosensor system, the optical biosensor systems have also been widely constructed along with the biosensors based on acoustic transduction that mainly dedicate to be employed for microbial contaminants detection. These biosensors have been employed for targeting the contaminating microorganisms presence on food such as *Salmonella typhimurium, Salmonella* group B, D, and E*, Staphylococcal* enterotoxin A and B*, E. coli* and *E. coli 0157: H7* (Kress-Rogers, 1993; Terry et al., 2004). Biosensors used for *Listeria monocytogenes* detection have been also developed (Liu, Chakrabartty, and Alocilja, 2007). Currently, a commercial biosensor called Food Sentinel System® (SIRA Technologies Inc.) has been marketed to detect pathogens in food. In this biosensor system as in the barcode, the specific antibodies have been linked to the membrane as bio-recognition part of the biosensor. The pathogens detected produce localized dark bar formation in the barcode, which causes the barcode unreadable anymore (Yam, Takhistov, and Miltz, 2005).

1.3.4   LEAK INDICATOR

In the case of leak indicator or sensor, it is needed particularly to monitor the freshness quality of some processed food, e.g., corn flakes, crackers, fruit bars, etc., including other processed fruits and vegetables, e.g., puree, jump, juice, etc., which their freshness quality are affected by the packaging integrity. In this context, leaks of packaging become a big problem to maintain its freshness quality. Therefore, the freshness quality could also be presented using a leak indicator or sensor as the integrity monitoring of the food packaging that directly associated with the freshness quality of food products. In this regard, the gas indicators are useful devices for monitoring of the toxic gases released from spoilage or decomposing food inside packaging, which can harmful to the consumers health (Matindoust et al., 2016). In addition, the color indicator using chemical or enzymatic reaction can be used as a control to measure a change occurs in food packaging (Fuertes et al., 2016).

The modified atmosphere packaging (MAP) and equilibrium MAP coupled with active packaging methods that employed to control quality of the food packaging freshness need this leak indicator (Shen, Kong, and Wang, 2006). The atmosphere in the MAP, is not air, but It contains a heightened level of CO_2 (20–80%) and a lowered level of O_2 (0–2%). Thus, leaks in the MAP could be an increase in the concentration of O_2 and a decrease in the concentration of CO_2, where, aerobic microbial growth able to occur. In the worst condition, the concentration of CO_2 will remain high despite leakage and thus make microbial growth due to food packaging deterioration was going faster. Thus, the leak indicators or sensors in the MAPs are more critical compare to in the active packaging. This is due to when it became smart packaging, it should depend on the O_2 detection than on the CO_2 detection (Smolander, Hurme, and Ahvenainen, 1997).

The O_2 sensors for MAP's main application are to make sure the proper functioning of O_2 absorption. Currently commercially available O_2 sensors, for instance, O_2 absorbing sachets under the trade name "Ageless" produced by Mitsubishi Gas Chemical Company (Japan) (http://www.mgc.co.jp/eng/products/abc/ageless/eye.html) (Figure 1.5). Other companies are also producing some commercial O_2 gas sensors to monitor properly removal of O_2 by O_2 absorbers (Hurme and Ahvenainen, 1996). In addition, a company called Cryovac-Sealed Air Ltd., has also produced the type of gas sensor for detecting of gas composition correctly (Anonym, 1996).

Commonly, a color leak indicator based on the gas sensor for O_2 contains a redox-dye, such as 2,6-dichloroindophenol (Nakamura, Nakazawa, and Kawamura, 1987), methylene blue (Nakamura, Nakazawa, and Kawamura, 1987; Goto, 1987), or N,N,N',N'-tetramethyl-p-phenylenediamine (Mattila-Sandholm et al., 1998), coupled with a reducing compound (reducing sugars) (Perlman and Linschitz, 1985) along with an alkaline compound, such as magnesium hydroxide (Yoshikawa et al., 1982), calcium hydroxide (Yoshikawa et al., 1979), sodium hydroxide (Shirozaki, 1990), and potassium hydroxide (Yamamoto, 1992). In addition, an oxygen gas sensor could be constructed based on oxidative enzymes (Gardiol et al., 1996). Moreover, these main components, such as a bulking agent (e.g., zeolite, silica gel, cellulose materials, polymers) and a solvent (usually water and/or alcohol), usually be added to the gas sensor as well. Herein, the gas sensor can be produced as a tablet (Goto, 1987; Nakamura, Nakazawa, and Kawamura, 1987), a laminated in a polymer film (Gardiol et al., 1996) or printed layer (Yamamoto, 1992; Davies and Garner, 1996).

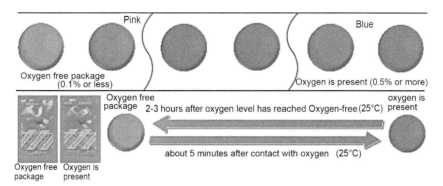

FIGURE 1.5   Schematic representation of the leak indicators of "ageless" oxygen indicator. *Source:* www.mgc.co.jp.

In the leak detection, where CO_2 use as indicators, it does not seem to be more reliable as leak detection over the gas sensor for O_2. Since after the packaging procedure and between 1–2 days, CO_2 will be dissolved into the product caused the head-space concentration of CO_2 increased, then could be reduced in the final concentration. After this time CO_2 concentrations decrease considerably and indicate the leakage in a package. The CO_2 indicators, other drawbacks are associated with the CO_2 production in the microbial metabolism. A package leak signed by CO_2 decreasing is often pushed by microbial growth that causes the CO_2 increase. Because this evident, commonly the CO_2 concentration will be constant, even when the package leakage or microbial spoilage occurs.

In the case of sensors for O_2 and CO_2 are employed as leak indicators in the MAP, there are some aspects that should be considered. The gas sensor with very low sensitivity of O_2 is not beneficial since this sensitive sensor could also detect the residual O_2 (typically 0.5–2.0%) that is often trapped inside the MAP (Raija and Hurme, 1997). Very low sensitivity of sensor could also be complicated in terms of the sensor handling, and need anaerobic conditions when the sensor constructed as well as the procedure in packaging. In addition, it has been stated that the O_2 gas sensors color change employed in MAPs also consists of an acidic CO_2 gas is not really clear yet (Ahvenainen and Hurme, 1997; Balderson and Whitwood, 1995). While, the reversibility sensor is not suitable when it is employed for leakage indicator and control, due to the fact that O_2 gas introducing the package via the leak can be consumed by the microbial growth as a result of the integrity loss in the food packaging (Balderson and Whitwood, 1995) that makes the indicator color or sensor will be same as or no change even if the packaged product already spoiled.

1.4 INDIRECT SENSORS

Principally, the indirect sensors or indicators were not interacted with the compound release of the packaged food, rather than mimics or follow the change of a certain freshness quality parameter of the food product undergoing the same exposure to temperature over time. Thus, they work based on the indirect detection of freshness markers. For it is used along with the food packaging as the monitoring devices for food freshness, the sensor or indicator change rate must correctly relate with the deterioration rate of the packaged food undergoing to the temperature variation over time during distribution, transportation, and storage as well as display. Generally, the indirect sensors or indicators will develop color as it exposed to the higher condition than recommended temperature conditions and will also develop to other colors when the end of product shelf-life has been achieved. Recently, the freshness indicators that can be grouped into the indirect sensor are temperature indicator (TI), TTIs, and RFID (radio frequency identification) that commercially available and used for food packaging, including for processed food, which originally derived from fruits and vegetables.

1.4.1 TIME INDICATOR (TI)

In order for packaging to make a self-cooling or self-heating, a sensor or indicator tells when the correct temperature has been achieved and the package becomes 'smart,' where currently such packaging is called smart packaging and commercially available. The TI that commonly marketed, it is consists of a thermochromic ink that uses to indicate when the product should be served at the correct temperature after refrigerated or heated. A plastic container of pouring syrup for pancakes can be bought in the UK and US where it is labeled using a dot of thermochromic ink to show that the syrup reaches the right temperature after heated using the microwave. Similarly, with orange juice pack labels, where it can be found on supermarket shelves that incorporate TI designed based on thermochromic ink to indicate when the orange juice is cold enough to drink following refrigeration as shown in Figure 1.6(a) (www.tetrapak.com).

In addition, another TI example is a technology produced by Smart Lid Systems, (Sydney, Australia) (www.smartlidsystems.com). The TI called smart lid is impregnated with an additive color development which able to change to a bright red color, from a coffee bean brown as it undergoing in the increasing temperature. The color develops running to start at 38°C, where it will reach the full intensity at 45°C. It indicates to consumers when the red color intensely emerges, meaning the cup of coffee is too hot for comfortable drinking (Figure 1.6(b)). However, when the lid is locked and not correctly positioned, where the brown color of the lit will not be evenly distributed, hence this will show that a potential spillage is exists. This color development additive is safe on food contact surfaces as it meets the US regulations associated with additive materials that has bed direct contact with food (www.smartlidsystems.com).

FIGURE 1.6 (a) Application of thermochromic ink, where the penguin appears when the temperature is right for the drink to be consumed (www.tetrapak.com). (b) Color changing disposable beverage lids showing increasing redness from left to right, when the red color appears intensely, it indicates that the coffee cup is too hot to drink.

Source: www.smartlidsystems.com.

1.4.2 TIME TEMPERATURE INDICATORS (TTIS)

One parameter which affects the food product quality is variations in temperature. Even though, the food products originated from fruits and vegetables are processed, packed, and distributed at the optimum temperature for their prolonging shelf life, during transportation and storage before retailing, the product quality may be reduced or even lost due to fluctuation in temperatures. In order to keep and control the temperature, so that the abuse did not occur, TTIs on packages were used. They can inform the product's quality with an assurance for consumers. They have the ability to monitor a time-temperature-dependent relationship as a small measuring device using irreversible color development (De Jong et al., 2005). Currently, marketed TTIs are shown in Table 1.1.

For example, 3M™ has two versions of TTIs. First is called MonitorMark™ (http:solutions.3m.com), this TTI is the threshold indicator for the industry, and intended for monitoring of product distribution, and second called the smart label, that intended for consumer information. The first one is an abuse indicator that works when a predetermined temperature has been exceeded otherwise will yield no response. The TTIs are based on a specific substance that has a blue dye with selected melting point. When they are activated, a film strip separates the wick from the reservoir is removed, which in turn, the porous wick, white in color, will show in the window. Undergoing a temperature exceeding the critical temperature, the substance starts to melt and diffuse through the porous wick, resulting appears of a blue color as shown in Figure 1.7(a). There are ALOS commercially available TTIs working at different critical temperatures, starting from −15°C to 26°C. On the other hand, the consumer label is working based on the blue dye melting point and diffusion as discussed previously. This is also an indicator for partial-history that develops color as it exposed to higher temperature than the recommended storage, and then it also develops color when the shelf-life of product has been reached.

Other smart labels are called Timestrips® (www.timestrip.com) that are monitoring how long a product has been open or it has been in use. These TTIs can monitor elapsed time, not only in minutes but also in a year, where it store in the refrigerator, in freezer, or at normal ambient temperature as well as at elevated conditions. The Timestrip® consists of a special porous membrane through, where a liquid with food-grade diffuses in a consistently and reproducible. By squeezing a start button of this TTIs, it is activated which makes the liquid inside moves into direct contact with the membrane. Then, the liquid diffuses via the membrane as the laws of physics take place. On the top of TTI surface, the markers have been printed-out and inform the all-important time since it is activated, as well as space for branding, including other graphics as shown in Figure 1.7(b). Since the Timestrip® is needed by the most applications to attach to a package or a product; therefore, they can be selected from a various range of adhesive tapes on the underside to fulfill the customer desired needs.

Other type of TTIs that can be used for freshness monitoring of fruits and vegetables is Fresh-Check® (www.lifelinestechnology.com). This freshness indicator is produced as self-adhesive labels that could be used to packages of perishable products to inform consumers the freshness of the product at point-of-purchase or at home as shown in Figure 1.7(c). The Fresh-Check have an active center circle that darkens irreversibly, where it is faster at higher temperatures and it is slower at lower temperatures. Thus, it is easy to control when the food product is use or not use within the product date codes. It gradually develops color, when its active center is exposed to temperature over time to show the food product freshness. The Fresh-Check as full history indicator that the working principle is based on the color development of a polymer formed diacetylene mono-mers. It contains a small circle, made of polymer, and surrounded with a printed ring as a reference for color change. When it starts, the polymer is lightly colored, then darken gradually depends on the cumulative exposure to temperature, as the color intendeds to reflect this condition. The rate of polymer color development is proportional to the rate of food quality loss; hence, the more rapidly the polymer develops in color, the higher the temperature exposed to the product.

Another TTIs that used to check for temperature abuse is CheckPoint (www.vitsab.com). It is a simple adhesive label attached to a food package that monitors a food package from the producer to the retailer and stays with the package until at the retail sale point. These TTIs react to temperature and time in similarly with the food product response; hence, they inform customers regarding the freshness state and shelf life of the product that still remaining. This indicator response is easy to read as it is in color dot as shown in Figure 1.7(d). The CheckPoint is also a full history indicator that works based on enzymatic reaction. The label consists of a bubble-like dot consisting of two elements: one for the substrate containing primarily of triglycerides and the other for the enzyme solution, i.e., lipase plus a pH indicator dye. At the beginning period of monitoring, the dot is activated by pressing the plastic bubble that breaks the seal between the two solutions. The solutions are mixed, and a color develops as the reaction precedes that caused a pH change in the dot. Initially, the dot is green, and then changes progressively to yellow when the product reaches the shelf-life end. The reaction in these TTIs is irreversible and will react slower as temperature reduced and faster as the temperature is increased.

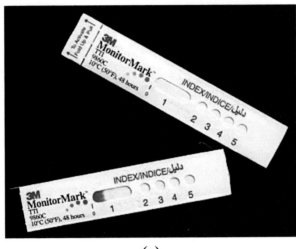

(a)

(b)

Fresh   Used soon   Should not be used

(c)

FIGURE 1.7   *(Continued)*

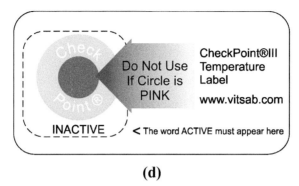

(d)

FIGURE 1.7 (a) MonitorMark™ TTIs product from 3M™ (solutions.3m.com); (b) Timestrips® TTIs product from Timestrip Plc (www.timestrip.com); (c) Fresh-Check® TTIs product from LifeLines (www.lifelinestechnology.com); (d) CheckPont® TTIs product from Vitsab (www.vitsab.com).

Source: www.vitsab.com.

1.4.3   RADIO FREQUENCY IDENTIFICATION (RFID)

An identification technology that works based on wireless sensors to identify the items and gather data automatically is called RFID. This device is employed with tags and readers (Tajima, 2007; Hong et al., 2011). Commonly, the tags store short of identification number, where the reader can retrieve information from a database related to the identification number and then response accordingly upon it (Todorovic, Neag, and Lazarevic, 2014) (Figure 1.8). The tags are classified into three classes, i.e., active, passive, and semi-passive, depending on the power supply used for communication and other functions. In active tags, to run the microchip's circuitry and to produce a signal to the reader is powered by an internal battery. While passive tags have no internal battery; and hence the tags are not able to communicate until the RFID reader emission is activated. The reader produces radio frequency (RF) field and provides enough power to the integrated circuit of the tags, so it is able to send energy to the reader. The semi-passive tags use a battery to power the electronics inside them and to maintain its memory or to modulate the RF field emitted by the reader antenna (Vanderroost et al., 2014). This technology has been employed successfully to control of traceability and processes management of supply chain, since to its capable to identify, detect, categorize, and manage the food products flow (Jones et al., 2004; Sarac, Absi, and Dauzère-Pérès, 2010; Ruiz-Garcia and Lunadei, 2011). This technology is much better compared to the barcode system for food traceability

(Jedermann, Ruiz-Garcia, and Lang, 2009). This is due to RFID serves supply chain visibility which capable automated processes rapidly at the chain for fast and automated processes at the supply chain levels, e.g., sharing of information and exception management (Tajima, 2007).

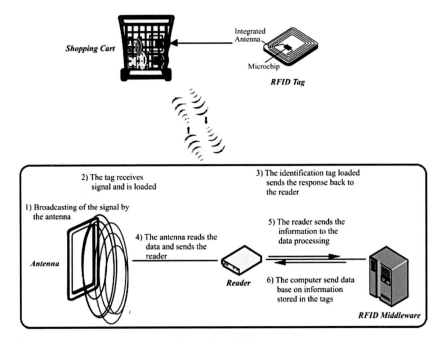

FIGURE 1.8   Schematic representation of the RFID system.

Recently, the RFID with tags are employed as freshness sensor that is marketed to monitor the temperature, relative humidity, pressure, and pH as well as light exposure, of the products. These tags detect the possible interruptions in the cold chain, which are harmful to food quality and safety (Vanderroost et al., 2014). In addition, many advance technologies have been produced in this technology, e.g., RFID tags employed for the real-time evaluation system of the packaged drink, for example, milk freshness during distribution and marketing (Potyrailo et al., 2012); The RFID tag also coupled with O_2 and CO_2 gas sensor for monitoring the vegetables freshness (Eom et al., 2014). In this context, the RFID can be categorized either as direct or indirect freshness sensors based on their function in direct or indirect monitoring of fruits and vegetables freshness. Therefore, various companies have produced various RFID solutions available on the market dependent on the needs of consumers as given in Table 1.1.

1.5   NANOSENSORS

Commonly, nanosensors can be defined as sensors that have structures and functions devices and fabricated in nanoscale size (1 nm = 10^{-9} m). As a sensing device, nanosensor is integrated in food packaging to monitor internal and external conditions of food (Ramachandraiah, Han, and Chin, 2015). The nanosensors, integrated into the smart packaging will have great advantages for the food industry, such as fruits and vegetables. Since this nanosensor is in the tiny chips form that invisible to our eye, therefore, it is can be incorporated with food or packaging. For example, it can be used as an electronic bar code that capable for the food monitoring in all stages (processing, production, distribution, marketing, and consumption). Currently, graphene, nanofibers, nanoparticles, and nanotubes as nanomaterials based carbon are employed in nanosensors, since they have excellent mechanical and electrical properties along with the surface area with high specificity (Vanderroost et al., 2014). This nanodevices can be used to monitor and detect spoilage, freshness, pathogens, track ingredients, chemical contaminants, product tampering, or products via the processing chain (Nachay, 2007; Azeredo, 2009; Liu, Chakrabartty, and Alocilja, 2007; Kuswandi, 2017). In industry 4.0, communication between these nanosensors are a future promising technology that makes the development of novel nanodevices able to perform simple and basic tasks at the nano level (computing, data storage, detection, and triggering) (Akyildiz and Jornet, 2010; Kuswandi, 2017). Due to a limited field of measurement, therefore the wireless nanosensor networks development is crucially important for the next smart packaging. These networks could be a set of nanosensors nodes that dynamically self-organizing in a wireless network with the possibility to be used with any kind of pre-existing infrastructure (Upadhayay and Agarwal, 2012). This nanosensor technology is still in its infancy of research and development for application in the smart or intelligent packaging.

Many works on the nanosensors have been developed and used in the monitoring of gases, molecules, microorganisms by surface-enhanced Raman spectroscopy (SERS) (Duncan, 2011); the fluorescent nanoparticles have been employed to monitor pathogens and toxins in crops and food (Burris and Stewart, 2012). For instance, the pathogenic bacteria detection (*Salmonella typhimurium, Shigella flexneri,* and *Escherichia coli O157: H7*), in food using functionalized quantum dots along with immunomagnetic separation in apple juice and milk (Zhao et al., 2008); the temperature changes have been detected by nanosensors (Iliadis and Ali, 2011; Lee et al., 2011); Escherichia coli detection in a food sample using nanosensor by detecting and measuring light scattering by cellular mitochondrial (Horner et al., 2006); Salmonella in foods has also been detected instantly by nanobiosensors (Fu et al., 2008). In addition of nanosensors, nano-biosensors have also been used to detect organophosphate pesticide residues in food (Liu et al., 2008); humidity or temperature changes due to moisture (Zhang et al., 2010); the food consumer profile (likes and dislikes), allergies, and nutritional deficiencies (Meetoo, 2011); CO_2 as a direct indicator of the food quality (Puligundla, Jung, and Ko, 2012); and the food pathogen, *Bacillus cereus* (Pal, Alocilja, and Downes, 2007). Moreover, e-tongue incorporated into packaging containing the nanosensors array that extremely sensitive to gases produced by spoiled food, as the food freshness indicator for a clear and visible sign for consumers (Bowles and Lu, 2014). The nanosensors research and development will lead to important scientific advances in the near future that capable for a new nanosensor generation to be integrated in food packaging smartly as truly smart packaging even it is still in the early development phase recently.

1.6　CONCLUSION AND FUTURE PROSPECT

The recent advance of ripeness and freshness sensor or indicators depend on the sensor technology development and materials that somehow directly detect or indirectly mimic the fruit and vegetable conditions to show its ripeness or freshness level that reflects the shelf-life, quality, and safety of fruits and vegetables. In the case of ripeness or freshness indicators or sensors to be incorporated in food packaging, they required to be reliable toward the current printing technology for the mass product, low-cost compare to the food product value, reliable, easy to use, simple, reproducible, and accurate, in the wide range of applications. Moreover, they should be eco-friendly, user-friendly, and safe for food contact.

The incorporation of an indicator or a sensor in food packaging has led to great benefits in smart or intelligent packaging. For example, enhancing shelf-life, quality, and safety of food products that resulted in significantly lowering food loss. Generally, the innovations in food packaging have been derived by consumer preferences, and global trends, e.g., ICT (Information and Communication Technology), including IoT (internet of thing) with their various devices and gadgets. In addition, some innovations have been derived from unexpected growths, e.g., the rise of nanosensor technologies that make the sensing technology at the nano-scale size. Perhaps, new ripeness and freshness indicator or sensor created in the future will be based on this nanotechnology, so that the on-package or on-food sensors could communicate intimately with the consumers to be more intelligent, not only with the color development, but also via the gadgets or smartphone in terms of ripeness, freshness, quality, and as safety, and will be more wider implementation, e.g., origin, authentication, convenience, tracking, online calorie and nutritional level, halalness as well as fruits and vegetables sustainability, including their processed food products and derivatives.

ACKNOWLEDGMENTS

The author thanks gratefully the DRPM, Higher Education, Ministry of Research, Technology, and Higher Education, the Republic of Indonesia for supporting this work via the Competency Grant Project 2018 (Hibah, 2018).

KEYWORDS

- **bromophenol blue**
- **information and communication technology**
- **intelligent packaging**
- **internet of thing**
- **modified atmosphere packaging**
- **radio frequency identification**
- **ripeness**

REFERENCES

Ahvenainen, R., & Eero, H., (1997). Active and smart packaging for meeting consumer demands for quality and safety. *Food Additives and Contaminants, 14*(6–7), 753–763. doi: 10.1080/02652039709374586.

Ahvenainen, R., & Hurme, E., (1997). Active and smart packaging for meeting consumer demands for quality and safety. *Food Addit. Contaminants, 14*, 753–763.

Ahvenainen, R., (2003). Active and intelligent packaging: An introduction. In: Ahvenainen, R., (ed.), *Novel Food Packaging Techniques* (pp. 5–21). Cambridge: Woodhead Publishing Ltd., Cambridge.

Akyildiz, I., & Josep, J., (2010). The internet of nano-things. *IEEE Wireless Communications, 17*(6), 58–63. doi: 10.1109/MWC.2010.5675779.

Alocilja, E. C., & Stephen, M. R., (2003). Market analysis of biosensors for food safety. *Biosensors and Bioelectronics, 18*(5–6), 841–846. http://www.ncbi.nlm.nih.gov/pubmed/12706600 (accessed on 23 May 2020).

Anonym, (1996). *Tufflex GS Product Information*. Sealed Air (FPD) Limited. Telford, UK.

Azeredo, & Henriette, M. C. D., (2009). Nanocomposites for food packaging applications. *Food Research International, 42*(9), 1240–1253. doi: 10.1016/j.foodres.2009.03.019.

Balderson, S. N., & Whitwood, R. J., (1995). *Gas Indicator for a Package*. US 5439648.

Burris, K. P., & Neal, S. C., (2012). Fluorescent nanoparticles: Sensing pathogens and toxins in foods and crops. *Trends in Food Science and Technology, 28*(2), 143–152. doi: 10.1016/j.tifs.2012.06.013.

Cameron, A. C., & Talasila, T., (1995). Modified-atmosphere packaging of fresh fruits and vegetables. In: *IFT Annual Meeting, Book of Abstracts* (p. 254).

Davies, E. S., & Garner, C. D., (1996). *Oxygen Indicating Composition*. UK Patent Application GB 2 298273.

De Cicco, B. T., & Keeven, J. K., (1995). *Detection System for Microbial Contamination in Health-Care Products*. US 5443987.

De Jong, A. R., Boumans, H., Slaghek, T., Van, V. J., Rijk, R., & Van, Z. M., (2005). Active and intelligent packaging for food: Is it the future? *Food Additives and Contaminants, 22*(10), 975–979. Taylor & Francis Group. doi: 10.1080/02652030500336254.

Duncan, & Timothy, V., (2011). Applications of nanotechnology in food packaging and food safety: Barrier materials, antimicrobials and sensors. *Journal of Colloid and Interface Science, 363*(1), 1–24. doi: 10.1016/j.jcis.2011.07.017.

Eom, K. H., Kyo, H. H., Sen, L., & Joo, W. K., (2014). The meat freshness monitoring system using the smart RFID tag. *International Journal of Distributed Sensor Networks*. doi: 10.1155/2014/591812.

Fu, J., Park, B., Siragusa, G., Jones, L., Tripp, R., Zhao, Y., & Cho, Y. Y. J., (2008). An Au/Si hetero-nanorod-based biosensor for Salmonella detection. *Nanotechnology, 19*(15), 155502.

Fuertes, G., Ismael, S., Raúl, C., Manuel, V., Jorge, S., Carolina, L., & Carolina, L., (2016). Intelligent packaging systems: Sensors and nanosensors to monitor food quality and safety. *Journal of Sensors*, pp. 1–8. Hindawi. doi: 10.1155/2016/4046061.

Gardiol, Alicia, E., Ruben, J. H., Bengt, R., & Bruce, R. H., (1996). Development of a gas-phase oxygen biosensor using a blue copper-containing oxidase. *Enzyme and Microbial Technology, 18*(5), 347–352. doi: 10.1016/0141-0229(95)00110-7.

Gardner, J. W., & Philip, N. B., (1994). A brief history of electronic noses. *Sensors and Actuators B: Chemical, 18*(1–3), 210–211. doi: 10.1016/0925-4005(94)87085-3.

Gómez, Antihus, H., Guixian, H., Jun, W., & Annia, G. P., (2006). Evaluation of tomato maturity by electronic nose. *Computers and Electronics in Agriculture, 54*(1), 44–52. doi: 10.1016/j.compag.2006.07.002.

Goto, M., (1987). *Oxygen Indicator*. JP 62-259059. Mitsubishi Gas Chemical Co., Inc., Tokyo, Japan, 1987.

Hogan, S. A., & Kerry, J. P., (2008). Smart packaging of meat and poultry products. In: Kerry, J. B., (ed.), *Smart Packaging Technologies for Fast Moving Consumer Goods* (pp. 33–59). West Sussex, England: John Wiley & Sons Ltd.

Hong, I-Hsuan, Jr-Fong, D., Yi-Hsuan, T., Chen-Shen, L., Wang-Tsang, L., Ming-Li, W., & Pei-Chun, C., (2011). An RFID application in the food supply chain: A case study of convenience stores in Taiwan. *Journal of Food Engineering, 106*(2), 119–126. Elsevier Ltd. doi: 10.1016/j.jfoodeng.2011.04.014.

Horner, Scott, R., Charles, R. M., Lewis, J. R., & Benjamin, L. M., (2006). A proteomic biosensor for enteropathogenic *E. Coli. Biosensors and Bioelectronics, 21*(8), 1659–1663. doi: 10.1016/j.bios.2005.07.019.

Hurme, E., & Ahvenainen, R., (1996). Active and smart packaging of ready-made foods. In: Ohlsson, T., Ahvenainen, R., & Mattila-Sandholm, T., (eds.), *Minimal Processing and Ready Made Foods* (pp. 169–182). Göteborg, SIK: SIK.

Iliadis, Agis, A., & Hasina, A. A., (2011). Properties of fast response room temperature ZnO-Si heterojunction Gas nanosensors. *IEEE Transactions on Nanotechnology, 10*(3), 652–656. doi: 10.1109/TNANO.2010.2089637.

Jedermann, Reiner, Ruiz-Garcia, L., & Walter, L., (2009). Spatial temperature profiling by semi-passive RFID loggers for perishable food transportation. *Computers and Electronics in Agriculture, 65*(2), 145–154. doi: 10.1016/j.compag.2008.08.006.

Jones, Peter, Clarke-Hill, C., Peter, S., Daphne, C., & David, H., (2004). Radio frequency identification in the UK: Opportunities and challenges. *International Journal of Retail and Distribution Management, 32*(3), 164–171. Emerald Group Publishing Limited. doi: 10.1108/09590550410524957.

Kerry, J. P., O'Grady, M. N., & Hogan, S. A., (2006). Past, current and potential utilization of active and intelligent packaging systems for meat and muscle-based products: A review. *Meat Science, 74*(1), 113–130. doi: 10.1016/j.meatsci.2006.04.024.

Kim, Jin-Ho, & Seimei, S., (2006). Fabrication of color changeable film to detect ethylene gas. *Japanese Journal of Applied Physics, 45*(5A), 4274–4278. Japan Society of Applied Physics. doi: 10.1143/JJAP.45.4274.

Kress-Rogers, E., (1993). The marker concept: Frying oil monitor and meat freshness sensor. In: Kress-Rogers, E., (ed.), *Instrumentation and Sensors for the Food Industry* (p. 52). Butterworth-Heinemann: Stoneham, Mass.

Kress-Rogers, E., (1998). Chemosensors, biosensors and immune sensors. In: Kress-Rodgers, E., (ed.), *Instrumentation and Sensors for the Food Industry* (pp. 581–669). Cambridge: Woodhead Publishing Ltd., Cambridge.

Kuswandi, B. (2017). Environmental friendly food nano-packaging. *Environmental Chemistry Letters, 15*(2), 205–221. Springer International Publishing. doi: 10.1007/s10311-017-0613-7.

Kuswandi, B., Maryska, C., Jayus, Abdullah, A., & Heng L. Y. (2013b). Real time on-package freshness indicator for guavas packaging. *Sensing and Instrumentation for Food Quality and Safety, 7*(1), 29–39. doi: 10.1007/s11694-013-9136-5.

Kuswandi, B., & Murdyaningsih, E. A., (2017). Simple on package indicator label for monitoring of grape ripening process using colorimetric pH sensor. *Journal of Food Measurement and Characterization*. doi: 10.1007/s11694-017-9603-5.

Kuswandi, B., (2013). Simple and low-cost freshness indicator for strawberries packaging. *Acta Manilana, 61*, 147–159.

Kuswandi, B., Andres, R., & Narayanaswamy, R., (2001). Optical fiber biosensors based on immobilized enzymes. *Analyst, 126*(8), 1469–1491. doi: 10.1039/b008311i.

Kuswandi, B., Jayus, Oktaviana, R., Abdullah, A., & Heng, L. Y., (2014). A novel on-package sticker sensor based on methyl red for real-time monitoring of broiler chicken cut freshness. *Packaging Technology and Science, 27*(1), 69–81. doi: 10.1002/pts.2016.

Kuswandi, B., Maryska, C., Jayus, Abdullah, A., & Heng, L. Y., (2013a). Real time on-package freshness indicator for guavas packaging. *Journal of Food Measurement and Characterization, 7*(1), 29–39. doi: 10.1007/s11694-013-9136-5.

Kuswandi, B., Wicaksono, Y., Jayus, Abdullah, A., Heng, L. Y., & Ahmad, M., (2011). Smart packaging: Sensors for monitoring of food quality and safety. *Sensing and Instrumentation for Food Quality and Safety, 5*(3–4), 137–146. doi: 10.1007/s11694-011-9120-x.

Lang, C., & Hübert, T., (2012). A color ripeness indicator for apples. *Food and Bioprocess Technology, 5*(8), 3244–3249. Springer US. doi: 10.1007/s11947-01-0694-4.

Lee, J., Mubeen, S., Hangarter, C. M., Mulchandani, A., Chen, W., & Myung, Y. N. V., (2011). Selective and rapid room temperature detection of H2S using gold nanoparticle chain arrays. *Electroanalysis, 23*(11), 2623–2628.

Liu, Shaoqin, Lang, Y., Xiuli, Y., Zhaozhu, Z., & Zhiyong, T., (2008). Recent advances in nanosensors for organophosphate pesticide detection. *Advanced Powder Technology, 19*(5), 419–441. doi: 10.1016/S0921-8831(08)60910-3.

Liu, Yang, Shantanu, C., & Evangelyn, C., A., (2007). Fundamental building blocks for molecular biowire based forward error-correcting biosensors. *Nanotechnology, 18*(42). IOP Publishing: 424017. doi: 10.1088/0957-4484/18/42/424017.

Matindoust, Samaneh, Baghaei-Nejad, M., Mohammad, H. S. A., Zhuo, Z., & Li-Rong, Z., (2016). Food quality and safety monitoring using gas sensor array in intelligent packaging. *Sensor Review, 36*(2), 169–183. Emerald Group Publishing Limited. doi: 10.1108/SR-07-2015-0115.

Mattila, T., Tawast, J., & Ahvenainen, R., (1990). New possibilities for quality control of aseptic packages: Microbiological spoilage and seal defect detection using head-space indicators. *Lebensm.-Wiss. Technol., 23*, 246–251.

Mattila-Sandholm, T., Ahvenainen, R., Hurme, E., & Järvi-Kääriäinen, T., (1998). *Oxygen Sensitive Color Indicator for Detecting Leaks in Gas-Protected Food Packages.* Espoo, Finland.

Meetoo, & Danny, D., (2011). Nanotechnology and the food sector: From the farm to the table. *Emirates Journal of Food and Agriculture, 23*(5), 387–403.

Nachay, K., (2007). Analyzing nanotechnology. *Food Technol., 61*(1), 34–36.

Nakamura, H., Nakazawa, N., & Kawamura, Y., (1987). *Food Oxidation Indicating Material-Comprises Oxygen Absorption Agent Containing Indicator Composed of Methylene Blue, Reducing Agent and Resin Binder.* JP 62-183834.

Nopwinyuwong, Atchareeya, Sudsai, T., & Panuwat, S., (2010). Development of a novel colorimetric indicator label for monitoring freshness of intermediate-moisture dessert spoilage. *Talanta, 81*(3), 1126–1132. Elsevier B.V. doi: 10.1016/j.talanta.2010.02.008.

Pal, Sudeshna, Alocilja, E. C., & Downes, F. P., (2007). Nanowire labeled direct-charge transfer biosensor for detecting bacillus species. *Biosensors and Bioelectronics, 22*(9–10), 2329–2336. doi: 10.1016/j.bios.2007.01.013.

Perlman, D., & Linschitz, H., (1985). *Oxygen Indicator for Packaging.* US 4526752.

Potyrailo, Radislav, A., Nandini, N., Zhexiong, T., Frank, J. M., Cheryl, S., & William, M., (2012). Battery-free radio frequency identification (RFID) sensors for food quality and safety. *Journal of Agricultural and Food Chemistry, 60*(35), 8535–8543. NIH Public Access. doi: 10.1021/jf302416y.

Puligundla, Pradeep, Junho, J., & Sanghoon, K., (2012). Carbon dioxide sensors for intelligent food packaging applications. *Food Control, 25*(1), 328–333. doi: 10.1016/j.foodcont.2011.10.043.

Ramachandraiah, K., Han, S. G., & Chin, Y. K. B., (2015). Nanotechnology in meat processing and packaging: Potential applications: A review. *Asian-Australasian Journal of Animal Sciences, 28*(2).

Robertson, G., (2006). *Food Packaging Principles and Practices* (1st edn.). Boca Raton, Florida: Taylor and Francis Group.

Rokka, Mervi, Susanna, E., Maria, S., Hanna-Leena, A., & Raija, A., (2004). Monitoring of the quality of modified atmosphere packaged broiler chicken cuts stored in different temperature conditions: B. biogenic amines as quality-indicating metabolites. *Food Control, 15*(8), 601–607. doi: 10.1016/j.foodcont.2003.10.002.

Ruiz-Garcia, L., & Lunadei, L., (2011). The role of RFID in agriculture: Applications, limitations and challenges. *Comput. Electron. Agr., 79*, 42–50.

Sarac, Aysegul, Nabil, A., & Dauzère-Pérès, S., (2010). A literature review on the impact of RFID technologies on supply chain management. *Integrating the Global Supply Chain, 128*(1), 77–95. doi: http://dx.doi.org/10.1016/j.ijpe.2010.07.039.

Seldman, J. D., (1995). Time-temperature indicators. In: Rooney, M. L., (ed.), *Active Food Packaging* (pp. 74–107). New York: Chapman & Hall.

Shen, Qun, Fanchun, K., & Qun, W., (2006). Effect of modified atmosphere packaging on the browning and lignifications of bamboo shoots. *Journal of Food Engineering, 77*(2), 348–354. doi: 10.1016/j.jfoodeng.2005.06.041.

Shirozaki, Y., (1990). *Oxygen Indicator, Rendering Reagent having Oxygen Detecting Capacity to Polyethylene.* JP 2-57975.

Smolander, M., (2003). The use of freshness indicators in packaging. In: Ahvenainen, R., (ed.), *Novel Food Packaging Techniques* (pp. 128–143). Cambridge: Woodhead Publishing Ltd.

Smolander, M., Hurme, E., & Ahvenainen, R., (1997). Leak indicators for modified-atmosphere packages. *Trends in Food Science and Technology, 8*(4), 101–106. doi: 10.1016/ S0924-2244(97)01017-0.

Tajima, M., (2007). Strategic value of RFID in supply chain management. *J. Purch. Supply. Manag., 13*, 261–273.

Taoukis, P. S., & Labuza, T. P., (2003). Time-temperature indicators (TTIs). In: Ahvenainen, R., (ed.), *Novel Food Packaging Techniques* (pp. 103–126). Cambridge: Woodhead Publishing Ltd., Cambridge.

Terry, L. A., Volpe, G., Palleschi, G., & Turner, A. P. F., (2004). Biosensors for the microbial analysis of food. *Food Sci. Technol., 18*(4), 22–24.

Todorovic, Vladimir, Marius, N., & Milovan, L., (2014). On the usage of RFID, tags for tracking and monitoring of shipped perishable goods. *Procedia. Engineering, 69,* 1345–1349. Elsevier B.V. doi: 10.1016/j.proeng.2014.03.127.

Upadhayay, V., & Agarwal, Y. S., (2012). Application of wireless nano sensor networks for wild lives. *International Journal of Distributed and Parallel Systems, 3*(4), 173–181.

Vanderroost, Mike, Peter, R., Frank, D., & Bruno, D. M., (2014). Intelligent food packaging: The next generation. *Trends in Food Science and Technology, 39*(1), 47–62. doi: 10.1016/j. tifs.2014.06.009.

Welt, B. A., Sage, D. S., & Berger, K. L., (2003). Performance specification of time-temperature integrators designed to protect against botulism in refrigerated fresh foods. *Journal of Food Science, 68*(1), 2–9. doi: 10.1111/j.1365-2621.2003.tb14105.x.

Yam, Kit, L., Paul, T. T., & Joseph, M., (2005). Intelligent packaging: Concepts and applications. *Journal of Food Science, 70*(1). Blackwell Publishing Ltd: R1–10. doi: 10.1111/j.1365-2621.2005.tb09052.x.

Yamamoto, H., (1992). *Laminated Oxygen Indicator Labels for Food Packages*. JP 4151557.

Yebo, Nebiyu, A., Sreeprasanth, P. S., Elisabeth, L., Christophe, D., Zeger, H., Johan, A. M., & Roel, B., (2012). Selective and reversible ammonia gas detection with nanoporous film functionalized silicon photonic micro-ring resonator. *Optics Express, 20*(11), 11855. Optical Society of America. doi: 10.1364/OE.20.011855.

Yoshikawa, Y., Nawata, T., Goto, M., & Fujii, Y., (1979). *Oxygen Indicator*. US 4169811.

Yoshikawa, Y., Nawata, T., Goto, M., & Kondo, Y., (1982). *Oxygen Indicator Adapted for Printing or Coating and Oxygen-Indicating Device*. US 4349509.

Zhang, H., Li, Z., Wang, W., Wang, C., & Liu, Y. L., (2010). Na$^+$-doped zinc oxide nanofiber membrane for high-speed humidity sensor. *Journal of the American Ceramic Society, 93*(1), 142–146.

Zhao, Y., Ye, M., Chao, Q., Jia, N., Ge, Y., & Shen, Y. H., (2008). Simultaneous detection of multi food-borne pathogenic bacteria based on functionalized quantum dots coupled with immunomagnetic separation in food samples. *Journal of Agricultural and Food Chemistry, 57*(2), 517–524.

CHAPTER 2

SENSORS FOR RIPENESS MEASUREMENT OF FRUITS AND VEGETABLES

SUCHETA[1] and KARTIKEY CHATURVEDI[2]

[1]Center of Innovative and Applied Bioprocessing, Mohali, Punjab, India

[2]Basic and Applied Sciences, National Institute of Food Technology and Entrepreneurship Management, Sonepat, Haryana, India

2.1 INTRODUCTION

Ripening of fruits and vegetables is a physiological process which brings about significant biochemical, textural, and color changes leading to consumer acceptability. The cells complete their morphological differentiation and continue to grow till senescence. The changes incurring during ripening leads to softening of texture of fruits and vegetables, taste becomes sweeter due to hydrolysis of polysaccharides to simple sugars and conversion of green-colored chlorophyll to various colored polyphenols. These changes can be detected only by either mechanical or subjective measurement methods which contribute to postharvest losses. Ripening can't be predicted accurately using single instrumental analysis or sensory evaluation but a combination of measurement methods like color, firmness, soluble solids, etc., can give reliable prediction. In addition, many commodities do not show a uniform significant change during ripening and may develop post-harvest disorder at the end of ripening if not harvested timely at particular maturity stages. Therefore, there is need to optimized or develop non-destructive (ND) sensors for ripeness measurement of fruits and vegetables to minimize post-harvest losses. The chapter provides details of such ND sensors for ripeness measurement of fruits and vegetables. ND sensors have been categorized in this chapter for firmness, color, and ethylene measurement as per their importance in ripeness measurement studies carried out by researchers in different fruits and vegetables.

2.2   MATURITY INDICES AND RIPENESS

A fruit or vegetable is considered mature if it reaches a stage after which it will continue to develop certain biochemical changes even after it is picked up from plants. Certain indices have been identified for different fruits and vegetables to enable growers know timely harvest of commodity. These are maturity indices and can be defined as measurements to determine maturity of a particular commodity and can contribute in design of better marketing strategies. Maturity indices can be physical attributes (size, shape, firmness, etc.) as well as certain biochemical changes (soluble solids content, sugar-to-acid ratio, starch content, etc.). Vegetables are evaluated for days from full bloom to harvest, desirable size, color, solidity, specific gravity, etc. (Sudheer and Indira, 2007). Vegetables generally are picked when they are mature or immature and may be allowed to ripen after harvest till few days. Carrots, melons, tomatoes, pumpkin, squashes are harvested when they are immature while cucumber, summer squash, eggplant, peas, okra are harvested after maturity. Fruits can be categorized into climacteric and non-climacteric depending upon their behavior after removal from plants. Climacteric fruits continue to ripen after harvesting and the duration of ripening varies with individuality, fertilizers applied, postharvest storage conditions, etc. Few examples are apple, banana, pear, guava, sapota, mango, papaya, etc. Eating the quality of such fruits improves after ripening due to the degradation of complex carbohydrates to simpler ones. These possess increased ethylene evolution and increased ethylene evolution until they are fully ripened. Therefore, postharvest management is necessary for such fruits to prolong their shelf life. Non-climacteric fruits do not ripen after harvest and include orange, litchi, grapes, pomegranate, cashew, etc. A thorough study of the interrelationship between maturity indices and subsequent ripening stages of fruits and vegetables and its application can help in eradicating postharvest losses. There is a need of ND methods to detect maturity indices and ripening on continuous grading-sorting lines. The development of ND sensors for ripening measurement can be the best alternative to ensure rapid application and have reliable results about the ripening stages of fruits and vegetables. Also, early-in field assessment of ripening using these sensors will greatly contribute to the reduction of postharvest losses, economic losses, and achieving food security.

2.3 NON-DESTRUCTIVE (ND) SENSORS FOR RIPENESS MEASUREMENT

The development of nondestructive sensors began with light transmittance techniques during the 1950's owing to the large sample size, variability in methods, and the sluggish process involved in the estimation of ripeness and maturity of fruits and vegetables. Though, the principle is universally applicable for all fruits and vegetables but with time, other techniques also involved based on texture and ethylene evolution rate and the use of a combination of such sensors might bring out the most reliable results of ripening indication. Texture and ethylene evolution has emerged out more accurate indication because the color may sometimes lead to erroneous conclusions about ripening as some fruits do not undergo chlorophyll degradation in concurrence with the progress of ripening stages.

2.3.1 FIRMNESS

2.3.1.1 IMPACT TEST-BASED SENSORS

Several researchers during 1980 have experimented for ND measurement of fruit firmness based on impact force of fruits like apples, pears, tomatoes (De Baerdemaeker et al., 1982; Delwiche, 1985). Firmness measurement of fruits and vegetables by impact testing was initially carried out to detect bruising (Homer et al., 2010) and is based on hertz's impact theory which is based on consideration of fruit as an elastic body and includes calculation of the compressions and tension in impact area by hitting the fruit with some element attached with the sensor or putting the fruit over a load cell and letting fruit fall over it or putting load cell beneath the plate containing fruit (Garcia-Ramos et al., 2005). The impact sensors can also be categorized as hammer impact or falling impact test sensors as reviewed by Mireei et al. (2015) and had been used for firmness studies of several fruits. The ones with piezoelectric accelerometer and impacting mass falls into the category of hammer impact sensors. In the case of falling impact test, the fruit is allowed to fall on force-sensitive surface; a load cell and energy of fall are transmitted onto the surface of the transducer and subsequent energy changes are correlated with the firmness of fruits (Lien et al., 2009).

ND impact test apparatus was developed by Lien et al. (2009) for assessment of tomato maturity. A pneumatic vacuum sucker holds the fruit and then releases it to fall on the load cell made of stainless steel. The height of fall and vacuum pressure may be standardized to not lead to bruising of fruit. A load cell is a piezoelectric transducer, which generates an analog signal proportional to force exerted by fruit. The analog signals may be amplified and recorded on digitizer (Figure 2.1.). A modified version of this for optimizing the online firmness determination of kiwifruits was developed by Ragni et al. (2010). It consisted of a conveyor belt which throws the fruit onto the load cell. Micro-accelerometers were fixed on the long and short edges of the plate. The signals from the load cell were then amplified and digitized by the acquisition board in PC. ND sensors have been commercialized by few companies like Sinclair International Ltd. of the United Kingdom which developed online firmness sorting tester based on low mass impact sensor under brand name Sinclair IQ^T firmness tester. The sensor head consists of a piezoelectric sensor surrounded by a rubber bellow activated by compressed air. The sensor moves vertically over the fruit (velocity 10 fruits/sec.).

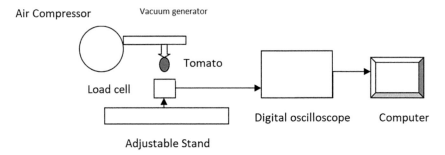

FIGURE 2.1   Experimental set up for impact force testing for tomatoes.

Geldermalsen, Netherlands-based GREEFA also developed intelligent firmness detector branded as iFD (Arana, 2012). This consisted of a large wheel with many sensor heads which impacts the fruits and rotates over the packaging line capturing up to 20 measurements at a time (velocity 7 fruits/sec.). Homer et al. (2010) evaluated the use of ND impact sensor for online fruit firmness measurement of peaches, apples, and pears. The impact sensor consisted of an optical sensor to detect the presence of fruit, an electromagnet, and spring to hold and release the impacting mass; and piezoelectric accelerometer to impact the fruit for firmness measurement. Fruit firmness index can be interpreted from the acceleration (m/s)-time (ms) curve (Figure 2.2) which shows higher acceleration for hard fruits and opposite for softer ones as per previous studies. The curve shown is a general curve not for any specific commodity. Ranatunga et al. (2009) evaluated firmness of tomatoes using Digital HPE II-FFF digital model firmness tester, Qualitest, USA. The anvil of the instrument does not penetrate the peel of the fruit. It consists of a flat-ended cylindrical test probe having a cross-sectional area of 0.25 cm². Firmness is indicated as units on the basis of conversion from kg/N; the maximum calibrating force 12.5 N corresponding to 100 units. Qualitest firmness sensor does not employ any piercing into a commodity and mainly pulp or peel determines the firmness.

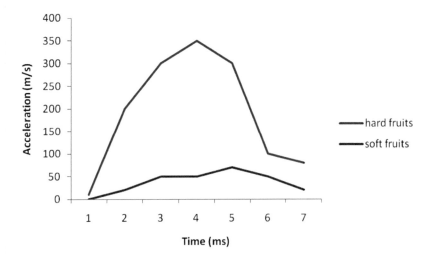

FIGURE 2.2   A typical acceleration (m/s) vs. time (ms) curve using an impact sensor.

Madrid Polytechnic University also developed a low-mass impact sensor consisting of spherical low mass (10 g) to impact the sample; a spring to release the impacting mass, an electromagnet to hold the impacting mass and a piezoelectric accelerometer of sensitivity 1 mV/ms^{-2}. Maximum acceleration in ms^{-2} is noted as a fruit firmness index.

2.3.1.2   ACOUSTIC IMPULSE SENSORS

Interests in using natural frequencies or impulse-response techniques of fruits for correlating with firmness begin after the introduction of ND acoustic impulse-response by Yamamoto et al. (1980). Fruit firmness may also be predicted by changes in the viscoelastic properties of the fruit surface. This kind of sensor does not require any contact or impose any pressure on the surface of fruits. Instead, the vibrators induce a response by the fruit at a specific frequency which is known as resonance frequency (Imran Al-Haq et al., 2004). Contactless transducers, i.e., microphones or the laser doppler vibrometer (LDV) or contact transducers, i.e., accelerometers may be attached to such sensing devices close to the fruits (Zhang et al., 2018) for measuring resonance frequencies which are correlated with the firmness of fruits. Several researchers have designated different terms like stiffness factor, elasticity index, transmission velocity, etc. for firmness prediction from the resonance frequency. In 1997, the acoustic impulse response technique was optimized by Langenakens and others. The modeling analysis showed that the best signals are recorded along the equator of fruit. The equipment designed consisted of a polystyrene block for holding the tomato and a small condenser microphone in the hole above the opening in the block. Best results were achieved by placing the tomatoes sideways along the equator. The firmness as first introduced by Abbott et al. (1968) was noted as stiffness factor derived from the equation below:

$$S = f^2 m^{2/3}$$

where, S is the stiffness coefficient ($kg^{2/3} s^{-2}$), f the dominant frequency where response magnitude is the greatest (Hz) and m the fruit mass (g).

Schotte et al. (1999) illustrated this using a frequency spectrum (amplitude vs. resonance frequency) for tomatoes. The highest peak amplitude was considered for the stiffness factor. For fruits of non-uniform size, transmission velocity has been found more effective firmness measurement method as the vibrations produced by the impact is transmitted in all directions on the surface of the fruit. Sugiyama et al. (2005) used a portable firm tester of Toyo Seiki Co. Ltd., Tokyo which consists of impact rod, two microphones, amplifier, and personal computer (PC). The impact rod induces vibrations in the sample which are then transmitted all over the surface of the fruit. Microphones then detect the vibrations as sound signals which are amplified using an amplifier and recorded in PC. Higher transmission velocity (m/s) entails proper ripening of fruit as observed in 'Keitt' mangoes (58 m/s) while lower velocities indicate any decay or spoilage in fruits. Similarly, the resonance frequencies decrease during maturation and subsequent ripening (Muramatsu et al., 2000). Another contactless transducer for measuring non-contact vibrations of fruit surface being used by researchers nowadays is LDV. The laser beam is directed at the fruit's surface and the vibration amplitude and frequency are extracted from the Doppler shift of the reflected laser beam frequency due to the motion of the surface. The output is directly proportional to the target velocity component along the direction of the laser beam. The amplitude and frequency shifts are extracted using a fast Fourier transform method. LDV has an advantage over accelerometer that it can target too small surfaces. The technique has been used to non-destructively evaluate the quality of Hosui pears, watermelons, pears, mandarin, and kiwifruit by Zhang et al. (2014); Abbaszadeh et al. (2015); Muramatsu et al. (2000); Taniwaki et al. (2009); Wang et al. (2006); and Terasaki et al. (2001).

2.3.1.3 MICRO DEFORMATION SENSORS

Small deformations using small plunger or jet of air aiming at deformation up to few mm's of fruit surface have been developed for ND firmness evaluation. The curve is produced by applying a small load for a fixed period of time (Macnish et al., 1997) or by calculating the force necessary to reach a pre-set deformation (Fekete and Felföldi, 2000). This ND technique (also known as micro-deformation) has led to the development of a number of force-deformation devices. Agro Technologie, UK developed microdeformation sensor under the trade name 'DUROFEL' based on flat-ended probe specific for fruits like tomato, cherries, and hard fruits. Firmness is determined by the penetration of probe into a fruit sample and is expressed as durofel index ranging from 0 to 100 (Agro-Technolgie, 2018). The ratio of the force applied to the fruit and resulting deformation is correlated with the fruit firmness. Both the analog (equipped with a spring and a gauge) and digital version (electronic sensor, display, and computer interface) have been widely used with apricots (Jay et al., 2000), tomatoes (Planton, 1991), cherries (Clayton et al., 1998) and other soft fruits. McGlone et al. (1999) established a non-contact laser air puff instrument for impact deformation of fruits. The instrument consisted of a compressed air tank, laser displacement sensor, glass plate, nozzle, and cradle. The compressed air from the tank flows through the tubing to the output nozzle having a diameter of 20 mm. The laser displacement sensor was located above the output nozzle and laser beam passes through the glass plate down to the nozzle and then onto the fruit along with puff of air. The cradle holds the fruit kept below the nozzle and prevents any sideways movement of fruit. A typical deformation curve observed is depicted in Figure 2.3.

The highest deformation indicates the hardness in fruits, i.e., unripe, and lowest deformation values indicate the softness in fruits which may be ripe or overripe. The principle has been utilized to determine harvest effects as well as firmness for several fruits including apples (Prussia et al., 1994), peaches (Maw et al., 2003), and blueberries (Changying et al., 2011).

Garcia-Ramos et al. (2005) reviewed certain micro deformation methods for firmness estimation like one developed by CEMAGREF, Agricultural equipment, and food process engineering division, France (Steinmetz et al., 1996) in collaboration with French enterprise, Caustier. A flexible positioning cup with a contact plunger (a probe with a sphere at the end) in the center helps the operator to slightly deform the fruit surface (maximum 2 mm approx.). A spring then shows the firmness index on a scale.

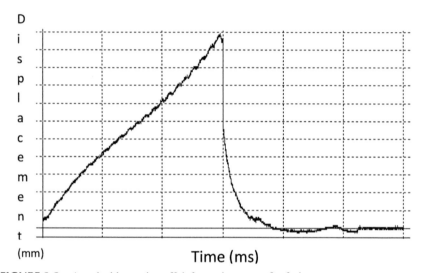

FIGURE 2.3   A typical laser air-puff deformation curve for fruits.

2.3.2   *COLOR*

2.3.2.1   *COLORIMETER*

In fruits and vegetables, ripening can be predicted by characteristic changes in color by the human eye which contains three kinds of color receptors corresponding to primary hues, i.e., red, blue, and green. Tristimulus colorimeters are widely accepted and applied for color measurement for quality or ripening prediction of fruits and vegetables. The most commonly used color systems are the CIE (Commission Internationale de' Eclairage) devised in 1931, the Hunter Lab developed in 1948 for photoelectric measurement to provide more uniform color differences in relation to human perception of differences (Abbott, 1999). The CIE system depicts the color- X, Y, and Z which are tristimulus values representing three primary colors- red, green, and violet. The chromaticity coordinates-x, y, and z is obtained by dividing each of the tristimulus values by the addition of three. These coordinates correspond to the total stimulus-specific to each primary color. The chromaticity diagram is a horseshoe-shaped curve from monochromatic light and for unripe fruits, chromaticity falls in the center of this curve. As the fruit ripens, the central point moves from hueless to reddish zone (Jha, 2010). In CIE lab color systems, L* indicates lightness which ranges from 0 (i.e., "black") to 100 (i.e., "white"). A + a* value indicates redness (a* is greenness) and + b* value yellowness (b* is blueness) on the hue-circle (Hutchings, 1994; Voss, 1992). In hunter colorimeter, L denotes lightness, a denote redness or greenness whereas b depicts yellowness or blueness. The L color stimulus function of hunter system is comparable to Y; a values are functions of X and Y; b values are functions of Z and Y (Pomeranz, 2000). CIE lab color systems and hunter colorimeter are similar in organization, with different numerical values. Both are mathematically derived from X, Y, Z values. Neither the hunter scale uniform as it is over expanded in blue region of neither color space nor CIE scale which is over expanded in yellow region. However, the current recommendation of CIE is to use L*, a*, b*. The instrument consists of light source, sample plane, filters, and detector. The light source should have sufficient radiant energy of visible spectrum. Mostly interference filters and gratings are used of bandwidth 10 nm. Tungsten filaments and xenon discharge lamps are generally used (Randall, 2010). Colorimeter may be having diffuse (sphere) geometry or directional one. The directional geometries have illumination at 45 and 0° angle (Jha, 2010).

The colorimeter CIE system and hunter values are widely used for determining the maturity and ripening stages or pigment changes during maturity. The L* value has been observed as a useful indicator of darkening during ripening of fruit on the tree and during storage which might resulted from either oxidative browning reactions or increasing pigment concentrations (Lysiak, 2012). The a* value is generally considered as a measure of greenness, and is observed to be highly correlated with color changes of apple flesh (Goupy et al., 1995). Gonclaves et al. (2007) measured the ripeness of four sweet cherry cultivars and observed that Burlat cherries which were redder and darker in color showed lower hue angle and L* whereas Summit fruits were less red hence showed higher L* and chroma values. In addition, L*, chroma, and hue angle of partially ripe cherries were always higher than in the ripe cherries, which means a less red fruit, and was correlated with lower anthocyanin content (Gonclaves et al., 2004). Golden delicious apples when evaluated for ripening by CIElab values resulted in an increase in L* values till green to bright green color but increased on the appearance of yellow color of apples or after complete ripening. Itle and

Kabelka (2009) also correlated increase in a* and b* values with decrease in chlorophyll content during ripening of climacteric fruits. Some fruits show degradation in pigments during ripening and there may be no significant changes in color of peel and pulp at the end of ripening. Ferrer et al. (2005) observed minimal difference in b* values of skin or pulp of peaches at the end of ripening because of increased content of yellowish pigments as compared to significant differences observed initially after harvest. Ruslan and Roslan (2016) observed steady increase in a* values during ripening of papaya.

The chroma can be calculated using the following equation:

$$c^* \sqrt{a^{*2} + b^{*2}}$$

where, c^* is the chroma; a^* and b^* are the color values in CIE L*a*b* color space.

The total change in color between samples on the first and the last day of ripening experiments can be calculated by ΔE (as reported by Nagle et al., 2016).

$$\Delta E = \Delta L^{*2} + \Delta a^{*2} + \Delta b^{*2}$$

where, L^*, a^*, and b^* are the color values in CIE L*a*b* color space.

2.3.2.2   FLUORESCENCE SENSORS

Fluorescence can be used to measure the degradation of chlorophyll which can be correlated to fruit ripening stages. Several researchers worked on the application of fluorescence since 1990's for monitoring the ripening process of different commodities like apricots, apples (Guidetti et al., 1998), lemons (Nedbal et al., 2000), oil palm (Hazir et al., 2012). Chlorophyll a molecules (photosystem II) possess fluorescence characteristics and when it is maximum referred to as F_m (a fraction of excited electrons are passed) and when it is minimum considered as F_o (all excited electrons are passed on). The difference among the two determines the chlorophyll activity (Bron et al., 2004). Fruit ripening may affect chlorophyll or lead to degradation which can be detected by fluorescence. However, certain other stress factors might be responsible for fluorescence effect, therefore, a combination of ripening measurement methods like color, size along with fluorescence can be a better prediction tool for ripening. Bodria et al. (2004) measured chlorophyll fluorescence (ChlF) of apples, peaches, and nectarines. The experimental set up consisted of a dark chamber, CCD monochrome camera to capture fluorescent images,

UV-visible light source or ultra-bright LED's, low pass or high pass filters and holder to keep the sample as shown in Figure 2.4. The fluorescence intensity as well as firmness of fruits decreases with an increase in ripening. FORCE-A, Orsay, France commercialized fluorescence-based optical sensor "Multiplex" (Ghozlen et al., 2010). The advantage of this sensor over the previous ones is its working even in daylight conditions. Its working resides in light-emitting diode excitation and filtered photodiode detection. The sensor possesses six UV-light sources and three Red-Blue-Green LED matrices emitting lights at wavelengths 470, 516, and 635 nm. Single measurement consisted of several flashes of four colors-UV, blue, green, and red, and the sensor results in a set of chosen ratios. The intensity of fluorescence emission depends upon the amount of excited light reaching the chlorophyll. The calculation regarding the measurement of simple chlorophyll fluorescence ration (SFR) is as follows:

$$SFR\_R = FRF\_R/RF\_R$$

where, FRF is far-red emission divided by (FR) red emission. The ration FRF/FR increases with decrease in chlorophyll content. Ghozlen et al. (2010) and Betemps et al. (2012) used this sensor for measuring grape and apple ripening and significant correlation between the SFR chlorophyll index and technical maturity for all three cultivars of grapes and apple fruits was observed. Hansatech, UK also commercialized pulse amplitude-modulated chlorophyll fluorometer which uses a rapid pulsing excitation light for pulsed fluorescence emission. It uses highly sensitive photodiode to detect the pulsed signals. The optical filters prevent the detection of non-fluorescence light wavelengths if ambient light is applied. Maximal, minimal, and variable ChlF (F_o, F_m, and F_v) is measured which indicates the ripening stage of fruits. With increase in ripening the values of F_v/F_m as also observed by Bron et al. (2004) in golden papaya fruits.

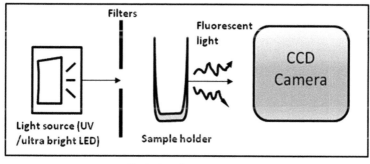

FIGURE 2.4 A laboratory set up for fluorescence sensor for fruits and vegetables.

2.3.2.3 HYPERSPECTRAL IMAGING SENSORS

Hyperspectral imaging has developed as an emerging process analytical tool for food quality and safety control. Hyperspectral imaging, also called imaging spectroscopy or imaging spectrometry has been introduced to integrate both spectroscopic and imaging techniques. Multispectral imaging is a simplified version of hyperspectral imaging, the difference is in the number of bands involved which is >100 for HSI, for MSI it is usually less than 10 (Elmasry and Sun, 2010). It is the simultaneous acquisition of spatial images in many spectrally contiguous bands measured from a remotely operated platform (Schaepman, 2007). A spectral image is a stack of images of the same object, each at a differential spectral narrow band. The light source interacts with the food sample. The detected portion contains both physical and chemical information of sample will be dispersed and projected on to the detector. The imaging spectrograph covers both visible and near-infrared (NIR) region. The images are then combined and form a three-dimensional (3D) hyperspectral cube also known as spectral cube, data cube, or spectral volume (Gowen et al., 2007) (Figure 2.5).

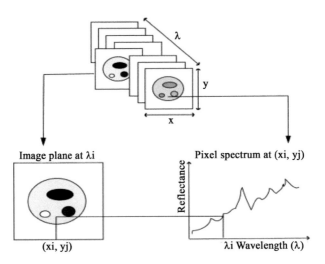

FIGURE 2.5 Combination of spectral and computer vision images to form hyperspectral cube. (Reprinted with permission from Gowen et al., 2007. © Elsevier.)

No single wavelength can describe the object completely, that is why hyperspectral imaging is useful in the analysis of an object (Elmasry et al.,

2012). Spectroscopy is a promising tool to detect the internal quality of fruits or vegetables and thus make it feasible for ripeness evaluation. HIS has been successfully used to determine the ripeness of several fruits like Astringent persimmon (Wei et al., 2014), strawberries (Tallada et al., 2006), pears (Khodabakhshian and Emadi, 2017), peaches (Lleo et al., 2011), etc. The most commonly used spectral ranges include: 400–1000 nm and 900–1700 nm. The superimposed spectral and computer vision images or HSI images contains many pixels and each pixel within this image shows the characteristic spectral characteristic of that pixel. The spectrum can be used to predict the physical, chemical, or specific information of the pixel by certain established models.

The set-up of acquisition of HIS images may be continuous and includes: conveyor belt for sample holding (moving speed and exposure time can be adjusted), spectrograph (for scanning wavelengths), CCD camera (to capture images) and tungsten halogen lamps (for illumination) (Zhang et al., 2016). The images can be acquired by scanning along the Y-axis of the sample moving along the X-axis at a certain speed. For optimum wavelength selection, principal component analysis (PCA), and certain models like support vector machine (SVM), chemometric can be used. Baiano et al. (2012) studied the spectra of a particular region of interest to predict the fruit quality of grapes. The average spectra from the whole wavelength scan (400–1000 nm) were used with a PLS model to predict the soluble solids content of grapes. Similarly, Liu et al. (2014) studied the same method to predict the TSS of strawberries and the firmness and SSC of blueberries (Leiva-Valenzuela, 2014). Selection of key wavelengths prior to modeling has also been studied, with different feature selection techniques, to reduce the redundancy of the whole spectral dataset.

2.3.3 ETHYLENE

Primary objective for shelf-life extension of fruits and vegetables is to control the atmospheric conditions. The shelf-life has been enhanced by many folds using transportation containers specifically designed and manufactured for the perishable goods (Saltveit, 1999). These containers offer temperature control, using refrigeration methods for optimum temperature maintenance during the transport. Recent developments involve the use of edible coatings and modified atmospheric packaging

(Tanner et al., 2003). However, these packaging methods are limited to specific fruits and vegetables and require technical expertise for implementation. Therefore, in the general typical supply chain of fruits and vegetables is still exposed to the inherent as well as extrinsic threats leading to spoilage and loss. The key factors to enhance shelf-life postharvest of fruits and vegetables along with low temperatures are maintained humidity, carbon dioxide, and ethylene gas. Ethylene is directly correlated with the freshness of fruits. Fruits consume oxygen and respirates, while releasing by product ethylene and carbon dioxide gas. It is essential to keep a track on levels of ethylene throughout the supply chain to insure the freshness of goods. Currently, most of detection methods are limited with requirements of stationery lab setups. However, the development of novel gas sensors has made it possible to detect the ethylene gas in real-time, which also offers their implementation throughout the supply chain. This section caters to a comprehensive discussion on methods of detection and quantification of ethylene sensors.

2.3.3.1 *GC-MS BASED SENSORS*

Ethylene gas is a phytohormone, associated with the ripening of fruits. 1 ppm ethylene gas by volume is sufficient enough to initiate the ripening process. Therefore, it is essential to use analytical tools which are capable enough to resolve concentration with a factor of 50 ppb at least. Gas chromatography (GC) becomes critical in separating ethylene gas from air and offers enhanced resolutions in quantification of concentrations when coupled with mass spectrophotometry (Zaidi et al., 2016). Despite the advantages of lower detection limits, GC-MS is non-mobile and expensive instrument. The main parts of GC system are carrier gas reservoir, pressure regulator, flow controller, rotameter, flow splitter, sample injector port, column in column oven, detector (in this case mass spectrophotometer), electrometer/bridge, analog-digital convertor, and a computer system to analyze the data, as illustrated in Figure 2.6. GC is in principle like column chromatography. The mobile phase is a carrier gas flowing with a fixed rate through the system. The flow rate is column specific. The two types of column (Figure 2.7) used in GC are:

1. Capillary (column thin film on the wall); and
2. Packed column it can be (small bead-like particles filled into the column).

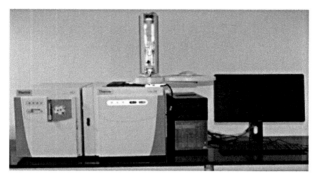

FIGURE 2.6   Gas-chromatography mass spectrophotometery system.

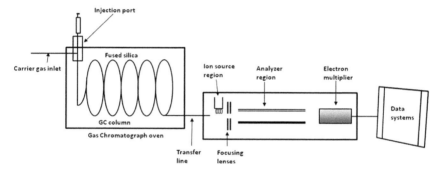

FIGURE 2.7   Block diagram of a typical capillary GC-MS system.

The sample gas mixture is injected through the sample injection port. In GC-MS, carrier gas should be inert to the sample gas for example Helium and Nitrogen. Separation of ethylene takes place because of the difference in passing time (retention time) in the chromatographic column. Specific gas has a specific retention time, thereof giving specific peak as an output signal which is used to identify the gas. The intensity and time at which the output signal is observed is used for quantification of ethylene concentration. The column is packed in a temperature-controlled oven and through this column; gas or mobile phase passes through. The concentration of a compound in the gas phase is solely a function of the vapor pressure of the gas. The injector can be used in one of two modes; split or spitless. Gas sample is introduced through a syringe in microliters into a heated small chamber to facilitate volatilization of the sample. The vaporized sample/carrier gas mixture then either passes through as such or as portion into the column. Former one is referred to as splitless while latter one is referred to as split mode. Split injection is preferred when working with samples with high analyte concentrations (>0.1%). Spitless injection is best suited for trace analysis with low amount of analyte (<0.01%) (Zaidi et al., 2016).

2.3.3.2   GAS IN SCATTERING MEDIA ABSORPTION SPECTROSCOPY (GASMAS)-BASED SENSORS

MS detector is a universal detector, which gives mass spectrum of the analyte. Mass spectrum enables analyst to obtain both structural a quantitative data. It can detect concentration as low as 1 ng per scan. The ions from the mass analyzer impinge on a surface of a detector where the charge is neutralized, either by collection or donation of electrons. Electric signals generate current which is further amplified and converted into a signal that is processed by computer. Detector can be selective 'selected ion monitoring (SIM)' or non-selective 'total ion current (TIC).' TIC is the sum of the detected current of all the ions being at any given moment. SIM is used to look for selective ions which are characteristic of a target compound or family of compounds. Certain algorithms are needed to set up to carry out tandem MS on components eluted from the GC column. Algorithms can be adapted to enable selective detection of many types of gases including ethylene gas (Zaidi et al., 2017). SIM selectively detects very low fractions of ethylene gas in sample gas mixture. The hurdle in the measurement of ethylene is lower v/v share in the mixture. Therefore, it shows similar retention times to vaporized water for several stationary phases. Therefore, for the separation, either a very long column should be used, or column must be heated to get the best separation. Recently, few μ-columns have been developed for mobile applications with onboard chips and platinum heater at the bottom of the column. A well-suited example is Carbosieve SII (CSII) system. Its significance and application are similar to GC-MS. However, for the efficiency of the system pre-concentrator device is used. Realization of recent advancements of the large-capacity-on-chip pre-concentrator device, the detection limit of ethylene measurement was found to be increased as compared to earlier devices.

2.3.3.3   ELECTROCHEMICAL SENSORS

Electrochemical sensors offer detection in the levels of ppm (0–10) with a resolution of 0.1 ppm. The basic setup of the electrochemical sensor is illustrated in Figure 2.8. It includes entrapment filters to filter out water vapor and dust. Passed gas through the filter reaches the sensor through small capillaries, using the principle of diffusion of a gas. The filter assembly includes an anti-condensation filter and hydrophobic membranes, through which gas diffuses to the capillaries to the detector chamber. Detector chamber is often a plastic housing, which has three electrodes present in the sensor assembly, i.e., a reference electrode, a sensing electrode, and a counter electrode.

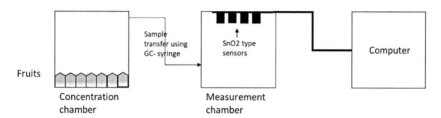

FIGURE 2.8   Electrochemical gas sensor.

Gas sensed is a measure of oxidation or reduction due to gas molecules producing or consuming electrons. The net ionic current produced reaches the counter electrode through the electrolyte present in the sensor. Conversion of current in voltage takes place during chemical reaction with ethylene. Transimpedance configuration is used to make measurements via op-amp IC1. To balance the current required by the sensing electrodes IC2 provides the current to counter electrode while C2 reduces high-frequency noises. JFET is used to ensure that both the sensing electrode and reference electrodes are running at the same potential. Before testing, calibration of the sensor is achieved using commercial ethylene gas of 10 ppm. The required concentration may be diluted by mixing the concentrated ethylene gas with synthetic air in a gas sample bag, in case of non-availability of 10 ppm concentration. The sample bag is a multi-layer foil composed of two outer layers of aluminum film providing a great barrier to gases permeating through walls of the bag offering moisture protection. It is vital that a gasbag should offer a shield from light and should be chemically inert with moisture (Barsan et al., 2013). Different concentrations of pure ethylene gas, viz., 10 ppm, 5 ppm, 2.5 ppm, 1.25 ppm is required for calibration. Once the gas is mixed and concentrations are achieved bags are stored for 12 h for homogeneity before the use in the system. The gas needs to be extracted and compressed from the gas bags for sensor calibration. The extraction system is shown in Figure 2.8. The extraction unit has a pump and a pressure chamber with two ports. One port generates pressure in the chamber while other port takes ethylene gas in the chromatographic system. The barometer and mass flow controller are used to maintain ambient flow. When the system is turned on, the pump draws air from the environment (Elmi et al., 2008). The drawn air is discharged into the chamber generating a positive pressure, as the chamber is already sealed. When the pressure reaches, 0.2 bar it squeezes the gas sample bag inside it and causes the ethylene to emerge from the gas sample bag (Janssen et al., 2014). Constant flow is maintained and controlled through the mass flow controller. Flow to the sensor is limited at a flow rate of 30 L/min. It takes about 15 min to reach a stable baseline with ambient air. Sensor electronics are calibrated with the sensor using the two potentiometers on the circuit board. 'Zero' potentiometer is adjusted to produce 40 mV as output signal against a known concentration of ethylene gas (Brezmes et al., 2000). Once the signal is, stable the second potentiometer 'span' is adjusted with the gain of 10 to produce 2 V as the output signal. The two potentiometers on the circuit board are used to calibrate the sensor electronics together with the sensor. The potentiometer "zero" is adjusted to produce an output signal of 40 mV. A maximum gas concentration of 40 mV was used and when a stable signal was achieved, the potentiometer "span" was adjusted with the gain of 10 to produce an output signal of 2 V.

2.3.3.4   FLEXIBLE TAG MICROLAB (FTM)

A visionary application of flexible tag microlab (FTM) coupled with chemical and physical sensors for monitoring fruit during transport and vending proposed and developed in the FP6 Integrated Project "Good-Food." This is a very useful integration of radio frequency identification (RFID) devices with sensors, in which detection data is gathered and processed using a wireless communication device. Abad and Estefania (2007) have defined it as "The idea for this device is a flexible label, hosting different sensor technologies and a RFID interface for wireless data exchange, capable of monitoring the quality of food during transport, storage, and vending." Basically, the analytical part is based on the sensors discussed is this book chapter whose usability, mobility, and applicability is enhanced by many folds with the help of wireless data exchange devices as RFID.

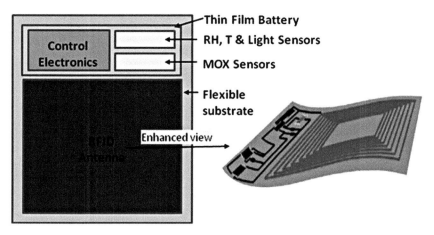

FIGURE 2.9   Functional block diagram of flexible tag microlab.

Currently, these FTM (Figure 2.9) devices allow monitoring of not only the ripening of fruits but also the factors affecting its freshness throughout the logistic supply chain. These factors may include physical variables such as light and temperature intensity and some chemical volatile variables like gases and ethylene (Opasjumruskit et al., 2006). The measurements can be automated to be done at regular intervals in parlance the data may further be stored in storage devices, which onset arrival of a shipment can be analyzed using computers. This data would help to understand and pinpoint the related events leading to the deterioration of the quality of fruit and its safety. Recently, several attempts are made to integrate gas sensors and temperature sensors (Besar et al., 2017) on commercial tags for monitoring of fruits. In general, a metal gas oxide sensor is sufficient in terms of stability and sensitivity as it falls due to the limits of detection. Initial developments of these sensors were restricted to the onboard application of the reader because of higher power consumptions but with low power requirements, battery-operated FTMs are recently developed. Four different coatings for the gas sensors consisted of WO3 + 1% wt Pt, SnO_2 + 1% wt Pt, SnO_2 + 1% wt Pd and SnO_2 + 1% wt Au are chosen depending on the requirements of heating temperature and gas selectivity (Espinosaa et al., 2010). IMM Bologna has developed ultra-low power consumption hotplates (ULPH) for metal oxide sensor integration to FTM substrates. With ULPH hotplates, only 8 mw is required for a 4-element sensor array of 1.5×1.0 mm^2 die to reach an operating temperature of 400°C. Since, we analysis is taking place at higher temperatures metal oxide sensor hotplates require certain assembling methods and protection of the chips from the environment.

2.3.3.5 E-NOSE SENSORS

Many methods for ethylene quantification as a relative measurement to the ripening index have been discussed in this chapter. However, the implementation of most of them to the farms and storage is limited because of the lack of mobility and cost required to meet the installation and running of facilities. E-nose sensors have started gaining a boom among the industrialist. This popularity is a result of the advantages over regular methods involving the compact form factor and inexpensive instruments required with ease in operation. This method is not only cost-effective and ND but also offers real-time screening (Brezmes et al., 2000). The prime components of e-nose sensors are sensor module in the housing of the measurement chamber and computer running data programs. It also requires a concentration chamber for the collection and concentration of headspace gases. A common example of an e-nose sensor used for detection and quantification of ethylene is tin oxide gas sensors. Fruits are placed in the concentration chamber, sampling is done in fixed intervals (1 h) and 150 ml of vapor sample is collected from the headspace, using a chromatographic syringe. The collected vapor gas sample is then injected into the sensor chamber, where modified conductance of tin oxide sensors is measured as a function of a sudden change in concentration (Vergara et al., 2007). The changes are recorded using programs in computer and data is analyzed. The changes are recorded till the time of stabilization (10 min or later). Pattern recognition techniques are further applied using conductance increment measured by the sensor. It is essential to flush both the collection chamber and the analytical chamber with synthetic air as a cleaning process (Simon, 1996). The concentration chamber enhances the limit of the detection of ethylene by manifolds and has a significant advantage. Recently, e-nose sensors have been successfully used for measurement of ripeness of peach, pear, grapes, and oranges. However, it is still an emerging technology, which requires further research work for universal applications.

2.4 CONCLUSION

Climacteric fruits continue to ripen even after harvest, so they must be picked up at the correct maturity stage to prevent loss of quality at the point of consumption. Looking towards the need of producers or retailers involved in the fruits and vegetables market for ND methods for ripeness measurement, several sensors have been developed, experimented, and commercialized by few companies. However, of all the above-described sensors, the best and reliable ripening measurement results could be achieved by integrating one or two as single quality characteristic can't predict accurate ripening stage. This can be achieved alternatively either by optimization of sensors for individual fruits category or development of sensor applicable to the large category of the fruits or vegetables.

KEYWORDS

- flexible tag microlab
- laser Doppler vibrometer
- radio frequency identification
- selected ion monitoring
- simple chlorophyll fluorescence ration
- ultra-low power consumption hotplates

REFERENCES

Abad & Estefania, (2007). Flexible tag micro lab development: Gas sensors integration in RFID flexible tags for food logistic. *Sensors and Actuators B: Chemical, 127*(1), 2–7.

Abbaszadeh, R., Rajabipour, A., Ying, Y., Delshad, M., Mahjoob, M. J., & Ahmadi, H., (2015). Nondestructive determination of watermelon flesh firmness by frequency response. *LWT-Food Science and Technology, 60*(1), 637–640.

Abbott, J. A., Bachman, R. F., Fitzgerald, J. V., & Matusik, F. J., (1968). Sonic techniques for measuring texture of fruits and vegetables. *Food Technol., 22,* 101–112.

Al-Haq, M. I., Sugiyama, J., Tomizawa, A., & Sagara, Y., (2004). Detection of manure effect on texture of muskmelon by non-destructive acoustic firmness tester. *Hort. Science, 39*(1), 142–145.

Arana, I., (2012). *Physical Properties of Foods: Novel Measurement Techniques and Applications* (1st edn., p. 57).

Baiano, A., Terracone, C., Peri, G., & Romaniello, R., (2012). Application of hyper spectral imaging for prediction of physico-chemical and sensory characteristics of table grapes. *Comput. Electron. Agric., 87,* 142–151.

Bârsan, N., & Weimar, U., (2013). Understanding the fundamental principles of metal oxide based gas sensors; the example of CO sensing with SnO_2 sensors in the presence of humidity. *J. Phys. Condens. Matter, 15,* R813.

Besar, K., Dailey, J., & Katz, H. E., (2017). Ethylene detection based on organic field-effect transistors with porogen and palladium particle receptor enhancements. *ACS Appl. Mater. Interfaces, 9*, 1173–1177.

Betemps, D. L., Fachinello, J. C., **Galarça, S. P.**, Portela, N. M., Remorini, D., Massai, R., & Agati, G., (2012). Non-destructive evaluation of ripening and quality traits in apples using a multiparametric fluorescence sensor. *Journal of the Science of Food and Agriculture, 92*(9), 1855–1864.

Bodria, L., Fiala, M., Guidetti, R., & Oberti, R., (2004). Optical techniques to estimate the ripeness of red-pigmented fruits. *Transactions of the ASAE, 47*(3), 815.

Brezmes, J., Fructuoso, L. L., Llobet, E., Vilanova, X., Recasens, I., Orts, J., Saiz, G., & Correig, X., (2000). Fruit ripeness monitoring using an electronic nose. *Sensors and Actuators B: Chemical, 69*(3), 223–229.

Bron, I. U., Ribeiro, R. V., Azzolini, M., Jacomino, A. P., & Machado, E. C., (2004). Chlorophyll fluorescence as a tool to evaluate the ripening of golden papaya fruit. *Postharvest Biol. Technol., 33*, 163–173.

Changyang, L., Luo, J., & Maclean, D., (2011). A novel instrument to delineate varietal and harvest effects on blueberry fruit texture during storage. *Journal of the Science of Food and Agriculture, 91*, 1653–1658.

Clayton, M., Biasi, B., & Mitcham, B., (1998). New devices for measuring firmness of cherries. *Perishables Handling Quarterly, 95*, 2–4.

De Baerdemaeker, J., Lemaitre, L., & Meire, R., (1982). Quality detection by frequency-spectrum analysis of the fruit impact force. *Trans. ASAE, 25*(l), 175–178.

Delwiche, M. J., Tang, S., & Mehlschau, J. J., (1989). An impact force response fruit firmness sorter. *Trans. ASAE, 32*(l), 321–326.

ElMasry, G., & Da-Wen, S., (2010). Chapter 1: Principles of hyper spectral imaging technology. *Hyper Spectral Imaging for Food Quality Analysis and Control* (pp. 3–43). 10.1016/B978-0-12-374753-2.10001–2.

ElMasry, G., Kamruzzaman, M., Sun, D. W., & Allen, P., (2012). Principles and applications of hyper spectral imaging in quality evaluation of agro-food products: A review. *Critical Review in Food Science and Nutrition, 52*, 999–1023.

Elmi, I., Zampolli, S., Cozzani, E., Mancarella, F., & Cardinali, G. C., (2008). Development of ultra low power consumption MOX sensors with ppb-level VOC detection capabilities for emerging applications, Sens. *Actuators, B: Chem., 135*, 342–351.

Espinosa, E., Ionescu, R., Zampolli, S., Elmi, I., Cardinali, G. C., Abad, E., Leghrib, R., Ramírez, J. L., Vilanova, X., & Llobet, E., (2010). Drop-coated sensing layers on ultra low power hotplates for an RFID flexible tag micro lab. *Sensors and Actuators B: Chemical, 144*(2), 462–466.

Fekete, A., & Felföldi, J., (2000). System for fruit firmness evaluation. *Proc. Intnl. Conf. Agricultural Engineering.* Warwick, UK. Paper 00-PH-034.

Ferrer, A., Remon, S., Negueruela, A. I., & Oria, R., (2005). Changes during the ripening of the very late season Spanish peach cultivar Calanda feasibility of using CIELAB coordinates as maturity indices. *Scientia Horticulture, 105*, 435–446.

García-Ramos, F. J., Valero, C., Homer, I., Ortiz-Cañavate, J., & Ruiz-Altisent, M., (2005). Non-destructive fruit firmness sensors: A review. *Spanish Journal of Agricultural Research, 3*(1), 61–73.

Ghozlen, N. B., Cerovic, Z. G., Germain, C., Toutain, S., & Latouche, G., (2010). Non-destructive optical monitoring of grape maturation by proximal sensing. *Sensors, 10*(11), 10040–10068.

Gonçalves, B., Silva, A. P., Moutinho-Pereira, J., Bacelar, E., Rosa, E. M., & Meyer, A. S., (2007). Effect of ripeness and postharvest storage on the evolution of color and anthocyannis in cherries (*Prunus avium* L.). *Food Chem., 103*, 976–984.

Goupy, P., Amiot, M. J., Richard-Forget, F., Duprat, F., Aubert, S., & Nicolas, J., (1995). Enzymatic browning of model solutions and apple phenolic substrates by apple polyphenoloxidase. *J. Food Sci., 60*(3), 497–501.

Gowen, A. A., Donnell, C. P. O., Cullen, P. J., Downey, G., & Frias, J. M., (2007). Hyper spectral imaging-an emerging process analytical tool for food quality and safety control. *Trends in Food Science and Technology, 18*, 590–598.

Guidetti, R., Mignani, I., & Oberti, R., (1998). Image analysis for evaluation of fruit quality: Measurement of fluorescence as an indicator of maturity stage. *Italius. Hortus., 5*(6), 23–26.

Hazir, M. H. M., Shariff, A. R. M., & Amiruddin, M. D., (2012). Determination of oil palm fresh fruit bunch ripeness—based on flavonoids and anthocyanin content. *Industrial Crops and Products, 36*(1), 466–475.

Homer, I., García-Ramos, F. J., Ortiz-Cañavate, J., & Ruiz-Altisent, M., (2010). Evaluation of nondestructive impact sensor to determine on-line fruit firmness. *Chilean Journal of Agricultural Research, 70*(1), 67–74.

Hutchings, J. B., (1994). *Food Color and Appearance*. Blackie, London.

Janssen, S., Tessmann, T., & Lang, W., (2014). High sensitive and selective ethylene measurement by using a large-capacity-on-chip preconcentrator device. *Sens. Actuators B Chem., 197*, 405–413.

Jay, M., Lancelin, N., Lichou, J., Chapon, J. F., Guinot, E., & Brunninck, M., (2000). Apricot quality: analysis of the Goldrich and Hargrand varieties. *Infos. Ctifl., 161*, 34–38.

Jha, S. N., (2010). Color measurements and modeling. *Non Destructive Evaluation of Food Quality* (1ˢᵗ edn., pp. 17–40).

Khodabakhshian, R., & Emadi, B., (2017). *Application of Vis/SNIR Hyper Spectral Imaging in Ripeness Classification of Pear* (pp. S3149–31463).

Langenakens, J. J., Vandewalle, X., & Baerdemaeker, J. D., (1997). Influence of global shape and internal structure of tomatoes on the resonant frequency. *J. Agric. Engng. Res., 66*, 41–49.

Leiva-Valenzuela, G. A., Lu, R., & Aguilera, J. M., (2014). Assessment of internal quality of blueberries using hyperspectral transmittance and reflectance images with whole spectra or selected wavelengths. *Innov. Food Sci. Emerg. Technol., 115*, 91–98.

Lien, C. C., Ay, C., & Tingh, C. H., (2009). Non-destructive impact test for assessment of tomato maturity. *J. Food Eng., 91*, 402–407.

Liu, C., Liu, W., Lu, X., Ma, F., Chen, W., Yang, J., & Zheng, L., (2014). Application of multispectral imaging to determine quality attributes and ripeness stage in strawberry fruit. *PLoS One, 9*, e87818.

Lleó, L., Roger, J. M., Herrero-Langreo, A., Diezma-Iglesias, B., & Barreiro, P., (2011). Comparison of multispectral indexes extracted from hyper spectral images for the assessment of fruit ripening. *Journal of Food Engineering, 104*, 612–620.

Llobet, E., (1999). Non-destructive banana ripeness determination using a neural network-based electronic nose. *Meas. Sci. Technol., 10*, 538–548.

Łysiak, G., (2012). The base color of fruit as an indicator of optimum harvest date for two apple cultivars (*Malus domestica* Borkh.). *Folia Horticulture, 24*(1), 81–89.

Macnish, A. J., Joyce, D. C., & Shorter, A. J., (1997). A simple non-destructive method for laboratory evaluation of fruit firmness. *Aust. J. Exp. Agr., 37*, 709–713.

Maw, B. W., Krewer, G. W., Prussia, S. E., Hung, Y. C., & Mullinix, B. G., (2003). Non-melting flesh peaches respond differently from melting flesh peaches to laser puff firmness evaluation. *Appl. Eng. Agri., 19*, 329.

McGlone, V., Ko, S., & Jordan, R., (1999). Non-contact fruit firmness measurement by the laser air-puff method. *Transactions of the ASAE, 42*(5), 1391.

Mireei, S. A., Sadeghi, M., Heidari, A., & Hemmat, A., (2015). On-line firmness sensing of dates using a nondestructive impact testing device. *Biosyst Eng., 129*, 288–297.

Muramatsu, N., Sakurai, N., Wada, N., Yamamoto, R., Tanaka, K., Asakura, T., Ishikawa-Takano, Y., & Nevins, D. J., (2000) Remote sensing of fruit textural changes with a laser Doppler vibro meter. *J. Am. Soc. Hortic. Sci., 125*(1), 120–127.

Nedbal, L., Soukupova, J., Whitmarsh, J., & Trilek, M., (2000). Postharvest imaging of chlorophyll fluorescence from lemons can be used to predict fruit quality. *Photosynthetica., 38*(4), 571–579.

Opasjumruskit, K., Thanthipwan, T., Sathusen, O., Sirinamarattana, P., Gadmanee, P., Pootarapan, E., Wongkomet, N., Thanachayanont, A., & Thamsirianunt, M., (2006). Self-powered wireless temperature sensors exploit RFID technology. *IEEE Pervasive Comput., 5*, 1268–1536.

Planton, G., (1991). Tomate-mesure de la fermeté au durofel 25. *Infos. Ctifl., 74*, 17–20.

Pomeranz, Y., (2000). Measurement of color. *Food Analysis: Theory and Practice* (3rd edn., pp. 87–97). Chapman & Hall, London.

Prussia, S. E., Astleford, J. J., Hung, Y. C., & Hewlet, R., (1994). *Non-Destructive Firmness Measuring Device*. U.S. Patent 5,372,030.

Ragni, L., & Berardinelli, A., (2010). Impact device for measuring the flesh firmness of kiwifruits, *Journal of Food Engineering, 96*, 591–597.

Ranatunga, C. L., Jayaweera, H. H. E., Suraweera, S. K. K., & Ariyaratne, T. R., (2009). *Effect of Measurement of Non-Destructive Firmness on Tomato Quality and Comparison with Destructive Methods, 25*, 29–35.

Randall., D. (1997). Instruments for the measurement of color, in Color Technology in the Textile Industry, *American Association of Textile Chemists and Colorists,* 2nd edn, pp. 9–17.

Ruslan, R., & Roslan, N., (2016). Assessment on the skin color changes of *Carica papaya* L. cv. Sekaki based on CIE L* a* b* and CIE L* C* h color space. *International Food Research Journal, 23*.

Saltveit, M. E., (1999). Effect of ethylene on quality of fresh fruits and vegetables. *Postharvest. Biol. Technol., 15*, 279–292.

Schaepman, M. E., (2007). Spectrodirectional remote sensing: From pixels to processes. *International Journal of Applied Earth Observation and Geoinformation, 9*, 204–223.

Schotte, S., De Belie, N., & Baerdemaeker, J. D., (1999). Acoustic impulse-response technique for evaluation and modeling of firmness of tomato fruit. *Postharvest Biol. Technol., 17*, 105–115.

Simon, J. E., (1996). Electronic sensing of aromatic volatiles for quality sorting of blueberries. *J. Food Sci., 61*, 967–970.

Steinmetz, V., Crochon, M., Bellon, M. V., Garcia, F. J. L., Barreiro, P., & Verstreken, L., (1996). Sensors for fruit firmness assessment: Comparison and fusion. *Journal of Agricultural Engineering Research, 64*, 15–27. 10.1006/jaer.1996.0042.

Sudheer, K. P., & Indira, V., (2007). *Post Harvest Technology of Horticultural Crops* (1st edn., pp. 34–36). New India Publishing, New Delhi.

Sugiyama, J., Al-Haq, M. I., & Tsuta, M., (2005). Application of portable acoustic firmness tester for fruits. *Sensor,* pp. 439–444.

Tallada, J. G., Nagata, M., & Kobayashi, T., (2006). Non-destructive estimation of firmness of strawberries (*Fragaria* × *ananassa* Duch.) using NIR hyper spectral imaging. *Environ. Control Biol., 44,* 245–255.

Taniwaki, M., Takahashi, M., Sakurai, N., Takada, A., & Nagata, M., (2009). Effects of harvest time and low temperature storage on the texture of cabbage leaves. *Post Harvest Biology and Technology, 54,* 106–110.

Tanner, D. J., & Amos, N. D., (2003). Heat and mass transfer: Temperature variability during shipment of fresh produce. *Acta Hortic., 599,* 193–204.

Terasaki, S., Sakurai, N., Wada, N., Yamanishi, T., Yamamoto, R., & Nevins, D. J., (2001). Analysis of the vibration mode of apple tissue using electronic speckle pattern interferometry. *Trans. ASAE, 44,* 1697–1705.

Vergara, A., Llobet, E., Ramírez, J. L., Ivanov, P., Fonseca, L., Zampolli, S., Scorzoni, A., Becker, T., Marco, S., & Wöllenstein, J., (2007). An RFID reader with onboard sensing capability for monitoring fruit quality. *Sens. Actuators, B: Chem., 127,* 143–149.

Voss, D. H., (1992). Relating colorimeter measurement of plant color to the royal horticultural society color chart. *Hort. Science, 27,* 1256–1260.

Wang, J., Gomez, A. H., & Pereira, A. G., (2006). Acoustic impulse response for measuring the firmness of mandarin during storage. *Journal of Food Quality, 29,* 392–404.

Wei, X., Liu, F., Qiu, Z., Shao, Y., & He, Y., (2014). Ripeness classification of Astringent persimmon using hyperspectral imaging technique. *Food Bioprocess Technol., 7,* 1371–1380.

Yamamato, H., Amoto, M. I., & Haginuma, S., (1980). Acoustic impulse response method for measuring natural frequency of intact fruits and preliminary applications to internal quality evaluation of apples and watermelons. *Journal of Texture Studies, 11,* 117–136.

Zaidi, N. A., Tahir, M. W., Vellekoop, M. J., & Lang, W., (2017). A gas chromatographic system for the detection of ethylene gas using ambient air as carrier gas. *Sensors, 17,* 2283.

Zaidi, N. A., Tahir, M. W., Vinayaka, P. P., Lucklum, F., Vellekoop, M., & Lang, W., (2016). Detection of ethylene using gas chromatographic system. In: *Proceedings of the Euro Sensors* (pp. 380–383). Budapest, Hungary;

Zhang, C., Guo, C., Liu, F., Kong, W., He, L., & Lou, B., (2016). Hyper spectral imaging analysis for ripeness evaluation of strawberry with support vector machine. *Journal of Food Engineering, 179.*

Zhang, H., Wu, J., Zhao, Z., & Wang, Z., (2018). *Nondestructive Firmness Measurement of Differently SHAPED Pears with a Dual-Frequency Index Based on Acoustic Vibration, 138,* 11–18.

Zhang, W., Cui, D., & Ying, Y., (2014). Non destructive measurement of pear texture by acoustic vibration method. *Postharvest Biology and Technology, 96,* 99–105.

CHAPTER 3

SENSORS FOR SORTING AND GRADING OF FRUITS AND VEGETABLES

NADIA SHARIF,[1] BURERA SAJID,[2] NEELMA MUNIR,[1] and
SHAGUFTA NAZ[1]

[1]*Department of Biotechnology, Lahore College for Women University,
Lahore, Pakistan, Tel.: +92 3237501948,
E-mails: nadiasharif.s7@gmail.com (N. Sharif),
neelma.munir@yahoo.com (N. Munir), drsnaz31@hotmail.com (S. Naz)*

[2]*Queen Mary College, Lahore, Pakistan,
E-mail: burerasajid123@gmail.com*

ABSTRACT

Grading is required for quality aspects like size, color, volume, texture, and hydration. A tremendous scope has arisen for grading the vegetables and fruits from farms, dispatch, and to consumers. More sophisticated robotic manipulators grading systems are needed for fruit and vegetables that are kept in piles and stock houses. Primarily based on the optical characteristics at near Infra-Red levels sensors are preferred for the grading of fruits and vegetables. Sensors assessed readings are utilized for algorithms and image processing. Advanced sensing technologies need to be addressed; some of these sensing systems are magnetic resonance, spectroscopy, chemical sensing, x-rays, wireless sensor networks (WSN), computer vision, mechanical contact, and radio-frequency identification sensors. This chapter focuses on the present eminence of dissimilar sensing systems in terms of commercial applications.

3.1 INTRODUCTION

Natural products quality is assessed by consumers essentially from their view of the worthiness of organic products in light of qualities including visual interest (absence of imperfections, shading, size, and surface), readiness, smell, and flavor. The nature of organic products (as estimated by fragrance, flavor, shading, and textural qualities) continually changes all through crop expansion as of pre-harvest through postharvest phases as fruit nurture and ripen, or for the duration of preservation among storage. Consumer proclivities for fruits and vegetables are reflected in their specific decisions of organic product assortments or cultivars chose for procurement. Natural product assortments fluctuate generally in aroma attributes because of contrasts in the piece of sweet-smelling volatiles show in organic product fragrances which are at last dictated by plant hereditary qualities (Ogundiwin et al., 2009). To make the fruits and vegetables selling great in the market, different techniques application is due to the requirement of maintaining their freshness and nourishment contents from harvest to postharvest. About 50 years ago, another way to transact with rendering different sustenance resources was prepared, in which foodstuffs were regarded as physical bodies to which different techniques and traditional manufacturing ideas were allied (Ruiz-Altisent et al., 2010). Consumers developing interest in health is directly involved in commercialization topping so is interest towards grading of vegetables and fruits. Affected by numerous ecological factors, fruits, and vegetables at the same tree differ in quality.

Predominantly, fruits of numerous orchards differ fundamentally in quality and size. Grading might institutionalize fruit product items similarly advance fruit quality management in orchards. The meticulousness of machine vision frameworks in restriction and identification of fruits is inclined by indefinite and capricious lighting conditions in the field condition, capricious and multifarious gloom constructions and divergent size, shape, and shading.

Additionally, the precision of organic product identification is generously constrained by the impediment of product in shade pictures by leaves, branches, and other fruits. In the past, numerous research endeavors have been accomplished to accurately recognize fruits and vegetables in outer conditions. Techniques deliberated in the past for fruit analysis used characteristic types of sensors, and diverse descriptions investigation and furthermore subtle calculation approaches. The significant advances engaged with organic product identification are: (i) picture securing; (ii) picture initial screening; (iii) fruit division; (iv) clamor sifting; (v) physical appearance commotion; (vi) marking; and (vii) include abstraction. To enhance accuracy, similarly in few examinations, categorizations were performed after element abstraction (Gongal et al., 2015).

Horticulture, floriculture nursery crops, dried fruits tree nuts particularly fruits and vegetables are defined as specialty crops. Non-damaging (ND) analysis for features and qualities of subject crops is basic for observing and governing item safety and excellence. Sensors perform the key part in recognition of items characteristics, and consequently, they have been a dynamic research region, as recognized by an enormous number of study reviews amid the previous 50 years. At several phases of the manufacture and commercializing manacle, excellence recognition is a prerequisite for nearly all vegetables and fruits. Fruit and vegetables cover different assortment items, which contrast significantly in physiology, morphology, and synthesis. Thus, it is customary to order them into various gatherings as per a particular measure. Fruits of temperate areas including papaya, pear, avocado, peach, and mango are collected manually for direct use, or mechanically to process, whereas machines are applied for nuts or dry fruits harvesting. World significance fruits e.g., grapes, and olives quality is assessed through sensors after harvesting through machines. Vegetables keep though a substantially more noteworthy figure of compounds; they are developed in various situations, together with green-house. For instance, there are green verdant vegetables (e.g., small greens, lettuce, cabbage, spinach) and foods grown from the ground (bell pepper, tomato, zucchini). Lastly, ornamentals pruned blossoms, cut blooms, and pruned green plants (Ruiz-Altisent et al., 2010).

Being transported to the consumer once harvesting, the forte crops principally fresh vegetables and fruits are better to progress all or part of these postharvest deeds: washing, pre-arranging, refrigeration, arranging, wrapping, evaluating (for quality classes), and bundling (setting into containers, little boxes, bushels, packs, nets, and so forth), storage for short period in cold storage or long term storage under altered or controlled atmosphere conditions. Moreover, a few stuffs need such extraordinary treatments; as the ripening of bananas, peaches, and citrus through temperature and gases treatments, wine grape and olive oil desolation, broccoli, bell peppers, cauliflower, and lettuce individual wrapping and cutting and fruits wrapping in small bags, lettuce, and mixed fruits salads (SwarnaLakshmi and Kanchanadevi, 2014).

Electromagnetic radiation-based advances have indicated incredible prospect; a lot are effectually used for discerning the nature of forte products.

Productive consumption of these inventions needs the alternations of persuasive sensors with complex scientific mockups and PC calculations to build up acquaintances amid physical/synthetic assets and eminence features of the fruits and vegetables. Hence, countless efforts are being made centered on ascertaining and using varied non-destructive (ND) optical systems for fruits and vegetables eminence grit. Implausible developments have been made in PC vision and spectroscopy, and these policies are generally employed for appraising the assets of the fruits and vegetables (Ruiz-Altisent et al., 2010).

As inventions in view of ultra-violet (UV), VIS, mid-infrared (MIR) and NIR are ending up more adequate and equipped with added easy to use facts cure and alteration aptitudes, they have nurtured hide their progress of credit practice for composition and qualities of fruits and vegetables. While currently, numerous advances in view of electromagnetic assets have ascended; nevertheless implausible progress moreover has been made on other prevailing machineries. They encompass magnetic resonant imaging (MRI), X-ray, nuclear magnetic resonance (NMR) or attractive thunderous imaging (MRI), fluorescence, and less consummate as of volatile sensing, microwave, and thermal sensing. Such advancements have unlocked innovative investigation zones additionally to diverse compliances for sensing the features of fruits and vegetables (Ruiz-Altisent et al., 2010; SwarnaLakshmi and Kanchanadevi, 2014).

3.2 SENSORS FOR GRADING AND SORTING THROUGH EXTERNAL CHARACTERS

Vegetables and fruits are checked for the ripeness, firmness, texture, and size. Different fruits or vegetables as elated over one place to another they must be patterned for excellence panels. Internal and external features need to be deliberated. The manual technique of handpicking the best fruit or vegetables amongst the stock is a time overwhelming procedure and the world is progressing headed for instinctive harvesting and eminence examination over sensors.

PC vision framework can recreate human vision to see the three dimensional components of spatial products and has fractional capacity of the human mind. The framework will exchange, interpret, dynamic, and distinguish the apparent data, and thus work out a choice and after that send an order to do expectant errand. The straightforward PC vision framework embraces of lighting up chamber, CCD camera, picture gathering card, and PC. The chamber retains up an ideal work condition for the camera, to be definite, observance a proportioned, and vague brightening in CCD vision constituency. CCD camera is a picture sensor for catching pictures. The picture collecting cards theorize the picture and make an elucidate the video signal into a digital image signal. The PC holds and differentiates the computerized flag to work out a supposition and illumination (Gao et al., 2009).

3.2.1 COLOR

Color is the utmost apprehension factor in the food industry and in the research arena. So well this factor may be determined through sensors. The computer vision technique is applied to analyze the correct color. In industry, a colorimeter is applied to measure the color. Through colorimeters, the pixel of an image can be restrained in three-coordinates; and digital imaging grounded on three primary colors red, green, and blue. Since diverse maneuvers, harvests diverse result on the same pixel study so for systematize the result the color space called standard Red Green Blue (sRGB) is used in this regard. An alternative device which is adjacent to human color discernment is the hue saturation intensity (HSI) value (Hoffman et al., 2000).

3.2.2 SIZE AND VOLUME

The size may portion from the characteristics of area, perimeter, length, and distance across. The size of globular or nearly orbicular fruits can be restrained certainly nevertheless with multifarious shapes fortitude in diverse angles amongst camera and object. The current procedures require the equatorial diameter to be used as the measurement of size. Through this image, analysis can be regulated; size measurement is done through equatorial diameter and the centroid (center of mass), and also being elated by rollers. The volume is a different way of formative the size of a fruit. For flat projection object divided into vertical section i.e., height along axis Y and horizontal section i.e., width along axis X from this we get the volume and for total volume add the entire flat surface volume (Moreda et al., 2012).

3.2.3 SHAPE

The shape is another grading factor for good quality fruits. It is easy to interpret the physical properties like shape from analyzing the images through

Computer vision. Some applications in computer vision like wavelengths outside human sensibility i.e., NIR or X-rays and also for resolution, the human attributes that we could not be easily viewed are automatically detected through computer vision systems.

Nowadays, U.S. and European countries, New Zealand, Korea, Australia, Japan, and Canada uses computer vision system in this manner hyperspectral imaging is highly used which cartels spectral and spatial computer vision systems with the use of the spectrograph (Kim et al., 2007). k-mean method used to determine the linear sequences (widths) length (size). Shape of the fruit is also viewed through circularity. To determine the standard of a fruit various characters are of concern including size, mass, average diameter, volume, density, and sphericity coefficient. Another method Fourier Harmonics is used for ellipse analysis shapes of the fruits using partial least squares (PLS) model (Costa et al., 2013).

3.2.4 TEXTURE

The texture is another tool to inspect fruits. It can be determined by texture-based separation which added intricate algorithms. To determine the texture and color defects in different varieties of citrus, principal component analysis (PCA) is an applied and best suit method. Detection of defects for about 91.5% is possible by applying this method. The co-occurrence matrix of color and its mathematical characteristics represent one of the most highly used methods to describe texture (Pydipati et al., 2006). A microscope is another way to detect the skin damages of fruits. Texture characteristics such as consistency, co-occurrence matrix correlation, contrast, and angular moment were studied by Menesatti et al. in 2009. For determination of texture, the visible spectrum and the NIR spectrum in the 1000–1700 nm range. The pear bruise was studied through automatic detection. The light values of R (red) and G (green) were rummage-sale to differentiate the defective surface from the non-defective surface. Another method is the fractal texture described by Quevedo et al. (2008) it can be determined from the Fourier spectral analysis.

3.2.5 DETECTION OF THE FRUIT AND VEGETABLES BRUISE AND DEFECTS ON ITS SURFACE

Surface defects and bruises severely affect the extrinsic and intrinsic characteristics of the fruits and vegetables. Defective or bruised fruit and vegetable removal is significant for the prevention of deterioration and rotting of fruits and vegetables along with their grading. Distinguishing the imperfections of vegetables and fruits is yet a hurdle to actualizing their continuous grading. A few discoveries demonstrated that there is an alternate spectral reflectivity in the range of visible light for the damaged locale or bruised vegetable or fruit surface contrasted with the normal. Thus, surface defects can be identified in the wavelength range of visible light. Also, fruits are affected frequently during picking, stacking, unloading, and transportation. Progress toward becoming criteria for evaluating in a PC vision framework is in advance.

3.3 SENSORS FOR SORTING AND GRADING THROUGH INTERNAL CHARACTERS

The main purpose of the fruits and vegetables grading is their commercialization. Many emerging techniques are presentation prodigious potential for evaluating the internal qualities of the fruits and vegetables i.e., physiological disorders, chemical constituents, flavor, and texture like sensory attributes, functional properties, and nutritive values.

3.3.1 NIR SPECTROSCOPY

NIR radiation covers the scope of the electromagnetic spectrum between 780 and 2500 nm. The spectrum of the transmitted or reflected radiation is estimated by way of the item is irradiated with NIR radiation. The spectral attributes of the incident beam are changed while it goes through the fruit or vegetables because of wavelength scattering and retention procedures. This change relies upon both the compound structure and the corporeal assets of the item. Progressed multivariate measurable procedures, for example, fractional least squares regression (PLS) are then connected to associate the NIR range to eminence properties, for example, the sugar substance or solidness of the item. The same connection can be utilized to ascertain the excellence properties of imminent samples from their NIR range (Nicolai et al., 2007). Rapid Protein, moisture, and fat analysis of the fruits and vegetables are possible by of the NIR spectroscopy. Earlier it has also been applied for the dry matter content estimation of onions and apples. It has been applied for the water contents of the mushrooms. Physical characteristics particularly microstructure related features such as sensory attributes and stiffness are easy to estimate through NIR (Liu et al., 2010).

In the NIR spectrum, results identified with chemical (absorption) and physical (scattering) properties of the natural tissue is consolidated. To isolate the absorption and scattering information, both space-settled and time-settled procedures have been utilized. For estimating the sugar substance and apple insistence, natural product, the space-resolved spectroscopy (SRS) has been utilized. A camera applied to procure a hyperspectral picture of a spot of light that is anticipated on the surface of the natural product in SRS. Contingent upon the wavelength and scattering, when contrasted with projection zone the measure of the reflected spot is bigger. The quality characteristics of the item at that point can be resolved from the hyperspectral picture. Time-domain reflectance spectroscopy (TRS) involves the illumination of fruit in a series of short NIR pulses through solid-state laser array or via tunable laser. At some distance from the incident photons entry, a detector is placed. The photons might follow a complex route in the tissue and it might take more or less time to range the detector contingent to the scattering characteristics of the tissue. The scattering and absorption coefficient is measured as the detector measure the photon time-of-flight (ToF) distribution. The internal quality of the fruits and vegetables is finally assessed from coefficient spectra.

As SRS and TRS there is an expanding concern about calibration strategies that depend on the physics of entrance of NIR radiation in fruit tissue to on empirically a measurable spectrum, for example, PLS on exactly preprocessed spectra. Propelled light transport models in light of the dissemination estimate, the including multiplying technique or the Monte Carlo strategy may direct these exploration endeavors, and make it likely to isolate the data identified with the physical (diffusing) and compound (retention) properties of the test samples of fruits and vegetables. NIR microscopy actualizing the innovation is currently came in to focus and it is predicted that they will give a good comprehension of microscope dispersion of, e.g., sugars or other intriguing biological parts at the histological or cell level. The accessibility of quick and generally inexpensive diode array spectrometers which permit to obtain a NIR range in as meager as 50 ms has helped in business and research applications, and evaluating lines furnished with NIR, sensors are presently industrially accessible from numerous producers (Nicolaï et al., 2008).

3.3.2 FLUORESCENCE SPECTROSCOPY

The longer wave-lengths light radiated commencing the fruit or vegetable as it is excited by the shorter wavelengths light is measured by fluorescence spectroscopy. Laser-induced fluorescence is also in use for more than 20 years to study the physiological and stress states in vegetative plants. Stresses, diseases, and nutrient deficiencies of fruits and vegetables in laboratory or field conditions are assessed by Fluorescence spectroscopy. Though industrial processing of this application fully needs to be developed. Two types of florescence are the chlorophyll (650–800 nm) fluorescence and blue-green (400–600 nm) fluorescence (Costa et al., 2017).

3.3.3 MID-IR SPECTROSCOPY

Electromagnetic radiation and interaction of matter could be determined via mid-infrared spectroscopy (IR) that works between 2.5 and 25 m wavelengths. Mid-IR light absorptions are through rotational and vibrational energy states of molecules. Fourier transform IR spectrometers (FTIR) are used to take the Mid-IR spectra, in comparison to NIR spectroscopy the Mid-IR spectrum is less intricate in comparison of NIR spectrum; it yields direct and sharp peaks of particular chemical components.

FTIR spectroscopy is consequently applied to illustrate consistent samples like juices, frequently combined with an attenuated total reflection (ATR) model appearance. ATR-FTIR is applied to estimate sugars and acids in apples and tomatoes, to find a modification in macromolecular components of hazelnut, and to quantify the acidity of olive oil. Mid-IR spectroscopy appears to be added apposite for qualitative determinations or to determine tarnishing moderately for the computable extent of eminence characteristics of food products.

3.3.4 NUCLEAR MAGNETIC RESONANCE (NMR) SPECTROSCOPY

Nuclear magnetic resonance (NMR) is utilized significantly to estimate the internal qualities of the fruits and vegetables. For quality classification of fruits and vegetables, it is necessary to have intrinsic information of the fruit and vegetable and this goal can be achieved using NMR technology which has the ability to collect information about the internal characteristics of the object. NMR is subtle to the suppleness, dispersal amid other spectacles concentration and chemical environment, related to certain nuclei which make it valuable ND intensive care procedures for a variety of solicitations. Magnetic resonance imaging (MRI), NMR relaxometry, and NMR spectroscopy are other NMR derived techniques. The resonance frequency in NMR spectroscopy encrypts the chemically corresponding nuclei populations at various chemical and electronic environments. MRI dedicated spatial codification of the signal force delivers 2D and 3D pictures. NMR relaxometry recognizes cores populaces discernable. MR pictures permit inward tomography where tissue differentiates chiefly emerges from contrasts in relaxation times and proton thickness, and are likewise utilized for classification.

Concentrates that have been led vegetables and fruits with NMR procedures mostly identify with the development and readiness, inside harm and deformity, and physiological issue showing up amid capacity and caused by pre-and postharvest environments. MRI and NMR spectroscopy are utilized to screen air spaces reduction and NMR relaxometry is as of late answered to distinguish diminishing unwinding times with readiness. Among various kinds of inner deformity, creepy-crawly harm, contagious disease, wound, and the nearness of pits and seeds is the major concern.

Larvae cavities were seen in MR pictures for peach, pear, and mango. Parasite contamination has been reviewed utilizing MRI in grape, nectarine, orange, coconut strawberry, and mandarin. MRI has likewise given differentiated pictures to the examination of bruises in potato, onions, apples, strawberry, peaches, pears, and guava. Multivariate picture examination of MR pictures of tomato turned out to be powerful to predict the conductivity score of pericarp tissue in tomatoes for bruises discovery. The nearness of pits or seeds was reviewed with MRI in various items, for example, mandarin peach, plum, olive, cherry, orange, and cucumber (Milczarek et al., 2009).

Earlier studies stated the feasibility of sensing freezing bruises in food items like kiwifruit, peaches, persimmons, and blue-berries through divergence alteration in MR pictures. MR relaxometry is good to study freezing bruises in apple tissues. It is possible to check the cavities and dehydrated tissue in all such fruit samples as well as in mandarine citrus fruit. Estimation of water content in apples is plausible as the decrease of the proton thickness and the unwinding times in influenced tissue include force differentiate in the MR pictures. The postharvest issues showing up amid capacity that have been completely assessed with NMR methods are inward browning and intrinsic breaking, wolliness, and coarseness and central breakdown in pears, apples, and peaches. Previously stated works have been encompassed with industrial NMR gear intended for therapeutic purposes. Such gear works at high attractive field quality of in excess of 2 T with superior and costly equipment requiring high beginning capital speculations. Also, they are hard to execute in an industrial atmosphere. Industrialized execution of NMR procedures for inner review needs agreeable execution under movement circumstances (Kim et al., 2008).

3.3.5 X-RAYS

X-rays cover the spectrum scope of 0.01–10 nm, which falls between gamma beams and UV beams. X-beams can enter through most fruit and vegetables and the level of X-beam vitality transmitted through the item relies upon the occurrence vitality and retention coefficient, thickness, and thickness of the item. X-beam imaging is hence valuable for assessing quality/development and inside imperfection of food items. Different X-beam imaging advancements are accessible; they incorporate X-beam radiography which examines layers of the item to make two-dimensional (2D) pictures and registered tomography or CT filtering which makes 3D pictures. X-beam imaging demonstrated the potential for assessing the development of peach, mango, and lettuce. Physiological disorder in vegetables and fruits and the presence of foreign objects or insects frequently cause changes in the thickness and water substance of the vegetables and fruits. Consequently, X-beam imaging is especially valuable for distinguishing different sorts of inside imperfections or outside items in agricultural items (Kotwaliwale et al., 2014).

3.3.6 VISIBLE/NEAR-INFRARED (NIR)

The reflectance properties of an item in the visible range (around 400–780 nm) are seen by people as shading, which gives color data about products. Skin shading has been viewed as demonstrative of development for some sustenance items, for example, mango, tomato, and banana. Color, in the human observation, specifically identifies with item appearance, and the relationship of colors, and in this manner the VIS reflectance unique finger impression, with weakening and development of organic products amid maturing has been built up. Numerous constituents of vegetables and fruit quality, including those that add to taste and fragrance and cancer prevention agent potential are integrated into chloroplasts or chromoplasts, and in the qualities. In the sustenance business, quality elements are frequently connected to item color or coloring highlights. VIS imaging sensors are in this way compelling methods for quality discovery of natural products, particularly for development and readiness. Comparable outcomes have been built up in fruits and vegetables.

Broadband pictures (i.e., dim scale and shading pictures) are unsuitable for distinguishing particular quality traits (other than shading properties or certain surface imperfections that are visible) on the grounds that numerous concoction segments (colors, sugar, starch, water, protein, and so forth) are delicate to particular tight wavebands in or past the obvious area. Thus otherworldly imaging innovation, which gains single or different pictures at chosen wavelengths, is utilized for recognition of particular quality characteristics of green items. Phantom imaging might be ordered into multispectral and hyper-otherworldly. Multispectral imaging procures phantom pictures at a couple of discrete thin wavebands (the transfer speed may extend in the vicinity of 5 and 50 nm). Hyperspectral imaging, then again, gets tens or several ghastly pictures at harmonious wavelengths or wavebands over a particular spectrum area.

Multispectral imaging has been utilized to distinguish quality-related synthetic parts, for example, natural product color focus, and supply of sugar (or solvent solids) in organic products. Since color estimations generally give poor outcomes for readiness appraisal and imperfection discovery, NIR multispectral imaging system has been created and is currently financially accessible for organic product flaw location. The innovation can accomplish up to 90% efficiency in distinguishing harmed vegetables and fruits with visible blemishes of more prominent than 5 mm in diameter.

Hyperspectral imaging coordinates the primary highlights of imaging and spectroscopy to get spatial and spectrum data from the product instantaneously, in this manner making it particularly appropriate and considerably more effective for investigating plant and food items, that's properties and attributes regularly, differ spatially. Hyperspectral imaging is ordinarily actualized in one of the two detecting modes: push-sweeper or line examining mode and channel-based imaging mode. In-line checking mode, the imaging framework line examines the moving food items, from which three-dimensional (3D) hyperspectral pictures, likewise called hypercubes, are made. In channel based imaging mode, spectral pictures are obtained from the stationary items for an arrangement of wavebands utilizing either liquid tunable channel (LCTF) or acousto-optic tunable filter (AOTF).

Line filtering mode is most regularly utilized in light of the fact that it is moderately simple to actualize, particularly when continuous, online applications are required. Channel-based hyperspectral imaging structures require more confused alignment and are not appropriate for online applications. An imaging spectrograph, which scatters line pictures into various wavelengths, is a basic part for a line examining hyperspectral imaging framework; it ought to have a fitting optical determination and spectral reaction effectiveness with negligible distortions.

3.3.7 FLUORESCENCE AND DELAYED-LIGHT EMISSION (DLE)

Phosphorescence and fluorescence and stated through photoluminescence. To enact a luminescence reaction in a body, excitation is proficient by short-wavelength light, customarily in the UV, yet in addition in the VIS range. Mostly fluorescence used in plants that are identified with the assurance of chlorophyll action. DLE fluorescence grows soon after excitation, because of intermediate reactions in photosynthesis. Uses of fluorescence announced in the writing have been identified with chlorophyll, as a backhanded marker for organic products' physiological status: chlorophyll debases in natural products, and furthermore in vegetables, as an outcome of aging, senescence, time, and medications. Applications are postharvest defects in apples, senescence in cucumber freshness of broccoli, and cold damage in oranges. A few chemicals in plants, not the same as chlorophyll, fluoresce, similar to phenolics. As fluorescence reaction is reliant on the excitation wavelengths and the reaction is likewise wavelength-dependent, the strategy demonstrates a potential high specificity of the reaction of various mixes.

Hyperspectral devices are utilized to describe fluorescence reactions, e.g., for the portrayal and separation of eatable oils of plant utilizing lasers as the excitation source. Fluorescence is a promising procedure, broadly utilized as a part of science, which could be the reason for new particular sensors. Though, considerable research is required in the field of fluorescence to propel conceivable applications, and also in the instrumentation and its pertinence, practically A few productions demonstrate the uses of fluorescence imaging for recognition of nature of organic products notwithstanding incorporating it with regular reflectance. More to depth information on the fluorescing operators of in place organic products are required, and their association with natural product quality before plausible applications might be created.

3.3.8 CHEMICAL SENSORS

Since the 1990s, a few researchers have done broad research on mimicking sonic vibrational conduct for various states of natural products. In these strategies, the item is energized by methods for a little effect, and the vibration (around 20–20,000 Hz) is estimated utilizing a receiver, piezoelectric sensors, or laser vibrometers. Minimal effort and nonstop checking of substance and microbiological quality (counting microbiological examination of sustenance: oxygen-consuming settlement tally, nearness as well as a number of pathogens), with quick reaction times, can be accomplished with compound sensors.

3.3.9 BIOSENSORS

Two prime modules of the biological acknowledged constituent of the biosensor are: lectins and nucleic acids, and antibodies like bioligands and microorganisms, tissue materials, and catalysts as biocatalysts. The customary transducers are ophthalmic, thermal, and electrochemical. The most recent biosensors (affinity biosensors) join the traditional estimation standards with piezoelectric and attractive transducers (Salinas-Castillo et al., 2008). Biosensors technologies include:

1. **Recognition of Pesticide Filtrates:** Compound bio-receptors are broadly utilized for this purpose. It was assessed by Amine et al. (2006) that around 71% of the solicitations depicted for these enzymatic biosensors are for the assurance of pesticides comprising organophosphorus blends and carbamates, while substantial metals location speaks to just 21%. These biosensors are constructed for the most part in light of the hindrance of Acetylcholinesterase (AChE), however, utilizing distinctive flag transduction techniques. As of late, Nagatani et al. (2007) have created pesticide buildup identification visual chips in light of AChE, which might separate concerning convergences of 0.1 and 0.2 ppm of diazinon oxon in genuine examples of orange juice and apple.

2. **Quality Regulation in Tidying Away Environment:** Assertion of ethanol is essential in agrarian and nature investigation as in adjusted environment bundles and regulated air stockpiling as a procedure for low O_2 damage location amid the treatment of vegetables and fruits. Avant gardebioenzymatic sensors are the most encouraging ethanol sensors in view of the co arrest of the chemicals alcohol oxidase with horseradish peroxidase, bienzymatic framework that are utilized to create commercial colorimetric sensors can distinguish low O_2 damage in altered troposphere sets keeping fresh-cut specialty crops.

3. Localization of toxins and foodborne pathogens in vegetables and fruits. As biosensors are the quickest developing innovation for pathogen recognition at present, being acceptable, this investigative strategy can supplant the reference test ELISA. This is an interesting

intrigue territory because of their potential toxin recognition, as well as because of determinants for spoilage or freshness of vegetables and fruits. There is moderately multiple researches on biosensors connected in postharvest innovation. An imperative comment is that most of the biosensors referred to will be to be utilized as a part of a fluid form, demonstrating that the non-damaging methodology is still far on account for solid samples (Velusamy et al., 2010).

3.3.10   HIGH SELECTIVITY VERSUS LOW SELECTIVITY SENSORS

The characterization of the chemical sensors might accord to the compound specificity-presenting instrument; biosensors that join an organic or biomimetic detecting component are a subgroup of substance sensors. The fundamental organic materials utilized as a part of biosensor innovation are the couples catalyst/substrate, antigen/correlative groupings/counteracting agent, and nucleic acids. The selectivity of the organic detecting component offers the open door for improvement of exceedingly particular gadgets for an ongoing investigation in complex blends. The most vital zone of enthusiasm for biosensor use is the location of poisons, microorganisms, pesticides, and pathogens. Commonly, a "lock-and-key" approach isn't appropriate for the recreation of human inclinations to the creation of multifarious blends as sustenance or drinks. A multi-part chemical media (MCM) not by a whole of their discrete segments (or relating reactions from the particular sensors) however by certain unique portrayal, a chemical image (CI) as a unique mark where an arrangement of constraints illustrates a given MCM. An "electronic nose (e-nose)," that imitates the human olfaction, is a case of "electronic detecting" in light of varieties of low-specific/cross-receptive sensors, that consolidated reactions and combined with a fitting example acknowledgment framework; permit describing a gaseous sample as a fingerprint or chemical picture.

3.3.11   ELECTRONIC NOSE (E-NOSE)

The idea of the e-nose was at that point figured by Gardner and Bartlett in 1993 as an apparatus equipped for perceiving basic or multipart aromas. It is probable to assemble a fingerprint that determines the impact of the aroma constituent test, with the sensor exhibit reaction for the association of the aroma segments with the receptor locales in the capacity of the compound specificity-giving component and the method of physic synthetic signal transduction (Snopok and Kruglenko, 2002). From the postharvest administration perspective, in natural product readiness, the most imperative changes in vegetables and fruit aroma are experienced amid the timeframe of realistic usability period, however various different components impact unpredictable discharge after natural product stockpiling, chiefly capacity treatment includes the period and O_2 and CO_2 fixations in the capacity air. Organic product smells, or some other volatiles identified with the procedure, are probable gages of the physical state of the natural product, which could be utilized to create steady and reproducible non-ruinous methods to assess natural product quality from producer to purchaser (Pathange et al., 2006).

The technology of aroma determination by EN has been viewed as extremely encouraging amid the most recent 15 years as a non-ruinous device to assess natural product quality: timeframe of realistic usability examination, apples quality amid time span of usability, to screen tomato organic product with various capacity time; harvest date assessing in mandarins, apples, and mango fruit, shortcomings location as to distinguish solidify harm in oranges or apples abandons as coarseness and skin harm in "Cox" and "Red Delicious" apples, and for blueberry natural product infection identification and grouping (Li et al., 2010).

3.3.12   WIRELESS SENSING IN SPECIALTY CROPS

The utilization of wireless sensors (WST) for fruits and vegetables quality assessment offers new highlights both as far as correspondences and sensor technologies that never have been accessible. The latest improvements in wireless sensor networking (WSN) innovation have prompted the improvement of ease, multifunctional sensor nodes, and low power. Sensor nodes empower condition detecting together with information handling. They can coordinate with different sensors frameworks and transfer information to external consumers. The utilization of this innovation for checking crops that are extensively cultivated is new, since the fundamental equipment has just as of late turned out to be accessible. In any case, a few applications have been shown for strength crops.

Motes (small single nodes) can shape organizes and coordinate as per different models and designs. They accompany scaled-down mounted sensors that permit, in a little space (2.5 cm × 5 cm × 5 cm), the social occasion of information pretty much temperature, as well as relative humidity, quickening, shock, and light. On account of RFID, there are commercial dynamic and semi-uninvolved labels that can gather temperature data. Another preferred standpoint for wireless sensor systems is the plausibility of establishment in places where cabling is unimaginable, for example, vast fields or implanted inside the apparatuses, which conveys their readings nearer to the valid *In situ* possessions of yields (Ruiz-Garcia and Lunadei, 2011).

In the greenhouse and other fields, many applications of the wireless sensors for vegetables and fruits are involved in growth and environmental conditions.

It is plausible just with a WSN to screen a complete plantation carrying 20 or 30 sensors for each hectare, similarly monitoring the temperature of each flowerpot of the nursery, these may be dispersed in plots averaging as 4 ha in analyzing up to 10 km. In fruit and vegetable crops, the use of WST is a vital field for pest and disease control. The greenhouse can incredibly profit by WST. Sensors assortment for assessing leaves wetness continuously, every day photosynthetic radiation, soil electrical conductivity, substrate water, and soil and air temperatures is the most developed frameworks.

Advantages that have been shown incorporate control for enhanced plant development, more effective water and compost applications, along with lessening illness issues identified with over-watering, and in this manner for upgrading the yields.

Assist vital applications demonstrating high potential are checking foods grown from the ground compartments and cold-storage facilities, with or without controlled air or observing quality and senescence of strength crops amid transport. Numerous other likely uses in the greenhouse field, and in connection to quality and security of items and to accuracy cultivating, can be imagined. Remote sensor innovations hold a vital specialty in food quality research labs.

3.3.13 OTHER TECHNOLOGIES

Terahertz imaging and near-infrared (NIR) tomography are the emergent technologies that are gaining ample contemplation in ND assessment of products. NIR tomography is helpful in gaining intrinsic images. NIR spectrum range is 750–1300 nm, as radiations are passed through the biological object; the sensors located at numerous positions from the event point are utilized to tackle the attenuation and scattering. To resolve the diffusion theory model, 3D or 2D descriptions are created through inverse algorithms.

Though technological challenges need to overcome to get the high-resolution pictures as light scattering is prevailing in 700–1300 nm.

As Terahertz imaging technology, a new frontline for the grading of vegetables and fruits has advanced. Ranging in microwaves and far-infrared, Terahertz radiation denotes the spectrum area of 0.1–3 mm. Because of vibrational activity protein, DNA, and correspond molecular rotations energy level, Terahertz imaging technology delivers biological tissues spectral fingerprints. Due to Terahertz wavelengths sensitivity to water, they give an accurate tissue condition overview. Although in early stages, nevertheless, it is probable to deliver auspicious outcomes in the near future (Pickwell and Wallace, 2006).

Thermal imaging is a new nascent system for non-invasive inquiry in the food industry (TI). Concerning food quality and safety, current developments and possible uses of TI are described. Though this technology is tranquil in development, yet it's been functional for food safety and quality for assessing citrus drying time and recognizing the tomato and apple bruises (Gowen et al., 2010).

Non-destructively ration techniques are established for certain eminence factors of fruit and vegetables, like steadfastness approximation, and acoustic retorts dimension to ambiances and sways. A piezoelectric device stood at the back of the compression plunger extent warp curve in fruit and vegetable devoid of any damage. It is identified as micro deformation applies and is the base of rising a number of associated practices.

CEMAGREF (French Agricultural and Environmental Engineering Research Institute) established a micro-deformation sensor that encompasses an elastic cup with a contact plunger in the center to distort the around 2 mm fruit surface. A new device was progressive by Copa Technology in association with CTIFL (Center Techniques Interprofessionnel des Fruits etLégumes). This instrument "Duro-fel" contains a metallic, flat-ended probe and keeps three (10, 25, 50 cm^2) possible contact areas. It is broadly applied for soft fruits, tomatoes, apricots, and cherries. Analog firmness meter (AFM) and digital firmness meter (DFM) are the techniques that are utilized to measure the fruits insistence predominantly in mangos and tomatoes. Fruits deformation is measured through the "laser air-puff" device imperiled to a short and strong air current (69 kPa in 100 ms). Firmness and deformation of apples, kiwifruit, and other fruits are tested through some related techniques (García-Ramos et al., 2005).

3.4 CONCLUSION

The development of computer technology along with electronic and mechanical technology is the prodigious advancement in postharvesting grading of vegetables fruits. Moreover, the coeval grading stratagem with purposes distinctive the characteristics of the fruits and vegetables is probable to be familiarized in upcoming times. So far, not several technologies have been practiced in industry; technology transfers need a close and more dynamic communications between food and vegetable industry and scientists. In this way, worthwhile and unswerving apparatus could be advanced with the capacity to satisfy industrial necessities.

KEYWORDS

- **chemical sensing**
- **magnetic resonance**
- **sensing technologies**
- **spectroscopy**
- **wireless sensor**
- **x-rays**

REFERENCES

Amine, A., Mohammadi, H., Bourais, I., & Palleschi, G., (2006). Enzyme inhibition-based biosensors for food safety and environmental monitoring. *Biosensors and Bioelectronics, 21*(8), 1405–1423.

Costa, A. G., De Carvalho, P. F. D. A., Unior, R. B. A., Motoike, S., Yoshimitsu, E., Gracia, L., & Navas, I. M., (2017). Determination of macaw fruit harvest period by biospeckle laser technique. *African Journal of Agricultural Research, 12*(9), 674–683.

Costa, C., Antonucci, F., Pallottino, F., Aguzzi, J., Sarriá, D., & Menesatti, P., (2013). A review on agri-food supply chain traceability by means of RFID technology. *Food and Bioprocess Technology, 6*(2), 353–366.

Gao, H., Cai, J., & Liu, X., (2009). Automatic grading of the post-harvest fruit: A review. *International Conference on Computer and Computing Technologies in Agriculture* (pp. 141–146). Springer.

García-Ramos, F. J., Valero, C., Homer, I., Ortiz-Cañavate, J., & Ruiz-Altisent, M., (2005). Non-destructive fruit firmness sensors: A review. *Spanish Journal of Agricultural Research, 3*(1), 61–73.

Gongal, A., Amatya, S., Karkee, M., Zhang, Q., & Lewis, K., (2015). Sensors and systems for fruit detection and localization: A review. *Computers and Electronics in Agriculture*, *116*, 8–19.

Gowen, A., Tiwari, B., Cullen, P., McDonnell, K., & O'Donnell, C., (2010). Applications of thermal imaging in food quality and safety assessment. *Trends in Food Science and Technology*, *21*(4), 190–200.

Hoffman, A., Egozi, H., Ben-Zvi, R., & Schmilovitch, Z., (2000). Machine for automatic sorting 'barhi' dates according to maturity by near infrared spectrometry. *IV International Conference on Postharvest Science* (Vol. 553, pp. 481–486).

Kim, D. E., Kim, M. H., Cha, J. E., & Kim, S. O., (2008). Numerical investigation on thermal-hydraulic performance of new printed circuit heat exchanger model. *Nuclear Engineering and Design*, *238*(12), 3269–3276.

Kim, K., Sun, Y., Voyles, R. M., & Nelson, B. J., (2007). Calibration of multi-axis MEMS force sensors using the shape-from-motion method. *IEEE Sensors Journal*, *7*(3), 344–351.

Kotwaliwale, N., Singh, K., Kalne, A., Jha, S. N., Seth, N., & Kar, A., (2014). X-ray imaging methods for internal quality evaluation of agricultural produce. *Journal of Food Science and Technology*, *51*(1), 1–15.

Li, C., Krewer, G. W., Ji, P., Scherm, H., & Kays, S. J., (2010). Gas sensor array for blueberry fruit disease detection and classification. *Postharvest Biology and Technology*, *55*(3), 144–149.

Liu, Y., Sun, X., & Ouyang, A., (2010). Nondestructive measurement of soluble solid content of navel orange fruit by visible-NIR spectrometric technique with PLSR and PCA-BPNN. *LWT-Food Science and Technology*, *43*(40), 602–607.

Menesatti, P., Zanella, A., D'Andrea, S., Costa, C., Paglia, G., & Pallottino, F., (2009). Supervised multivariate analysis of hyper-spectral NIR images to evaluate the starch index of apples. *Food and Bioprocess Technology*, *2*(3), 308–314.

Milczarek, R. R., Saltveit, M. E., Garvey, T. C., & McCarthy, M. J., (2009). Assessment of tomato pericarp mechanical damage using multivariate analysis of magnetic resonance images. *Postharvest Biology and Technology*, *52*(2), 189–195.

Moreda, G., Muñoz, M., Ruiz-Altisent, M., & Perdigones, A., (2012). Shape determination of horticultural produce using two-dimensional computer vision: A review. *Journal of Food Engineering*, *108*(2), 245–261.

Nagatani, N., Takeuchi, A., Hossain, M. A., Yuhi, T., Endo, T., Kerman, K., Takamura, Y., & Tamiya, E., (2007). Rapid and sensitive visual detection of residual pesticides in food using acetyl cholinesterase-based disposable membrane chips. *Food Control*, *18*(8), 914–920.

Nicolai, B. M., Beullens, K., Bobelyn, E., Peirs, A., Saeys, W., Theron, K. I., & Lammertyn, J., (2007). Nondestructive measurement of fruit and vegetable quality by means of NIR spectroscopy: A review. *Postharvest Biology and Technology*, *46*(2), 99–118.

Nicolaï, B. M., Verlinden, B. E., Desmet, M., Saevels, S., Saeys, W., Theron, K., Cubeddu, R., et al., (2008). Time-resolved and continuous wave NIR reflectance spectroscopy to predict soluble solids content and firmness of pear. *Postharvest Biology and Technology*, *47*(1), 68–74.

Ogundiwin, E. A., Peace, C. P., Gradziel, T. M., Parfitt, D. E., Bliss, F. A., & Crisosto, C. H., (2009). A fruit quality gene map of Prunus. *BMC Genomics*, *10*(1), 587.

Pathange, L. P., Mallikarjunan, P., Marini, R. P., O'Keefe, S., & Vaughan, D., (2006). Non-destructive evaluation of apple maturity using an electronic nose system. *Journal of Food Engineering*, *77*(4), 1018–1023.

Pickwell, E., & Wallace, V., (2006). Biomedical applications of terahertz technology. *Journal of Physics D: Applied Physics*, *39*(17), R301.

Pydipati, R., Burks, T., & Lee, W., (2006). Identification of citrus disease using color texture features and discriminant analysis. *Computers and Electronics in Agriculture*, *52*(1–2), 49–59.

Quevedo, R., Mendoza, F., Aguilera, J., Chanona, J., & Gutiérrez-López, G., (2008). Determination of senescent spotting in banana (*Musa cavendish*) using fractal texture Fourier image. *Journal of Food Engineering*, *84*(4), 509–515.

Ruiz-Altisent, M., Ruiz-Garcia, L., Moreda, G., Lu, R., Hernandez-Sanchez, N., Correa, E., Diezma, B., et al., (2010). Sensors for product characterization and quality of specialty crops: A review. *Computers and Electronics in Agriculture*, *74*(2), 176–194.

Ruiz-Garcia, L., & Lunadei, L., (2011). The role of RFID in agriculture: Applications, limitations and challenges. *Computers and Electronics in Agriculture*, *79*(1), 42–50.

Salinas-Castillo, A., Pastor, I., Mallavia, R., & Mateo, C. R., (2008). Immobilization of a trienzymatic system in a sol-gel matrix: A new fluorescent biosensor for xanthine. *Biosensors and Bioelectronics*, *24*(4), 1053–1056.

Snopok, B., & Kruglenko, I., (2002). Multisensor systems for chemical analysis: State-of-the-art in electronic nose technology and new trends in machine olfaction. *Thin Solid Films*, *418*(1), 21–41.

SwarnaLakshmi, R., & Kanchanadevi, B., (2014). A review on fruit grading systems for quality inspection. *International Journal of Computer Science and Mobile Computing*, *3*(7), 615–621.

Velusamy, V., Arshak, K., Korostynska, O., Oliwa, K., & Adley, C., (2010). An overview of food borne pathogen detection: In the perspective of biosensors. *Biotechnology Advances*, *28*(2), 232–254.

CHAPTER 4

FLUORESCENCE TOOLS FOR SENSING OF QUALITY-RELATED PHYTOCHEMICALS IN FRUITS AND VEGETABLES

GIOVANNI AGATI,[1] WOLFGANG BILGER,[2] and ZORAN G. CEROVIC[3]

[1]*Institute of Applied Physics "Nello Carrara" (IFAC) – National Research Council (CNR), 50019 Sesto Fiorentino, Florence, Italy*

[2]*Botanical Institute, Christian-Albrechts-University Kiel, Olshausenstraße 40, 24098 Kiel, Germany*

[3]*Paris-Saclay University, CNRS, AgroParisTech, Ecology Systematics and Evolution, 91405, Orsay, France*

ABSTRACT

The quality of fruits and vegetables can be related to particular classes of compounds possessing peculiar optical properties. This aspect has been exploited by developing specific spectroscopic techniques for the non-destructive detection of such compounds, which are also suitable for the product quality monitoring. Among these, fluorescence spectroscopy was preferentially employed for a long time in the labs of the authors. By using various tools, partially developed by the authors, this technique proved to be suitable not only in basic research, but also for applied purposes. In this chapter, we are going to recall basic concepts of fluorescence and fluorescence measurements, give a brief description of different fluorescence methods employed in the acquisition of parameters describing the quality of fruits and vegetables and define what are the quality-related phytochemicals detectable by fluorescence. A report on portable fluorescence sensors available for in-field measurements is presented in subsection 4.3. Finally, some recent applications of the various fluorescence techniques and sensors on quality assessment of both fruits and vegetables are outlined.

4.1 INTRODUCTION

Since the first scientific approach to the fluorescence phenomenon in the middle of the 19th century (Valeur and Berberan-Santos, 2011), fluorescence spectroscopy became a useful analytical tool to characterize molecules. It is amazing that the first application of fluorescence, dating back to 1565, concerned the identification of a water extract, from a Mexican tree wood, with curative properties (Valeur and Berberan-Santos, 2011). Nowadays, fluorescence is still largely employed to investigate the plant's physiological status or to typify phytochemical products.

In principle, several plant constituents could be detected in vivo by measuring their fluorescence (Duval and Duplais, 2017; Garcia-Plazaola et al., 2015). However, the overlapping of various emission bands, the low intensity of fluorescence, and the localization of compounds into the plant organs significantly restricts the number of detectable phytochemicals.

Because of the limited penetration of the excitation wavelengths into the sample tissues, fluorescence methods can give information only on compounds present in the superficial layers. They are, therefore, useful to detect compounds mostly concentrated on the skin. This may represent a limitation of the technique; however, often the detectable superficial compounds are well correlated to the total compounds. Moreover, an indirect correlation between the measured signals and different quality parameters can be found, even with the help of chemometric analyzes.

Chlorophyll is the main fluorophore present in vegetable and green fruit samples. Chlorophyll fluorescence (ChlF) has been largely employed to assess the photosynthetic activity of plants (Baker, 2008; Kalaji et al., 2017; Murchie and Lawson, 2013). Senescence processes accompanying fruit ripening include chlorophyll degradation and a decrease of photosynthetic activity. Therefore, ChlF can be correlated to fruit quality parameters, such as sugar content and firmness, which change with maturity.

Polyphenols represent an interesting class of compounds that should be worthy to estimate because of their antioxidant health-promoting and disease-preventing properties (Chiva-Blanch and Badimon, 2017; George et al., 2017; Nabavi et al., 2017; Vanamala, 2017). Some of them possess significant fluorescence quantum yields allowing their direct *in situ* detection by fluorescence techniques. Others, present in leaf epidermises and fruit skins, can be determined indirectly by analyzing the ChlF from the underlying cell layers.

This last technique, called the chlorophyll fluorescence excitation screening (ChlFES) method, is relatively new, being developed within the last 20 years. It is based on the filtering effect of phenolic compounds that reduces the incoming light impinging on chlorophyll molecules.

The description of the method and its application to vegetables and fruits will be the main issue of this chapter, although other suitable fluorescence methods will be mentioned.

Due to the importance of extending the techniques from the lab to the field, particular emphasis is given to the description of portable fluorescence sensors and their use directly on cultivations.

Fluorescence spectroscopy has been also applied to the analysis of products and extracts from fruits and vegetables. We do not consider this application here, limiting to the description of *in situ* determinations.

4.2   FLUORESCENCE METHODS

4.2.1   BASIC CONCEPTS OF FLUORESCENCE AND FLUORESCENCE MEASUREMENTS

The term *fluorescence* refers to the physical phenomenon whereby a molecule excited by the absorption of electromagnetic radiation releases part of its energy as photons (Lakowicz, 2006). The molecules that take part in this process are then named fluorophores. As depicted in Figure 4.1A, following the absorption of one photon, occurring within 10^{-15} s, the fluorophore reaches an excited state of higher energy. This is an unstable and temporary condition lasting for about 10^{-9} s. The interaction of the excited molecule with its environment leads to different processes of energy relaxation such as: (i) thermal dissipation, (ii) production of other molecules (photochemistry), and (iii) emission of photons, i.e., fluorescence. Because of the presence of different minor competing relaxation processes already in the excited state, the energy of the emitted photon is usually only a fraction of that of the absorbed photon. Consequently, the wavelength of fluorescence is always longer than that of the excitation light. Such an aspect is depicted in Figure 4.1B where the absorption and fluorescence spectra of malvidin 3-O-glucoside in solution are reported. The shape and width of the spectra are determined by the structure of the vibrational and rotational energy levels, within each electronic level, allowing for different, more or less probable, absorption and emission transitions.

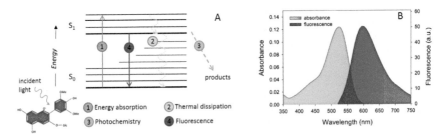

FIGURE 4.1 (A) Schematic representation of the absorption-relaxation processes occurring during the light-fluorophore interaction. (B) Absorbance and emission spectra of malvidin 3-O-glucoside (see structure drawn at the left of panel A) in aqueous/EtOH solution at pH 1.0.

The efficiency of a molecule to emit fluorescence is measured by the fluorescence quantum yield, which is defined as the number of photons emitted per number of photons absorbed:

$$\Phi_F = \frac{\#\,of\ emitted\ photons}{\#\,of\ absorbed\ photons} \qquad (1)$$

or alternatively,

$$\Phi_F = \frac{k_F}{k_F + k_{NR}} \qquad (2)$$

as the ratio between the fluorescence deactivation rate (k_F) and the total of the deactivation rates, that is the sum of k_F and all the non-radiative decays (k_{NR}).

This value can range from almost 0 to close to 1, depending on the molecule's structure and the environment around it. For example, when increasing molecule rigidity or reducing the temperature the fluorescence quantum yield increases, since non-radiative processes, as due to rotational-vibrational motion and collisions, are reduced.

The organic molecules employed in the laser technology possess rather high Φ_F, so that their fluorescence can be observed by the naked eye, as shown in Figure 4.2.

In order to properly measure the fluorescence intensity of a sample, attention must be paid that the detector is not reached by the measuring light strayed by the sample. Stray light can have intensity comparable to or even higher than that of the fluorescence, affecting significantly the signal-to-noise ratio of the measurement.

FIGURE 4.2   (A) Bright yellow fluorescence from coumarin in ethanol under blue excitation, to be compared to the pale-yellowish appearance under ambient light (B). A stilbene ethanolic solution emits blue fluorescence under UV irradiation (C), while it appears transparent under ambient light (D).

Figure 4.3 shows the basic set-up for fluorescence measurements. The excitation source should preferably emit in a narrow spectral band, to reduce tails overlapping to the detector spectral window. A short pass filter in the excitation beam and a long pass filter in front of the detector, which avoids any spectral overlap of stray light, guarantee that the detector senses only fluorescence. The 90° detection with respect to the irradiation direction further reduces the entrance of excitation into the detector device.

Yet, a narrow-wavelength excitation can have the advantage of a more selective excitation of specific fluorophores of interest.

Fluorescence is usually weaker than ambient light; therefore, its detection must be performed under darkness. Alternatively, under natural conditions, the fluorescence signal can be derived by using pulsed or modulated excitation sources and synchronized detectors. Electronic filtering will remove the noisy contribution of non-modulated ambient light.

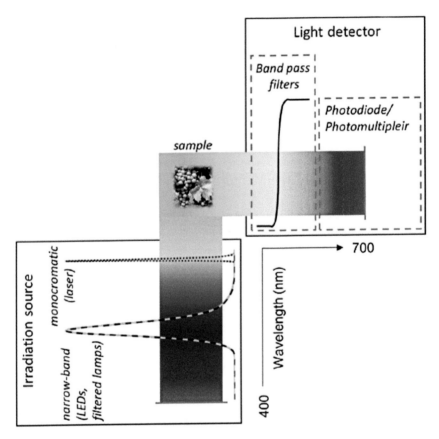

FIGURE 4.3 Basic set-up for fluorescence measurements. Samples are excited by monochromatic (laser) or narrow-band radiation. Fluorescence is detected by a photodiode or a photomultiplier at 90° with respect to the irradiation direction. The detector is preceded by a band-pass filter to limit stray light.

4.2.2 PULSE-AMPLITUDE-MODULATION (PAM) METHOD

Particularly in the measurement of fluorescence of chlorophyll, the main fluorophore in green vegetables, techniques using a modulated excitation beam have been applied. In 1986, the so-called Pulse-Amplitude-Modulation (PAM) technique was introduced (Schreiber, 1986). While other fluorophores present in plants such as coumarin emit fluorescence with a constant yield, ChlF emitted from photosynthesizing chloroplasts is highly variable. Since photosynthesis is consuming the absorbed excitation energy photochemically (process 3 in Figure 4.1A) it directly affects fluorescence yield. Furthermore, processes regulating the photosynthetic light reactions are affecting ChlF yield non-photochemically (process 2 in Figure 4.1A). Combining the PAM technique with short pulses of strong light saturating the photochemical reactions allows to monitor photosynthetic activity in intact plant organs, even under natural conditions in full sunlight. A deeper explanation of this technique is beyond the scope of this article and the reader is referred to exhaustive reviews (Maxwell and Johnson, 2000; Schreiber, 2004). However, as the function of photosynthesis is intimately related to the function of the whole plant cells, ChlF measurements are a potent means to report the progress of senescence or to indicate the effects of stress on leaves or green fruits.

4.2.3 SPECTRAL MEASUREMENTS

The concentration and number of different phytochemical compounds in fruits and vegetables affect the size of the fluorescence signal. Their identity or at least the class of chemical compounds to which they belong can be elucidated by recording their fluorescence emission and excitation spectra.

As mentioned before, the dominating fluorophore in plants is Chl. The typical *in situ* ChlF emission spectrum consists of two bands with maxima in the red (685–690 nm) and far-red (730–740 nm) regions. In intact plant organs, its shape depends on the Chl content. In fact, as depicted in Figure 4.4, the Chl absorption spectrum is partially overlapping the fluorescence spectrum. Therefore, the photons at around 690 nm emitted inside the leaf tissues, in their path to the detector will be partially re-absorbed by Chl itself. The re-absorption is proportional to the Chl concentration.

Because of this, the red/far-red Chl fluorescence ratio has been adopted as a non-destructive (ND) index of Chl content (Buschmann, 2007). The use of the inverted far-red/red Chl fluorescence ratio (Gitelson et al., 1999) is even more practical because it is linearly related to Chl content and increasing with the increase in Chl. It is named a simple fluorescence ratio (SFR) in the recent literature.

The re-absorption effect explains also the observed dependence of the ChlF emission spectrum on the excitation wavelength (Lichtenthaler and Rinderle, 1988). Radiation around the absorption maxima of Chl (blue and red) penetrates less into the leaf tissues in comparison to radiation at wavelengths (green-orange) which are less efficiently absorbed. In the first case, excited ChlF originates mainly from the upper mesophyll cells, therefore only little re-absorption takes place.

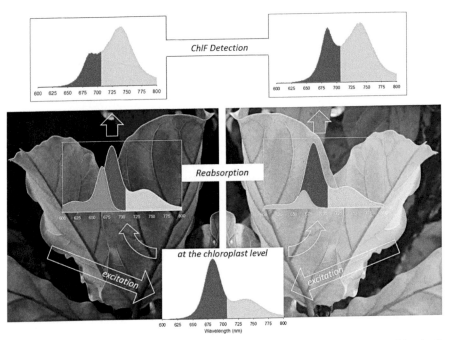

FIGURE 4.4   Schematic representation of the fluorescence reabsorption effects inside a leaf. On the right, the case of lower chlorophyll concentration is represented: the fluorescence emitted (red area) inside the leaf (*at the chloroplast level*) undergoes a moderate reabsorption (green area) leading to a reduced red/far-red ChlF ratio in the detected spectrum. On the left, the Chl concentration is higher, determining a greater reasbsorption of ChlF and then a larger change in the ChlF spectrum.

When UV excitation is used, blue-green fluorescence between 400 and 600 nm is recorded, in addition to the red/far-red Chl bands.

The assignment of blue-green fluorescence is complex since many different fluorophores can contribute to it. Lists of possible natural compounds showing UV-induced blue-green fluorescence are reported elsewhere (Cerovic et al., 1999; Lagorio et al., 2015). It is important to recall that localization of fluorophores in the sample tissues and the fluorophore's micro-environment are two fundamental factors determining the *in situ* detection of compounds. In fact, compounds located into the deeper sample tissues are likely not receiving a sufficient intensity of radiation to be significantly excited. On the other hand, the interaction of the fluorophores with their environment (viscosity, pH, presence of quenchers) can activate more or less deactivation channels (Figure 4.1A) and then significantly change their fluorescence quantum yield.

For this, phenolic, and flavonoid compounds bound to cell walls and present in the vacuoles of epidermal cells are the most likely molecules to contribute to the measurable blue-green fluorescence in plants.

Spectral measurements can be performed by rather different devices. Bench laboratory spectrofluorimeter provide both emission and excitation spectra by using a xenon lamp as an excitation source and monochromators to select wavelengths. Solid samples can be measured on front-face geometry, also with the help of bifurcated optical fiber bundles (Pfündel et al., 2006).

Pulsed laser-excitation systems coupled with synchronized high-sensitivity detectors allowed to improve the signal-to-noise ratio even under ambient light. They have been used in proximal sensing laboratory measurements as well as in the outdoors remote sensing of vegetation, embedded in a LIDAR device (Ounis et al., 2001; Rascher et al., 2009). On the opposite microscopic scale, fluorescence spectra can be recorded at the cellular and sub-cellular levels. This is possible by using a standard wide-field epifluorescence microscope connected to a diode-array or charge-coupled-device spectrometer (Raimondi et al., 2009). Most of confocal microscopes can also record fluorescence spectra, but usually at low spectral resolution (≥ 5 nm). In Figure 4.5, three examples of fluorescence spectra recorded by three different devices is reported.

4.2.4   FLUORESCENCE IMAGING

The spatial distribution of fluorophores over the sample can be obtained by fluorescence imaging. This technique first appeared as an application in epifluorescence microscopy to image leaf tissues by high-pressure mercury lamps coupled with band-pass interference filters. Later on, it was employed in confocal microscopy mainly for the detection of exogenous fluorophore

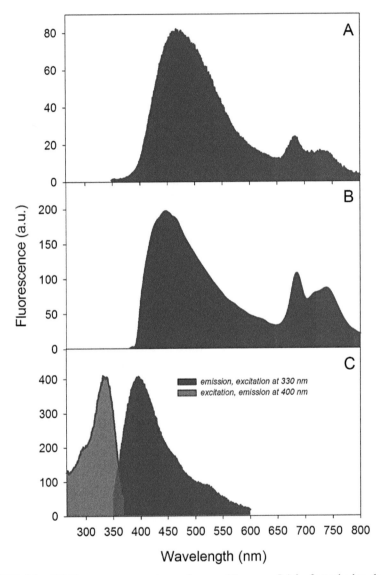

FIGURE 4.5 (A) Fluorescence spectrum of a corn (*Zea mays* L.) leaf attached to the plant recorded in the field by means of a hyperspectral Fluorescence LIDAR (FLIDAR) imaging system with a pulsed Nd YAG laser excitation at 355 nm. (B) Fluorescence spectra, under 365 nm excitation, of the adaxial layers from a 70 mm thick *Olea europaea* L. leaf cross-section recorded by a diode array multichannel spectral analyzer connected to an inverted epi-fluorescence microscope. (C) Emission and excitation spectra of an intact plum (*Prunus domestica* L.) fruit recorded by means of an optical fiber bundle connected to a bench spectrofluorimeter. All spectra were measured by the research group of the first author.

markers. In this way, intact leaf or fruit parts can be observed down to a depth of about 100 μm and 3D images of the fluorophore distribution can be obtained.

Multispectral fluorescence imaging is aimed to acquire several pictures of the sample at different emission bands. The elaboration of the digitized images is used to improve the discrimination and sensitivity of the technique. Co-localization of different compounds can be presented merging separate pictures in a single image.

At the macroscopic level, the technique employs a pulsed excitation source (laser, LED, or flash lamp) and an intensified CCD monochromatic camera with a filter wheel to select the acquisition spectral band. Usually for leaves and fruits, under UV excitation four emission bands in the blue, green, red, and far-red regions are acquired.

A comprehensive description of the technique, the needed instrumentation, and applications was reported by Buschmann et al. (2008).

Imaging-PAM systems have been developed to study the light-induced ChlF kinetics at the entire leaf level. This is the extension of the PAM technique applied as punctual measurements by using as detector a CCD camera synchronized with the pulse modulated excitation light.

The application of ChlF imaging in horticultural research to detect plant stresses and assess the postharvest quality of fruits and flowers has been reviewed by Gorbe and Calatayud (2012).

4.2.5 THE CHLOROPHYLL FLUORESCENCE EXCITATION SCREENING (CHLFES) METHOD (DETECTION OF NON-FLUORESCING COMPOUNDS)

Non-fluorescing compounds present on the sample surface can be still indirectly detected by measuring ChlF. This technique developed for the *in vivo* detection of polyphenols in leaves (Bilger et al., 1997) uses chlorophyll located below the compounds of interest as a sensor of light transmitted through the upper layers. Compounds absorbing fluorescence excitation light will cause a reduced emission of ChlF.

The *in vivo* ChlF excitation spectrum represents the convolution of the absorption spectra of Chl *a* and of the accessory pigments, carotenoids, and Chl *b*, which transfer the absorbed energy to Chl *a*. It contains also the negative contribution of compounds located in the epidermal layers above Chl that attenuate the excitation light reaching the chloroplasts (Figure 4.6, left side) (The absorption spectra reported in Figure 4.6 refers to *in vitro* conditions, while *in vivo* band broadening and a shift to longer wavelength are expected). Therefore, the intensity of ChlF at each excitation wavelength is given by:

$$ChlF = I T_{ep} A_{chlo} \Phi_F \tag{3}$$

where, I is the intensity of the excitation beam, T_{ep} is the transmittance of the epidermal layers, A_{chlo} is the absorbance of chloroplasts and Φ_F is the ChlF quantum yield.

Let's consider two excitation wavelengths, λ_1 inside the absorption band of an epidermal compound and λ_2 outside it so that $T_{ep}(\lambda_2)=1$, with the same intensity $[I(\lambda_1) = I(\lambda_2)]$ and similar chloroplast absorbance. According to Eq. (3), the ratio of the two ChlF will be:

$$\frac{ChlF(\lambda_1)}{ChlF(\lambda_2)} = T_{ep}(\lambda_1) \tag{4}$$

The fluorescence excitation method can then provide a quantitative estimate of epidermal transmittance on intact samples (Markstädter et al., 2001).

And, according to Beer-Lambert's law ($A = log\ T^{-1}$):

$$\log \frac{ChlF(\lambda_1)}{ChlF(\lambda_2)} = A_{ep}(\lambda_1) \tag{5}$$

that is proportional to the concentration of the epidermal compound taken into account (see Figure 4.6, right side).

On the other hand, comparing the ChlF excitation spectrum from two samples with different epidermal compound concentrations, *1* and *2*, allows to derive the spectrum of the difference in the epidermal absorbance between samples *2* and *1*:

$$\Delta A_{ep,2-1}(\lambda) = \log \frac{ChlF_1}{ChlF_2}(\lambda) \tag{6}$$

The ChlFES method works properly when the compound to be detected is spatially well separated from the first Chl layer. In leaves, the sub-epidermal localization of Chl allows for the detection of transmittance (absorbance) of compounds in the epidermis. This is the case reported in Figure 4.7, where the 3-D dual-band fluorescence image of a purple basil (*Ocimum basilicum*) leaf adaxial surface shows the epidermal distribution of anthocyanins (purple color) above the first Chl layer of the mesophyll (green color). Although phenols present in the epidermis have the strongest influence on chlorophyll excitation, it has been shown that also compounds present in sub-epidermal layers can be detected by the ChlFES method (Nichelmann and Bilger, 2017).

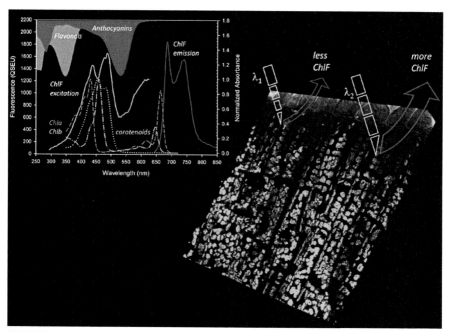

FIGURE 4.6   Left side, excitation, and emission spectra of a Chl containing sample (full lines) along with the positive spectral contributions due to absorbance of Chl and carotenoids located into the chloroplasts (dashed lines) and the negative spectral contributions due to absorbance of compounds (Flavonoids, Anthocyanins) (top of the figure) located into the leaf epidermis. The reported absorption spectra refer to *in vitro* conditions, while *in vivo* band broading and a shift to longer wavelength are expected. Right side, schematic representation of the chlorophyll fluorescence excitation screening (ChlFES) method using the three-dimensional view of the chloroplast distribution (green color) in the adaxial portion of a *Phillyrea latifolia* leaf cross-section. Sixty Chl fluorescence images were recorded (at 0.3-mm-steps) along the z-axis in a Confocal Laser Scanning Microscope, with $\lambda_{exc} = 488$ nm and λ_{em} over the 687–756 nm waveband. The yellow coloration of the epidermis indicates the presence of compounds absorbing at λ_1, but not at λ_2. The attenuation at λ_1 leads to less chlorophyll fluorescence (less ChlF) with respect to excitation at λ_2 that fully penetrates to the mesophyll resulting in higher chlorophyll fluorescence (more ChlF).

In grapevine leaves, the localization of phenolic compounds assessed by means of multispectral fluorescence microscopy showed that flavonols but not hydroxycinnamic acids can be detected non-destructively by the fluorimetric method (Agati et al., 2008a).

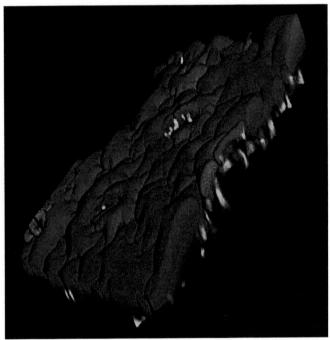

FIGURE 4.7   Three-dimensional view of a purple basil (*Ocimum basilicum*) leaf adaxial surface built with 127 fluorescence images recorded (at 0.3-mm-steps) along the z-axis in a Confocal Laser Scanning Microscope, with $\lambda_{xc} = 488$ nm and λ_{em} over the 571–646 nm and 676–732 nm wavebands for the anthocyanins and Chl acquisition, respectively.

In fruits, flavonoids are mostly localized above Chl (Hagen et al., 2006; Kolb et al., 2003) or they are partially co-localized with Chl, as is the case of anthocyanins in olives (Agati et al., 2005) or grapevine berries (Agati et al., 2007). Since the co-localized pigments will screen chlorophyll excitation with lower efficiency than the epidermal pigments the ChlFES technique will quantify the total pigment content in a non-linear manner.

In the practical situation, it is difficult to achieve an identical intensity of the measuring beam at λ_1 and λ_2 (Eq. 4). In addition, chloroplast absorption at both wavelengths may not be the same. In that case, referring the data to values determined with an epidermis free leaf will allow calculating UV transmittance and absorbance (Bilger et al., 1997; Cerovic et al., 2002). Alternatively, the lower epidermis of bifacial leaves or basal segments of monocotyledonous leaves, which have usually much reduced contents of screening compounds, can be used as a reference (Cerovic et al., 2002). The identification of the UV-screening compounds present in the leaf epidermis can be facilitated by the spectral application of the method (Cerovic et al., 2002).

TABLE 4.1 Commercially Available Sensors for the Detection of Chlorophyll Fluorescence Parameters

Based on	Type	Company/Web Site	Descriptive Features
ChlF kinetics parameters			
	Plant efficiency analyzer (PEA); FMS1; FMS2	Hansatech, King's Lynn, UK www.hansatech-instruments.com	Mouget and Tremblin, 2002; Öquist and Wass, 1988
	FluorPen	Photon Systems Instruments, Brno, Czech Republic, www.psi.cz	
	OS-500	Opti-Sciences Inc., Tyngsboro, MA, USA www.optisci.com	
	MINI-PAM; PAM-2000; PAM-2500	Walz, Effeltrich, Germany www.walz.com	Walz, Effeltrich, Germany www.walz.com
	HarvestWatch™	Satlantic Inc., Halifax, NS, Canada www.harvestwatch.net	DeLong et al., 2004; Prange et al., 2002
Imaging			
	Handy FluorCam FC 1000-H	Photon Systems Instruments, Brno, Czech Republic, www.psi.cz	Nedbal et al., 2000a
	IMAGING-PAM	Walz, Effeltrich, Germany www.walz.com	Ralph et al., 2005
Fluorescence excitation ratios			
	DUALEX®	Force-A, Orsay, France www.force-a.com	Goulas et al., 2004
	MULTIPLEX®	Force-A, Orsay, France www.force-a.com	Ben Ghozlen et al., 2010a
	UVA-PAM*	Gademann Messgeräte, Würzburg, Germany	Kolb et al., 2005
	Xe-PAM	Walz, Effeltrich, Germany	Bilger et al., 1997

* no longer available.

4.3 SENSORS DESCRIPTION

Several fluorescence sensors for applications on vegetables and fruits have been developed. Table 4.1 reports a list of the most common commercially available devices for ChlF detection. Most of them are portable and accordingly usable under *in-field* conditions. The description of the sensors can be found at the producer web page and in the reported references.

The portable handy plant efficiency analyzer (PEA) (Hansatech, UK) sensor uses a continuous excitation by an array of ultra-bright red LEDs at 650 nm (22 nm half-width) with NIR short pass cut-off filters. The maximal intensity can be up to 3,500 μmol m^{-2} s^{-1}. The detector consists of a fast response PIN photodiode with a RG9 long pass filter (fluorescence measuring range above 700 nm). The Hansatech Fluorescence Monitoring System FMS2 is a field portable pulse modulated (at 4 different frequencies) device exciting with a 594 nm amber LED or optionally with a 470 nm blue LED. Actinic, up to 3,000 μmol m^{-2} s^{-1}, and saturating, up to 20,000 μmol m^{-2} s^{-1}, lights are provided by a halogen lamp. A PIN photodiode detects fluorescence above 700 nm.

The FluorPen (PSI, Czech Republic) sensor uses a pulsed optically filtered blue LED at 470 nm and detects fluorescence between 667 to 750 nm, through a bandpass filter, by a PIN photodiode. Actinic and saturating lights are up to 1,000 and 3,000 μmol m^{-2} s^{-1}, respectively. The Handy FluorCam FC 1000-H (PSI, Czech Republic) imaging system acquires Chl fluorescence by a CCD camera with a RG695 optical filter (measuring range above 700 nm). The pulsed measuring LED emits at 620 nm. Three different actinic/saturating lights, white, blue (455 nm) or orange (620 nm) can be set up.

The Opti-Sciences chlorophyll fluorimeter has a modulated LED light probe at 660 nm (optional at 450 nm) with a 690 nm short-pass filter. A PIN photodiode with a 700–750 nm bandpass filter acts as detector. Saturation pulses and actinic light with white LEDs provide 15,000 and 5,800 μmol m^{-2} s^{-1}, respectively.

The MINI-PAM (Walz, Germany) fluorimeter exists as a blue (470 nm) and a red (655 nm) LED version measuring fluorescence above 630 and 700 nm, respectively. Actinic and saturation light reaches 3,000 and 6,000 μmol m^{-2} s^{-1}, respectively. In the PAM-2500 (Walz, Germany), detection by a PIN photodiode occurs above 715 nm with a measuring red LED at 630 nm. Actinic light can reach 4,000 μmol m^{-2} s^{-1}, while saturation pulses can be up to 25,000 μmol m^{-2} s^{-1}.

The light sources for the Imaging PAM (Walz, Germany) consist of 44 blue (450 nm) or red (650 nm) 3 W LEDs providing an actinic intensity of 1,900 µmol m^{-2} s^{-1} and saturation intensity of 4,000 µmol m^{-2} s^{-1}. Detection is through a CCD camera protected by a long-pass filter (RG 645) and a short-pass filter (<780 nm).

Multispectral fluorescence imaging is a common tool present in commercial confocal or deconvolution microscopes. For wide-scale detection, it is not easy to find complete multispectral fluorescence imaging systems on the market. Often, researchers adopt self-assembled instruments (Buschmann et al., 2008). One may mention the customized versions of FluorCam (PSI, Czech Republic) that allow for a different combination of LED panel excitation wavelengths, from UVA to red, and up to 7 acquisition fluorescence bands.

Several instruments applying the ChlFES method to detect UV transmittance of leaf epidermis have been developed during the last decades. The Xe-PAM fluorimeter (Xe discharge lamp and Pulsed Amplitude Modulation, Walz, Effeltrich, Germany) provides three bands of excitation, in the UV-B (λ_{max} at 314 nm), UV-A (λ_{max} at 366 nm), and blue-green (400–550 nm) spectral ranges, and detects ChlF above 700 nm (Bilger et al., 1997).

While the Xe-PAM fluorimeter operates on detached leaves, the UV-A PAM fluorimeter (Gademann Instruments, Würzburg, Germany) can perform in situ measurements for assessing epidermal compounds on attached leaves in the field (Bilger et al., 2001). It provides two bands of excitation in the UV-A (peak 375 nm) and blue (peak 470 nm) spectral regions and detects ChlF above 650 nm.

The use of the blue-excited ChlF as a reference signal limits the application of both the Xe-PAM and UV-A-PAM fluorometers to sample free of blue-absorbing compounds in the epidermis, as evidenced by Barnes et al. (2000) and Pfündel et al. (2007) in anthocyanin-containing leaves. Recently, it has been shown that the blue beam of the UV-A-PAM can be used to detect changes in carotenoid contents, when referenced to the ChlF excited with a red beam of a Mini-PAM (Nichelmann et al., 2016).

The Dualex Scientific+ (Force-A, Orsay, France) is a leaf-clip sensor, providing excitation in the UV-A (375 nm), in the green (520 nm), and in the red (655 nm) and detecting ChlF above 700 nm. The Dualex and the UV-A-PAM fluorometers were compared on leaves of a wide range of different species (Pfündel et al., 2007).

Applications of the ChlFES technique to relatively large (up to 8-cm in diameter) and thick samples is possible by the Multiplex sensor (Force-A,

Orsay, France; Ben Ghozlen et al., 2010a). It provides excitation in the UV-A (375 nm), blue (450 nm), green (520 nm) and red (630 nm) spectral regions and detects over three channels in the blue-green (or yellow), red, and far-red spectral regions, these two lasts at 680–690 nm and 730–780 nm, respectively, corresponding to the two emission peaks of chlorophyll.

As an example of a self-assembled instrument to apply the ChlFES method as an imaging technique, it is worthy to mention the system used by Lenk et al. (2007; Lenk and Buschmann, 2006). The fluorescence was excited by a Xenon flash lamp filtered at 340 nm (half bandwidth of 75 nm) or at 430 nm (half bandwidth of 70 nm) for UV- and blue-excitation, respectively. Fluorescence images were taken using a gated intensified video camera with a filter wheel housing 4 narrow-bandwidth (10 nm) interference filters in the blue (440 nm), green (520 nm), red (690 nm) and far red (740 nm) spectral regions.

4.4   APPLICATIONS ON VEGETABLES

Some recent applications of the various fluorescence techniques and sensors are outlined in the following. They are limited to the assessment of quality-related features of products and grouped according to the method used. Detection of the plant's physiological status, stress effects, nutrient deficiency, and plant diseases are not covered here.

4.4.1   CHLF KINETICS

The ND monitoring of the ChlF dark-light kinetics parameters was related to the quality of broccoli during storage under a modified atmosphere (DeEll and Toivonen, 2000). It was also applied to cabbage, green pepper, and lettuce (Prange et al., 2003).

Schofield et al. (2005) showed that several ChlF kinetics parameters (Possibly: $F_v/F_m = F_m - F_0/F_m$) were strongly correlated to the overall visual quality and decay of the iceberg lettuce during storage. ChlF measured at harvest can predict the storage potential of lettuce.

The storability at 4°C and 10°C of lamb's lettuce *Valeriana locusta* was evaluated by the analysis of the fast rise of the ChlF kinetics (Ferrante and Maggiore, 2007). The authors highlighted the potential of different ChlF indices of the health status of vegetables. Analogously, Qiu et al. (2017) found a correlation between ChlF kinetics parameters and the freshness of spinach, lettuce, and cabbage leafy vegetables.

4.4.2 IMAGING

ChlF imaging was proved to be useful for monitoring the tissue alteration induced by minimal processing (cutting and washing) of romaine lettuce (*Lactuca sativa* L. var. longifolia) and endive (*Cichorium endivia* L.) during storage (Hägele et al., 2016). Simko et al. elaborated a lettuce decay index based on ChlF imaging for measuring product deterioration and freezing injuries (Simko et al., 2015).

The imaging of the ratio between two fluorescence bands in the red spectral region was able to detect bovine fecal contaminants on Romaine lettuce and baby spinach leaves (Lee et al., 2014). Other applications to the horticulture of ChlF imaging can be found in the review by Gorbe and Calatayud (2012). The use of 4-bands multispectral fluorescence imaging to the assessment of quality on green bell peppers (*Capsicum annuum* L.) was reported (Buschmann et al., 2008).

4.4.3 CHLFES METHOD

A self-assembled apparatus was used to detect non-destructively flavo- noids on marketable broccoli heads (Bengtsson et al., 2006). UV-A-PAM measurements on curly kale leaves were found to be well correlated with the quercetin content determined by HPLC (Hagen et al., 2009). Other recent applications of the ChlFES method by the UV-A-PAM sensor comprised the evaluation of pre-harvest UV-B treatments at 22°C and 9°C and controlled storage conditions on epidermal UV-A absorbance in pak choi (*Brassica campestris* L. ssp. *chinensis* var. *communis*) leaves (Harbaum-Piayda et al., 2010).

Zivcak et al. (2017) applied the Multiplex-3 sensor to study the differ- ences in the accumulation of phenolic compounds between genotypes and growing conditions in lettuce (*Lactuca sativa* L.) (Zivcak et al., 2017).

The Multiplex sensor was also used to control the synthesis of flavonoids induced by UV-B radiation and white fluorescent lamps in greenhouse produced lettuce (*Lactuca sativa* L.) cv. Lollo Rosso, in order to improve their antioxidant properties (Rodriguez et al., 2014). Furthermore, the kinetics of flavonol accumulation during postharvest temperature and radia- tion treatments of Broccoli (*B. oleracea* L. var. *italica*) inflorescences was evaluated (Rybarczyk-Plonska et al., 2016).

4.5 APPLICATIONS ON FRUITS

4.5.1 CHLF KINETICS

The F_m and F_0 ChlF were found to decline with ripening in papaya fruits, due to chlorophyll degradation. They were correlated to fruit firmness and skin color and represent a complementary parameter to evaluate the overall fruit quality (Bron et al., 2004).

Kolb et al. (2006) found a negative association between F_0 ChlF and the concentrations of fructose, glucose, and total sugars, and the fructose/glucose ratio in white grapes (*Vitis vinifera* L. Cv. Bacchus and Silvaner) at various degrees of ripeness (Kolb et al., 2006).

The HarvestWatch system, based on an estimate of F_0 ChlF, has been applied to the control of quality retention in stored apples (DeLong et al., 2004) and other Chl-containing fruits such as avocado, pears, kiwifruits, banana, and mango (Prange et al., 2002, 2003).

In an *infield* study, ChlF was measured directly on mango fruits attached to the tree. F_0, F_m, and $F_v = F_m - F_0$ were significantly lower on fruits at the top of the canopy with respect to those within the canopy and correlated to fruit maturity (Lechaudel et al., 2010).

ChlF kinetics parameters combined with chemometrics represent a useful nondestructive tool to predict the sugar content in figs at different maturity stages (Jiang et al., 2013).

In mulberry (*Morus alba* L.) fruits, a high correlation was found between ChlF and total flavonoids, total phenols, sugars, and antioxidant activity (Lou et al., 2012). ChlF was also used to evaluate the quality and the optimal harvest time of jujube fruit (*Zizyphus jujuba* Mill. cv.) (Lu et al., 2012).

All of the primary ChlF parameters (F_0, F_m, F_v, and F_v/F_m) were found to be correlated to the grape mass loss during postharvest dehydration of 'L'Acadie' and 'Thompson Seedless' grape clusters (Wright et al., 2009).

4.5.2 LASER-INDUCED FLUORESCENCE

Laser-induced fluorescence spectroscopy with excitation at 337 nm was applied to detect phenolic compounds in strawberry fruits (Wulf et al., 2008) and polyphenols in apples (Wulf et al., 2005).

Laser excitation of ChlF in the blue or the red spectral region allowed monitoring of ripening in Riesling and Cabernet Sauvignon grapevine berries (Navrátil and Buschmann, 2016), as well as senescence-induced changes in peel chlorophyll of 'Jonagold' and 'Golden Delicious' apples during shelf-life (Kuckenberg et al., 2008).

4.5.3 IMAGING

Multicolor fluorescence yield and fluorescence ratio imaging over four distinct bands, from the blue to the far-red, was applied to monitor storage and ripening of Braeburn apples up to 6 months (Buschmann et al., 2008; Lichtenthaler et al., 2012).

ChlF imaging can identify damages to the lemon rind before the appearance of visual symptoms (Obenland and Neipp, 2005). Nedbal et al.(2000b) showed the potential of ChlF imaging in detecting injured areas or infected spots on lemon fruits in postharvest (Nedbal et al., 2000b).

Hyperspectral fluorescence imaging was used to detect defects on cherry tomato (*Lycopersicon esculentum* Mill. cv. Yoyo) fruits (Cho et al., 2013). Blue-green fluorescence on cracked cuticle spots was significantly higher than that of the sound surfaces.

The intensity of ChlF images of fresh apples correlated with the fruit firmness of fresh apples (*Malus domestica* cv. Red Delicious), fresh peaches, and nectarines (*Prunus persica* cv. Elegant Lady and Sweet Lady, respectively) (Bodria et al., 2004).

Integrated techniques of multispectral reflectance and fluorescence imaging were found to be accurate tools for recognition of various disorders on apples (Ariana et al., 2006).

Hyperspectral laser-induced fluorescence imaging of apple fruits with the combination of a principal component analysis (PCA) and neural networks modeling showed the potential for assessing selected quality parameters of apple fruits (Noh and Lu, 2007). An excellent prediction of apple skin hue, relatively good predictions for fruit firmness, skin chroma, and flesh hue, and poorer correlations for soluble solids content, titratable acid, and flesh chroma were found.

4.5.4 CHLFES METHOD

Most of the ChlFES method applications on fruits concerned the use of the Multiplex sensor. Among these, a large number of papers have been published in the viticulture area, with the aim to define a new ND tool for the assessment of the wine grape (*Vitis vinifera* L.) phenolic maturity directly in the vineyards. The Multiplex sensor was used for the measurement of the temporal accumulation of anthocyanins in the grape bunches to predict the best harvest time (Agati et al., 2013; Ben Ghozlen et al., 2010a, b; Diago et al., 2013; Tuccio et al., 2011). It was possible to evaluate the spatial heterogeneity of the anthocyanin berry content and to graph the results on vineyard maps by using the sensor either manually (Agati et al., 2013; Baluja et al., 2012a, b) or mounted on a harvester for on-the-go sensing (Bramley et al., 2011). Calibration of the Multiplex sensor against the wet-chemical destructive analysis of berry extracts is a critical issue and is still in progress (Ben Ghozlen et al., 2010a; Ferrandino et al., 2017; Pinelli et al., 2018).

These in-field studies were preceded by in-lab investigation defining the basic protocol of the ChlFES method at the single grape berry level (Agati et al., 2007) and whole bunch level (Cerovic et al., 2008).

Lately, attention has been devoted to the ND Multiplex detection of flavonols in white wine grape cultivars (mainly Vermentino, Chardonnay, and Nascetta) (Agati et al., 2013; Ferrandino et al., 2017).

Berry skin UV absorbance in the berries of the Bacchus white grape cultivar were also investigated by the Xe-PAM and UV-A-PAM fluorimetric methods. A comparison of ND data with those from wet chemistry on skin extracts suggested that UV screening in berries depends almost exclusively on flavonol content (Kolb et al., 2003).

Other applications of the Multiplex sensor consisted in the assessment of flavonols in apple (Betemps et al., 2012) and kiwifruit (Pinelli et al., 2013) exocarps, as well as of anthocyanins in olive fruits (Agati et al., 2005) and anthocyanins and flavonoids in sun-exposed mango fruits (Sivankalyani et al., 2016).

The Multiplex fluorimeter was as well able to: (i) follow the ripening of Thompson Seedless grapes and the change in the color of Crimson Seedless grapes (Bahar et al., 2012); (ii) provide complementary information on the effect of cytokinin analog forchlorfenuron on the ripening of Thompson Seedless table grape (Maoz et al., 2014); (iii) classify the maturity stage of oil palm bunches (Hazir et al., 2012a, b); and (iv) monitor the pre- and postharvest ripening of tomato fruits (Hoffmann et al., 2015).

The combination of a UV-A-PAM and a PAM-2000 fluorimeters was used to assess the content of anthocyanins and flavonoids in the apple skin of the Aroma red cultivar (Hagen et al., 2006) and their kinetics during postharvest treatments by visible light and UV-B radiation (Hagen et al., 2007).

The application of the ChlFES method as an imaging technique for assessing flavonols in white grape berries (Lenk et al., 2007) and anthocyanins in whole red grape bunches (Agati et al., 2008b) deserves a mention.

4.6 CONCLUDING REMARKS

The above described optical techniques have several major advantages. They can be applied to intact produce saving time-consuming and costly destructive analyzes, thereby providing as a side effect also a positive environmental impact. Repeated analyzes of identical samples are possible, which enhances the precision of analyzes when time courses need to be followed. A further strong advantage is the rapidity of the measurements allowing large amounts of samples tested in a short time. This combined with almost non-existing costs for consumables makes optical techniques very well suited for routine analyzes.

The compounds, which can be detected, as e.g., polyphenolic antioxidants, have attained a high importance as so-called nutraceuticals in recent years. A group of the flavonoids, the flavanones have been found to protect vitamin C and were even termed for some time vitamin P (Rice-Evans and Packer, 2003).

The usefulness and the spreading of these tools within the agro-food sector have been proven by a large number of applications of the fluorescence techniques for the assessment of quality features in vegetables and fruits. Whereas the most sophisticated spectroscopic techniques are mainly suitable to quality control laboratories, portable sensors can be used directly in the field (or in greenhouses) to forecast the harvest date of products with the highest level of beneficial phytochemicals. This practice would improve fruit and vegetable qualities, considering their large dependence on the variability of climatic conditions such as rainfall, temperature, and irradiance, even within short periods.

The ChlFES method for the assessment of non-fluorescing superficial compounds is particularly useful to detect the UV-absorbing antioxidants, for which detection by reflectance spectroscopy is limited by the very low reflectance signals in the UV region. These detected compounds are also indicative of plant growth conditions. Flavonoids are readily formed under field conditions but are strongly reduced in plants grown under greenhouse conditions. Hence, the detection of flavonoids may also serve to distinguish field-grown plants from those grown in the greenhouse, which not only have different flavonoid contents but contain also different amounts of vitamins (E and C). Due to the dependence of flavonoid biosynthesis on nutrient conditions, they can serve as indicators of nutrient deficiency.

Most of the described techniques are based on the detection of ChlF. Much less, information has been derived from the *in situ* measurement of blue-green fluorescence from beneficial phytochemicals belonging to the group of hydroxycinnamic acid derivatives. Therefore, it would be appealing if the relationship between the blue-green fluorescence and the sample quality could be better investigated in the future.

During postharvest, the rapid optical sensing could improve the control of the whole supply chain from production, through storage and distribution, to consumers in order to guarantee the highest value of products. Fluorescence sensors can be embedded into sorting machines to improve the automated grading system of produce. On-line quality monitoring of postharvest treatments can predict shelf-life of fresh produce, increasing profitability.

Thanks to the ongoing progress in the field of electronic components and to the increasing power of LEDs, miniaturization of devices for ChlFES measurement is possible, and is currently underway. This may lead in the future to a higher spectral resolution also in portable instruments, enabling the determination of more complex indices providing higher information.

It is worth to recall that the ChlFES technique is limited to the detection of compounds located at the sample's superficial layers and that it is as poorly specific as any spectrophotometric method. When the detailed compound composition is sought, the standard extraction and chromatographic procedures are still needed. However, the fluorescence sensors can still be useful for the intentional selection of plants with the most interesting antioxidant content.

There are three major competing techniques for nondestructive sensing of intact fruits and vegetables: colorimetry, near-infrared (NIR) spectroscopy (NIRS, coupled to chemometrics), and ChlFES. The major problem of the first is the limitation to chromophores absorbing visible light and its saturation at physiological concentration; of the second the omnipresence of water in the sample and the changing matrix effect; and of the third the loss of chlorophyll in very mature fruits. Only samples retaining at least minimal chlorophyll content can be measured.

The absolute intensity of fluorescence signals may be disturbed by unwanted contributions, for example by reflecting surfaces like waxes or hairs, dust, or residues of phytosanitary treatments. Fluorescence imaging can be affected by non-uniform contributions due to non-flat surfaces, as in curled leaves or rounded fruits. For this, fluorescence signal ratios should be preferred to minimize the above-mentioned problems.

When the mentioned limitations are kept in mind and the available instruments are employed wisely, they have a high potential to lead to further considerable progress in agriculture.

KEYWORDS

- chlorophyll fluorescence
- chlorophyll fluorescence excitation screening
- near-infrared spectroscopy
- plant efficiency analyzer
- pulse-amplitude-modulation
- simple fluorescence ratio

REFERENCES

Agati, G., Cerovic, Z. G., Dalla, M. A., Di Stefano, V., Pinelli, P., Traversi, M. L., & Orlandini, S., (2008a). *Optically-Assessed Preformed Flavonoids and Susceptibility of Grapevine to Plasmopara Viticola under Different Light Regimes., 35*, 77–84.

Agati, G., D'Onofrio, C., Ducci, E., Cuzzola, A., Remorini, D., Tuccio, L., Lazzini, F., & Mattii, G., (2013). Potential of a multiparametric optical sensor for determining in situ the maturity components of red and white *Vitis vinifera* wine grapes. *J. Agr. Food Chem., 61*, 12211–12218.

Agati, G., Meyer, S., Matteini, P., & Cerovic, Z. G., (2007). Assessment of anthocyanins in grape (*Vitis vinifera* L.) berries using a non-invasive chlorophyll fluorescence method. *J. Agric. Food Chem., 55*, 1053–1061.

Agati, G., Pinelli, P., Cortes-Ebner, S., Romani, A., Cartelat, A., & Cerovic, Z. G., (2005). Nondestructive evaluation of anthocyanins in olive (*Olea europaea*) fruits by in situ chlorophyll fluorescence spectroscopy. *J. Agric. Food Chem., 53*, 1354–1363.

Agati, G., Traversi, M. L., & Cerovic, Z. G., (2008b). Chlorophyll fluorescence imaging for the non-invasive assessment of anthocyanins in whole grape (*Vitis vinifera* L.) bunches. *Photochem. Photobiol., 84*, 1431–1434.

Ariana, D., Guyer, D. E., & Shrestha, B., (2006). Integrating multispectral reflectance and fluorescence imaging for defect detection on apples. *Comput. Electron. Agr., 50*, 148–161.

Bahar, A., Kaplunov, T., Zutahy, Y., Daus, A., Lurie, S., & Lichter, A., (2012). Auto-fluorescence for analysis of ripening in Thompson seedless and color in crimson seedless table grapes. *Aust. J. Grape Wine Res., 18*, 353–359.

Baker, N. R., (2008). Chlorophyll fluorescence: A probe of photosynthesis *in vivo*. *Annu. Rev. Plant Biol., 59*, 89–113.

Baluja, J., Diago, M. P., Goovaerts, P., & Tardaguila, J., (2012b). Assessment of the spatial variability of anthocyanins in grapes using a fluorescence sensor: Relationships with vine vigor and yield. *Precis. Agric., 13*, 457–472.

Baluja, J., Diago, M., Goovaerts, P., & Tardaguila, J., (2012a). Spatio-temporal dynamics of grape anthocyanin accumulation in a Tempranillo vineyard monitored by proximal sensing. *Aust. J. Grape Wine Res., 18*, 173–182.

Barnes, P. W., Searles, P. S., Ballaré, C. L., Ryel, R. J., & Caldwell, M. M., (2000). Non-invasive measurements of leaf epidermal transmittance of UV radiation using chlorophyll fluorescence: Field and laboratory studies. *Physiol. Plant., 109*, 274–283.

Ben Ghozlen, N., Cerovic, Z. G., Germain, C., Toutain, S., & Latouche, G., (2010a). Non-destructive optical monitoring of grape maturation by proximal sensing. *Sensors, 10*, 10040–10068.

Ben Ghozlen, N., Moise, N., Latouche, G., Martinon, V., Mercier, L., Besancon, E., & Cerovic, Z., (2010b). Assessment of grapevine maturity using a new portable sensor: Non-destructive quantification of anthocyanins. *J. Int. Sci. Vigne. Vin., 44*, 1–8.

Bengtsson, G. B., Schöner, R., Lombardo, E., Schöner, J., Borge, G. I. A., & Bilger, W., (2006). Chlorophyll fluorescence for non-destructive measurement of flavonoids in broccoli. *Postharvest Biol. Technol., 39*, 291–298.

Betemps, D. L., Fachinello, J. C., Galarca, S. P., Portela, N. M., Remorini, D., Massai, R., & Agati, G., (2012). Non-destructive evaluation of ripening and quality traits in apples using a multiparametric fluorescence sensor. *J. Sci. Food Agr., 92*, 1855–1864.

Bilger, W., Johnsen, T., & Schreiber, U., (2001). UV-excited chlorophyll fluorescence as a tool for the assessment of UV-protection by the epidermis of plants. *J. Exp. Bot., 52*, 2007–2014.

Bilger, W., Schreiber, U., & Bock, M., (1995). Determination of the quantum efficiency of photosystem II and of non-photochemical quenching of chlorophyll fluorescence in the field. *Oecologia, 102*, 425–432.

Bilger, W., Veit, M., Schreiber, L., & Schreiber, U., (1997). Measurement of leaf epidermal transmittance of UV radiation by chlorophyll fluorescence. *Physiol. Plant., 101*, 754–763.

Bodria, L., Fiala, M., Guidetti, R., & Oberti, R., (2004). Optical techniques to estimate the ripeness of red-pigmented fruits. *Trans. ASAE, 47*, 815.

Bramley, R. G. V., Le Moigne, M., Evain, S., Ouzman, J., Florin, L., Fadaili, E. M., Hinze, C. J., & Cerovic, Z. G., (2011). On-the-go sensing of grape berry anthocyanins during commercial harvest: Development and prospects. *Aust. J. Grape Wine Res., 17*, 316–326.

Bron, I. U., Ribeiro, R. V., Azzolini, M., Jacomino, A. P., & Machado, E. C., (2004). Chlorophyll fluorescence as a tool to evaluate the ripening of 'Golden' papaya fruit. *Postharvest Biol. Technol., 33*, 163–173.

Buschmann, C., (2007). Variability and application of the chlorophyll fluorescence emission ratio red/far-red of leaves. *Photosynth. Res., 92*, 261–271.

Buschmann, C., Langsdorf, G., & Lichtenthaler, H. K., (2008). Blue, green, red, and far-red fluorescence signatures of plant tissues, their multicolor fluorescence imaging, and application for agro-food assessment. In: Zude, M., (ed.), *Optical Methods for Monitoring Fresh and Processed Food: Basics and Applications for a Better Understanding of Non-destructive Sensing* (pp. 272–319).Taylor & Francis Group, CRC Press: Boca Raton.

Cerovic, Z. G., Moise, N., Agati, G., Latouche, G., Ben, G. N., & Meyer, S., (2008). New portable optical sensors for the assessment of wine grape phenolic maturity based on berry fluorescence. *J. Food Compos. Anal., 21*, 650–654.

Cerovic, Z. G., Ounis, A., Cartelat, A., Latouche, G., Goulas, Y., Meyer, S., & Moya, I., (2002). The use of chlorophyll fluorescence excitation spectra for the non-destructive in situ assessment of UV-absorbing compounds in leaves. *Plant Cell Environ., 25*, 1663–1676.

Cerovic, Z. G., Samson, G., Morales, F., Tremblay, N., & Moya, I., (1999). Ultraviolet-induced fluorescence for plant monitoring: Present state and prospects. *Agronomie, 19*, 543–578.

Chiva-Blanch, G., & Badimon, L., (2017). Effects of polyphenol intake on metabolic syndrome: Current evidences from human trials. *Oxid. Med. Cell. Longev.* https://doi.org/10.1155/2017/5812401.

Cho, B. K., Kim, M. S., Baek, I. S., Kim, D. Y., Lee, W. H., Kim, J., Bae, H., & Kim, Y. S., (2013). Detection of cuticle defects on cherry tomatoes using hyper spectral fluorescence imagery. *Postharvest Biol. Technol., 76*, 40–49.

DeEll, J. R., & Toivonen, P. M., (2000). Chlorophyll fluorescence as a nondestructive indicator of broccoli quality during storage in modified-atmosphere packaging. *Hort. Science, 35*, 256–259.

DeLong, J. M., Prange, R. K., Leyte, J. C., & Harrison, P. A., (2004). A new technology that determines low-oxygen thresholds in controlled-atmosphere-stored apples. *Hort. Technology, 14*, 262–266.

Diago, M. P., Guadalupe, Z., Baluja, J., Millan, B., & Tardaguila, J., (2013). Appraisal of wine color and phenols from a non-invasive grape berry fluorescence method. *J. Int. Sci. Vigne Vin., 47*, 55–64.

Duval, R., & Duplais, C., (2017). Fluorescent natural products as probes and tracers in biology. *Nat. Prod. Rep., 34*, 161–193.

Ferrandino, A., Pagliarani, C., Carlomagno, A., Novello, V., Schubert, A., & Agati, G., (2017). Improved fluorescence-based evaluation of flavonoid in red and white wine grape cultivars. *Aust. J. Grape Wine Res., 23*, 207–214.

Ferrante, A., & Maggiore, T., (2007). Chlorophyll a fluorescence measurements to evaluate storage time and temperature of Valeriana leafy vegetables. *Postharvest Biol. Technol., 45*, 73–80.

Garcia-Plazaola, J. I., Fernandez-Marin, B., Duke, S. O., Hernandeza, A., Lopez-Arbeloa, F., & Becerril, J. M., (2015). Auto fluorescence: Biological functions and technical applications. *Plant Sci., 236*, 136–145.

George, V. C., Dellaire, G., & Rupasinghe, H. P. V., (2017). Plant flavonoids in cancer chemoprevention: Role in genome stability. *J. Nutr. Biochem., 45*, 1–14.

Gitelson, A. A., Buschmann, C., & Lichtenthaler, H. K., (1999). The chlorophyll fluorescence ratio F 735/F 700 as an accurate measure of the chlorophyll content in plants. *Remote Sens. Environ., 69*, 296–302.

Gorbe, E., & Calatayud, A., (2012). Applications of chlorophyll fluorescence imaging technique in horticultural research: A review. *Sci. Hort., 138*, 24–35.

Goulas, Y., Cerovic, Z. G., Cartelat, A., & Moya, I., (2004). Dualex: A new instrument for field measurements of epidermal ultraviolet absorbance by chlorophyll fluorescence. *Appl. Opt., 43*, 4488–4496.

Hägele, F., Baur, S., Menegat, A., Gerhards, R., Carle, R., & Schweiggert, R. M., (2016). Chlorophyll fluorescence imaging for monitoring the effects of minimal processing and warm water treatments on physiological properties and quality attributes of fresh-cut salads. *Food Bioprocess Tech., 9*, 650–663.

Hagen, S. F., Borge, G. I. A., Bengtsson, G. B., Bilger, W., Berge, A., Haffner, K., & Solhaug, K. A., (2007). Phenolic contents and other health and sensory related properties of apple fruit (*Malus domestica* Borkh., cv. Aroma): Effect of postharvest UV-B irradiation. *Postharvest Biol. Technol., 45*, 1–10.

Hagen, S. F., Borge, G. I. A., Solhaug, K. A., & Bengtsson, G. B., (2009). Effect of cold storage and harvest date on bioactive compounds in curly kale (*Brassica oleracea* L. var. acephala). *Postharvest Biol. Technol., 51*, 36–42.

Hagen, S. F., Solhaug, K. A., Bengtsson, G. B., Borge, G. I. A., & Bilger, W., (2006). Chlorophyll fluorescence as a tool for non-destructive estimation of anthocyanins and total flavonoids in apples. *Postharvest Biol. Technol., 41*, 156–163.

Harbaum-Piayda, B., Walter, B., Bengtsson, G. B., Hubbermann, E. M., Bilger, W., & Schwarz, K., (2010). Influence of pre-harvest UV-B irradiation and normal or controlled atmosphere storage on flavonoid and hydroxycinnamic acid contents of pak choi (*Brassica campestris* L. ssp. chinensis var. communis). *Postharvest Biol. Technol., 56*, 202–208.

Hazir, M. H. M., Shariff, A. R. M., & Amiruddin, M. D., (2012a). Determination of oil palm fresh fruit bunch ripeness-based on flavonoids and anthocyanin content. *Ind. Crops Prod., 36*, 466–475.

Hazir, M. H. M., Shariff, A. R. M., Amiruddin, M. D., Ramli, A. R., & Saripan, M. I., (2012b). Oil palm bunch ripeness classification using fluorescence technique. *J. Food Eng., 113*, 534–540.

Hoffmann, A. M., Noga, G., & Hunsche, M., (2015). Fluorescence indices for monitoring the ripening of tomatoes in pre-and postharvest phases. *Sci. Hort., 191*, 74–81.

Jiang, L., Shen, Z., Zheng, H., He, W., Deng, G., & Lu, H., (2013). Noninvasive evaluation of fructose, glucose, and sucrose contents in fig fruits during development using chlorophyll fluorescence and chemometrics. *J. Agr. Sci. Tech., 15*, 333–342.

Kalaji, H. M., Schansker, G., Brestic, M., Bussotti, F., Calatayud, A., Ferroni, L., Goltsev, V., et al., (2017). Frequently asked questions about chlorophyll fluorescence, the sequel. *Photosynth. Res., 132*, 13–66.

Kolb, C. A., Kopecky, J., Riederer, M., & Pfündel, E. E., (2003). UV screening by phenolics in berries of grapevine (*Vitis vinifera*). *Func. Plant Biol., 30*, 1177–1186.

Kolb, C. A., Wirth, E., Kaiser, W. M., Meister, A., Riederer, M., & Pfündel, E. E., (2006). Noninvasive evaluation of the degree of ripeness in grape berries (*Vitis vinifera* L. cv. Bacchus and Silvaner) by chlorophyll fluorescence. *J. Agric. Food Chem., 54*, 299–305.

Kolb, C., Schreiber, U., Gademann, R., & Pfundel, E., (2005). UV-A screening in plants determined using a new portable fluorimeter. *Photosynthetica, 43*, 371–377.

Kuckenberg, J., Tartachnyk, I., & Noga, G., (2008). Evaluation of fluorescence and remission techniques for monitoring changes in peel chlorophyll and internal fruit characteristics in sunlit and shaded sides of apple fruit during shelf-life. *Postharvest Biol. Technol., 48*, 231–241.

Lagorio, M. G., Cordon, G. B., & Iriel, A., (2015). Reviewing the relevance of fluorescence in biological systems. *Photochem. Photobiol. Sci., 14*, 1538–1559.

Lakowicz, J. R., (2006). *Principles of Fluorescence Spectroscopy* (3rd edn.). Springer: New York.

Lechaudel, M., Urban, L., & Joas, J., (2010). Chlorophyll fluorescence, a nondestructive method to assess maturity of mango fruits (Cv.'Cogshall') without growth conditions bias. *J. Agric. Food Chem., 58*, 7532–7538.

Lee, H., Everard, C. D., Kang, S., Cho, B. K., Chao, K., Chan, D. E., & Kim, M. S., (2014). Multispectral fluorescence imaging for detection of bovine faces on *Romaine lettuce* and baby spinach leaves. *Biosyst. Eng., 127*, 125–134.

Lenk, S., & Buschmann, C., (2006). Distribution of UV-shielding of the epidermis of sun and shade leaves of the beech (*Fagus sylvatica* L.) as monitored by multi-color fluorescence imaging. *J. Plant Physiol., 163*, 1273–1283.

Lenk, S., Buschmann, C., & Pfündel, E. E., (2007). *In vivo* assessing flavonols in white grape berries (*Vitis vinifera* L. cv. Pinot Blanc) of different degrees of ripeness using chlorophyll fluorescence imaging. *Func. Plant Biol., 34*, 1092–1104.

Lichtenthaler, H. K., & Rinderle, U., (1988). *The role of Chlorophyll Fluorescence in the Detection of Stress Conditions in Plants* (Vol. 19, pp. S29–S85).

Lichtenthaler, H. K., Langsdorf, G., & Buschmann, C., (2012). Multicolor fluorescence images and fluorescence ratio images of green apples at harvest and during storage. *Isr. J. Plant Sci., 60*, 97–106.

Lou, H., Hu, Y., Zhang, L., Sun, P., & Lu, H., (2012). Nondestructive evaluation of the changes of total flavonoid, total phenols, ABTS and DPPH radical scavenging activities, and sugars during mulberry (*Morus alba* L.) fruits development by chlorophyll fluorescence and RGB intensity values. *LWT - Food Sci. Technol., 47*, 19–24.

Lu, H., Lou, H., Zheng, H., Hu, Y., & Li, Y., (2012). Nondestructive evaluation of quality changes and the optimum time for harvesting during jujube (*Zizyphus jujuba* Mill. cv. Changhong) fruits development. *Food Bioprocess Tech., 5*, 2586–2595.

Maoz, I., Bahar, A., Kaplunov, T., Zutchi, Y., Daus, A., Lurie, S., & Lichter, A., (2014). The effect of the cytokinin forchlorfenuron on the tannin content of Thompson seedless table grapes. *Am. J. Enol. Vitic. Ajev.*, 13095.

Markstädter, C., Queck, I., Baumeister, J., Riederer, M., Schreiber, U., & Bilger, W., (2001). Epidermal transmittance of leaves of *Vicia faba* for UV radiation as determined by two different methods. *Photosynth. Res., 67*, 17–25.

Maxwell, K., & Johnson, G. N., (2000). Chlorophyll fluorescence: A practical guide. *J. Exp. Bot., 51*, 659–668.

Mouget, J. L., & Tremblin, G., (2002). Suitability of the fluorescence monitoring system (FMS, Hansatech) for measurement of photosynthetic characteristics in algae. *Aquat. Bot., 74*, 219–231.

Murchie, E. H., & Lawson, T., (2013). Chlorophyll fluorescence analysis: A guide to good practice and understanding some new applications. *J. Exp. Bot., 64*, 3983–3998.

Nabavi, S. F., Tejada, S., N Setzer, W., Gortzi, O., Sureda, A., Braidy, N., Daglia, M., Manayi, A., & Mohammad, N. S., (2017). Chlorogenic acid and mental diseases: From chemistry to medicine. *Curr. Neuropharmacol., 15*, 471–479.

Navrátil, M., & Buschmann, C., (2016). Measurements of reflectance and fluorescence spectra for nondestructive characterizing ripeness of grapevine berries. *Photosynthetica, 54*, 101–109.

Nedbal, L., Soukupová, J., Kaftan, D., Whitmarsh, J., & Trtílek, M., (2000a). Kinetic imaging of chlorophyll fluorescence using modulated light. *Photosynth. Res., 66*, 3–12.

Nedbal, L., Soukupova, J., Whitmarsh, J., & Trtilek, M., (2000b). Postharvest imaging of chlorophyll fluorescence from lemons can be used to predict fruit quality. *Photosynthetica, 38*, 571–579.

Nichelmann, L., & Bilger, W., (2017). Quantification of light screening by anthocyanins in leaves of *Berberis thunbergii*. *Planta, 246*, 1069–1082.

Nichelmann, L., Schulze, M., Herppich, W. B., & Bilger, W., (2016). A simple indicator for non-destructive estimation of the violaxanthin cycle pigment content in leaves. *Photosynth. Res., 128*, 183–193.

Noh, H. K., Lu, R. F., (2007). Hyperspectral laser-induced fluorescence imaging for assessing apple fruit quality. *Postharvest Biol. Technol., 43*, 193–201.

Obenland, D., & Neipp, P., (2005). Chlorophyll fluorescence imaging allows early detection and localization of lemon rind injury following hot water treatment. *Hortscience, 40*, 1821–1823.

Öquist, G., & Wass, R., (1988). A portable, microprocessor operated instrument for measuring chlorophyll fluorescence kinetics in stress physiology. *Physiol. Plant., 73*, 211–217.

Ounis, A., Cerovic, Z., Briantais, J., & Moya, I., (2001). Dual-excitation FLIDAR for the estimation of epidermal UV absorption in leaves and canopies. *Remote Sens. Environ., 76,* 33–48.

Pfündel, E. E., Agati, G., & Cerovic, Z. G., (2006). Optical properties of plant surfaces. In: Reiderer, M., & Müller, C., (eds.), *Biology of the Plant Cuticle* (Vol. 23, pp. 216–249). Blackwell Publishing Ltd: Oxford-Ames-Victoria.

Pfündel, E. E., Ghozlen, N. B., Meyer, S., & Cerovic, Z. G., (2007). Investigating UV screening in leaves by two different types of portable UV fluorimeters reveals in vivo screening by anthocyanins and carotenoids. *Photosynth. Res., 93,* 205–221.

Pinelli, P., Romani, A., Fierini, E., & Agati, G., (2018). Prediction models for assessing anthocyanins in grape berries by fluorescence sensors: Dependence on cultivar, site, and growing season. *Food Chem., 244,* 213–223.

Pinelli, P., Romani, A., Fierini, E., Remorini, D., & Agati, G., (2013). Characterization of the polyphenol content in the kiwifruit (*Actinidia deliciosa*) exocarp for the calibration of a fruit-sorting optical sensor. *Phytochem. Anal., 24,* 460–466.

Prange, R. K., DeLong, J. M., Harrison, P. A., Leyte, J. C., & McLean, S. D., (2003). Oxygen concentration affects chlorophyll fluorescence in chlorophyll-containing fruit and vegetables. *J. Am. Soc. Hortic. Sci., 128,* 603–607.

Prange, R. K., DeLong, J. M., Leyte, J. C., & Harrison, P. A., (2002). Oxygen concentration affects chlorophyll fluorescence in chlorophyll-containing fruit. *Postharvest Biol. Technol., 24,* 201–205.

Qiu, Y., Zhao, Y., Liu, J., & Guo, Y., (2017). A statistical analysis of the freshness of postharvest leafy vegetables with application of water based on chlorophyll fluorescence measurement. *Inform. Process Agric., 4,* 269–274.

Raimondi, V., Agati, G., Cecchi, G., Gomoiu, I., Lognoli, D., & Palombi, L., (2009). *In vivo* real-time recording of UV-induced changes in the auto fluorescence of a melanin-containing fungus using a micro-spectrofluorimeter and a low-cost webcam. *Opt. Express, 17,* 22735–22746.

Ralph, P. J., Schreiber, U., Gademann, R., Kühl, M., & Larkum, A. W., (2005). Coral photobiology studied with a new imaging pulse amplitude modulated fluorometer. *J. Phycol., 41,* 335–342.

Rascher, U., Agati, G., Alonso, L., Cecchi, G., Champagne, S., Colombo, R., Damm, A., et al., (2009). CEFLES2: The remote sensing component to quantify photosynthetic efficiency from the leaf to the region by measuring sun-induced fluorescence in the oxygen absorption bands. *Biogeosciences, 6,* 1181–1198.

Rice-Evans, C. A., & Packer, L., (2003). *Flavonoids in Health and Disease* (2nd edn.). Marcel Dekker, Inc.: New York.

Rodriguez, C., Torre, S., & Solhaug, K. A., (2014). Low levels of ultraviolet-B radiation from fluorescent tubes induce an efficient flavonoid synthesis in *Lollo Rosso lettuce* without negative impact on growth. *Acta Agric. Scand. Sect. B. Soil Plant Sci., 64,* 178–184.

Rybarczyk-Plonska, A., Wold, A. B., Bengtsson, G. B., Borge, G. I. A., Hansen, M. K., & Hagen, S. F., (2016). Flavonols in broccoli (*Brassica oleracea* L. var. italica) flower buds as affected by postharvest temperature and radiation treatments. *Postharvest Biol. Technol., 116,* 105–114.

Schofield, R. A., DeEll, J. R., Murr, D. P., & Jenni, S., (2005). Determining the storage potential of iceberg lettuce with chlorophyll fluorescence. *Postharvest Biol. Technol., 38,* 43–56.

Schreiber, U., (1986). Detection of rapid induction kinetics with a new type of high-frequency modulated chlorophyll fluorometer. *Photosynth. Res., 9*, 261–272.

Schreiber, U., (2004). Pulse-amplitude-modulation (PAM) fluorometry and saturation pulse method: An overview. In: George C. Papageorgiou and Govindjee, (eds.), *Chlorophyll a Fluorescence: A Signature of Photosynthetis* (pp. 279–319). Springer: The Netherlands.

Simko, I., Jimenez-Berni, J. A., & Furbank, R. T., (2015). Detection of decay in fresh-cut lettuce using hyperspectral imaging and chlorophyll fluorescence imaging. *Postharvest Biol. Technol., 106*, 44–52.

Sivankalyani, V., Feygenberg, O., Diskin, S., Wright, B., & Alkan, N., (2016). Increased anthocyanin and flavonoids in mango fruit peel are associated with cold and pathogen resistance. *Postharvest Biol. Technol., 111*, 132–139.

Tuccio, L., Remorini, D., Pinelli, P., Fierini, E., Tonutti, P., Scalabrelli, G., & Agati, G., (2011). A rapid and nondestructive method to assess in the vineyard grape berry anthocyanins under different seasonal and water conditions. *Aust. J. Grape Wine R., 17*, 181–189.

Valeur, B., & Berberan-Santos, M. N., (2011). A brief history of fluorescence and phosphorescence before the emergence of quantum theory. *J. Chem. Educ., 88*, 731–738.

Vanamala, J., (2017). Food systems approach to cancer prevention. *Crit. Rev. Food Sci. Nutr., 57*, 2573–2588.

Wright, H., DeLong, J., Lada, R., & Prange, R., (2009). The relationship between water status and chlorophyll a fluorescence in grapes (Vitis spp.). *Postharvest Biol. Technol., 51*, 193–199.

Wulf, J. S., Geyer, M., Nicolai, B., & Zude, M., (2005). Non-destructive assessment of pigments in apple fruit and carrot by laser-induced fluorescence spectroscopy (LIFS) measured at different time-gate positions. In: Mencarelli, F., & Tonutti, P., (eds.), *Proceedings of the 5th International Postharvest Symposium* (Vol. 1–3, pp. 1387–1393). International Society Horticultural Science: Leuven 1.

Wulf, J., Rühmann, S., Rego, I., Puhl, I., Treutter, D., & Zude, M., (2008). Nondestructive application of laser-induced fluorescence spectroscopy for quantitative analyses of phenolic compounds in strawberry fruits (Fragaria x ananassa). *J. Agric. Food Chem., 56*, 2875–2882.

Zivcak, M., Bruckova, K., Sytar, O., Brestic, M., Olsovska, K., & Allakhverdiev, S. I., (2017). Lettuce flavonoids screening and phenotyping by chlorophyll fluorescence excitation ratio. *Planta, 245*, 1215–1229.

CHAPTER 5

ELECTROCHEMICAL SENSORS FOR QUALITY DETERMINATION OF FRUITS AND VEGETABLES

ESMAEIL HEYDARI-BAFROOEI,[1] ALI A. ENSAFI,[2] and MARYAM AMINI[1]

[1]*Department of Chemistry, Faculty of Science, Vali-e-Asr University of Rafsanjan, P.O. Box 518, Rafsanjan, Iran*

[2]*Department of Chemistry, Isfahan University of Technology, Isfahan – 84156–83111, Iran*

ABSTRACT

Sensing and biosensing methods are very broad fields of research that have a great impact on health, environmental quality control, and food quality. The aim of this chapter is limited to (bio)sensor technologies developed in the very last year exactly for screening quality control of vegetables and fruits. The chapter covers the fundamental principles and types of electrochemical (bio)sensors designs that are testified for vegetables and fruits-specific applications. Furthermore, the integration of nanotechnology into electrochemical biosensors design was also investigated.

5.1 ELECTROCHEMICAL BIOSENSORS

Electrochemistry provides many methods for measuring the quality of fruits and vegetables so that it can be considered as one of the pioneers in providing sensors for food technology (Rapini and Marrazza, 2017; Liu et al., 2017; Ontiveros et al., 2017; Ensafi, Heydari-Bafrooei, and Amini, 2012; Ensafi et al., 2014, 2016; Sadrabadi et al., 2016). Moreover, the combination of electrochemistry and nanotechnology with molecular biology is made possible to fabricate miniaturized devices for chemical and biological sensing (Cui et al., 2018; Felix and Angnes, 2018; Puiu and Bala, 2018; Heydari-Bafrooei and Askari, 2017; Heydari-Bafrooei and Shamszadeh, 2017; Heydari-Bafrooei, Amini, and Ardakani, 2016). For instance, electrochemical biosensors can offer quick, sensitive, ease of use, long-term reliability, and reproducibility, and low-cost detection of target analytes. They are sensitive and selective technologies for the monitoring of fresh and processed crops (Manzanares-Palenzuela et al., 2015; Verma and Bhardwaj, 2015). Electrochemical biosensors utilize a wide variety of chemistries. However, all use nanoscale interactions between the analyte in unknown solution, the bioreceptor layer, and a solid electrode surface (Silva et al., 2018; Sánchez-Paniagua, Redondo-Gómez, and López-Ruiz, 2017; Kang et al., 2017). The bioreceptor layer contains the biomolecules that react with the target analyte in the solution and change the electrochemical characteristics of the solid electrode surface. Electrochemical biosensors can detect a variety of target analytes by simply varying the bio-recognition molecules used. Electrochemical biosensors can detect pesticides, veterinary drugs, steroids, pathogenic bacteria, and toxins and have the potential to be established for the simultaneous determination of several target analytes. For food applications, the bioreceptor can be an aptamer, antibody, nucleic acid, enzyme, and molecularly imprinted polymer (MIP) (Figure 5.1). Among them, traditionally, enzymes, and antibodies are the most popular for quality determination of fruits and vegetables due to their specific binding affinities and catalytic activities.

In electrochemical enzymatic biosensors, the selective reaction between the enzyme (recognition element) and its substrate (target analyte) occurs, and then electrochemical transducer detects the activity of enzyme inhibitors or formation and degradation of an electroactive compound (substrate detection) (Kurbanoglu, Ozkan, and Merkoçi, 2017; Etienne et al., 2015). It is clear that the enzymes that belong to the oxidoreductases family such as oxidases, laccase, peroxidases, di-oxygenases, and dehydrogenases should be used in the case of substrate detection. In this case, detection is based on the electrochemical analysis of hydrogen peroxide or nicotinamide adenine dinucleotide (NADH) in the reduced form as electroactive compounds (Alagiri, Rameshkumar, and Pandikumar, 2017; Huang et al., 2018; Zappi et al., 2017; Korani, Salimi, and Karimi, 2017; Pilas et al., 2017). In the case of determination of the inhibitors, the enzyme inhibits by the analyte and by obtaining the difference between the enzymes activity in the presence and the absence of inhibitor, the inhibition grade is associated with the concentration of the analyte (Malvano et al., 2017; Nalini et al., 2016; Attar et al., 2015).

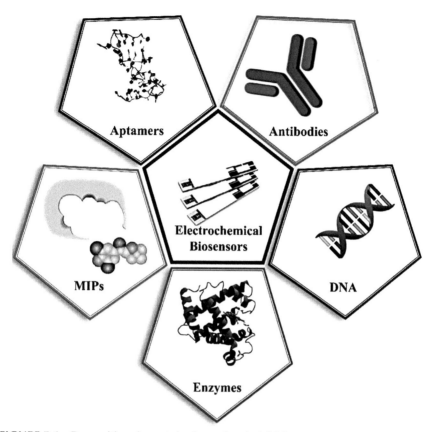

FIGURE 5.1 Recognition elements in electrochemical (bio)sensors.

In the antibody-based electrochemical biosensors or immunosensors, immobilization of antigen corresponding to the target antibody on the electrode surface without changing their specificity and immunological activity is one of the most key steps in the construction of an electrochemical immunosensor. After sample incubation, an antibody labeled with the electroactive compound is added in order to allow the detection or a direct detection base on the affinity event is performed (known as label-free detection) such as in impedimetric-based biosensors. Impedimetric biosensors measure any change of the impedance at the electrode-solution interface that occurs after antigen-antibody binding (Figure 5.2) (Felix and Angnes, 2018; Patris, Vandeput, and Kauffmann, 2016; Farka et al., 2017; Zamora-Gálvez et al., 2017).

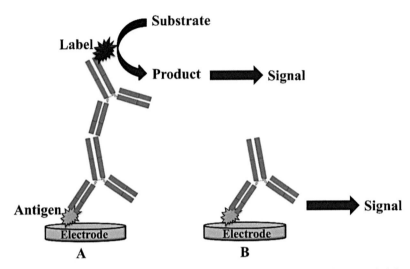

FIGURE 5.2 Some plans at antibody-based biosensors. A. detection by a sandwich-type immunosensor in multi-steps, the reaction between the label and reagent leads to the signal. B: Label-free immunosensor detection in one step, the antigen-antibody binding makes the signal.

The construction of the biosensor that is used nucleic acid as biore-ceptor layer is simple, fast, and cheap and therefore it is extensively utilized in food safety. One of the basic advantages of nucleic acid over enzyme or antibodies biorecognition elements is its easily synthesis and regenerability. An electrochemical DNA biosensor is consisted of a DNA probe layer immobilized on an electrode surface and is used to recognize its complementary DNA target to form a DNA double helix formation (hybridization) or detect the DNA structure changes induced by interaction with DNA-binding molecules (the analyte). This phenomenon changes electrochemical properties of the sensor, which are further translated into an electrochemical signal (Figure 5.3). Like other biorecognition elements, immobilization of DNA on the surface of solid electrode surfaces is also a critical step in the development of DNA-bases biosensors (Minaei et al., 2015; Chang, Deng, and Chen, 2015; Manzano et al., 2018).

Whole-cells of living organisms such as algae, fungi, bacteria, yeast, and plant and animal cells have also been utilized as bioreceptor element in biosensing devices. Whole-cell biosensors normally show multi-receptor behavior, but they have numerous advantages over enzyme-based sensors, such as high stability, less purification steps, low cost of preparation, and effective layer regeneracy. In whole-cell biosensors, the general metabolic status of the cells records in response to the samples to be analyzed (Hansen and Unruh, 2017). For instance, a sensitive method for determination of hydrocarbons is the level of oxygen consumption and catechol production during the metabolism of hydrocarbons (Di Gennaro et al., 2011).

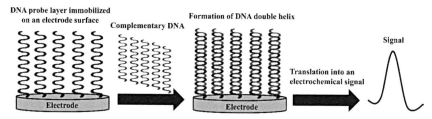

FIGURE 5.3   A schematic of general design of DNA biosensor.

Aptamer-based biosensors (aptasensors) have attracted much attention in the quality determination of fruits and vegetables. Aptamers are short-chain single-stranded DNA and RNA oligonucleotides with a stable three-dimensional (3D) structure synthesized in vitro with no need for animal or cell cultures. These single-stranded sequences have the ability to twist and can bind specifically to the target species (e.g., proteins, drugs, and other organic or inorganic analytes), that can be considered oligonucleotide analogs of antibodies as biorecognition layer in biosensors. The values of the dissociation constant for aptamer-target complexes are in the range of nanomolar to picomolar that this shows high affinity and specificity of this bioreceptors (Seo and Gu, 2017). As listed in Table 5.1, aptamers have numerous advantages over other bioreceptors such as they can be selected by in vitro technologies for their target molecules and they are thermally stable and reusable under high temperatures, and they are greatly flexible and easily modifiable with functional groups. Therefore, in recent years, the use of aptamers as biological recognition layer in electrochemical biosensors has increased sharply (Vasilescu and Marty, 2016; Ansari et al., 2017; Malekzad et al., 2017).

The technique of molecular imprinting makes it possible to form specific recognition layers in synthetic polymers through the use of templates or imprint molecules. The synthesized MIPs mimic specific recognition in a manner similar to natural biorecptors such as antibodies and enzyme, which are famous as low stability biomaterials. Therefore,

MIPs called artificial antibodies or plastic antibodies that researchers have a strong interest in their use of electrochemical sensors due to stability, ease of preparation, reusability, and less process cost (Table 5.1). A typical MIP system usually contains a template molecule, a functional monomer, cross-linking reagents, and an initiator in a porogenic solvent. Basically, MIPs are synthesized by three-step processes involving: (1) interaction of functional monomers with templates to form a complex through covalent or non-covalent bonds in solution; (2) polymerization of cross-linkers and initiators with the complex under photo-/thermal conditions; and (3) the removal of the template molecules through extraction. Active sites are formed in the building of the polymer after template removal, which is matching in size, functional groups, and shape to the template molecule (Figure 5.4). Mimicking the biological activity of antibodies, MIPs can bind the target analyte (with the near-shape and microstructure of templates) in the molecularly imprinted sites (Gui et al., 2018; Turco et al., 2018; Ansari, 2017).

TABLE 5.1   Advantages and Disadvantages of Receptors in Electrochemical Sensors and Biosensors

Receptor	Sensor Name	Advantages	Disadvantages
Antibodies	Immunosensor	High specificity	Sensitive to the temperature
		High affinity	Irreversible denaturation
			Limited number of targets (proteins) usually detected in an distinct assay
			Laborious generation (in vivo synthesis)
			Poor stability
			Limited number of substrates for which enzymes
			Complexity of the assays
Enzymes	Enzymatic biosensor	Simple instruments	Laborious and time-consuming assay
		Modulation of enzyme activity	Low stability of materials
		Easy to modify the catalytic properties	Heavy dependence on physiological conditions
		High Specificity	

TABLE 5.1 *(Continued)*

Receptor	Sensor Name	Advantages	Disadvantages
Whole cells	Whole-cell biosensor	Possibility to study the samples via processes in which several enzymes are involved	Low specificity Slow response/recovery times
		Cost-effectiveness	
Aptamers	Aptasensor	Not using animals for the production of the molecules (in-vitro synthesis)	
		High-temperature stability	
		Reversible denaturation	
		Easy modifications for enhancing specify	
		Discriminate analytes with different functional groups	
		Reusability	
Nucleic acids	Genosensor	High specificity	Limited target (complementary nucleic acid)
		High stability	
		Inexpensive	
		in-vitro synthesis	
Molecularly imprinted polymers (MIPs)	MIP sensor	Reusability	Complexity of the assays
		Inexpensive	Laborious and time consuming assay
		High temperature chemical, and mechanical stability	Inconsistency with aqueous solutions
			Possibility to leak the template molecules

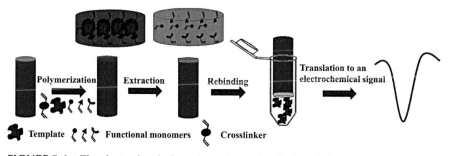

FIGURE 5.4 The electrochemical sensing using molecular imprinting process.

5.2　NANOMATERIALS IN ELECTROCHEMICAL BIOSENSORS

The investigation of the literature indicates that the ultra-sensitive electrochemical biosensors for the quality determination of fruits and vegetables combined nanomaterials in the biosensor architecture. The nanomaterials were used in the electrochemical biosensing for two reasons: to increase the electrical conductivity and electrochemical response of the sensor and as the immobilization matrices for the biorecognition element. As shown in Figure 5.3, nanomaterials can be used as support materials for aptamer, DNA, enzyme, or antibody immobilization, or as labels for electrochemical signal amplification. Both factors improve detection limits. Carbonaceous nanomaterials, metal nanomaterials, and magnetic nanoparticles have been successfully used to construct numerous electrochemical biosensors for the quality determination of fruits and vegetables. In this part, the fundamental properties of these three categories of nanomaterials utilized in food safety electrochemical biosensors are argued.

5.2.1　CARBON NANOMATERIALS

Carbon nanomaterials, including single- and multi-walled carbon nanotubes, nanodiamonds, fullerene, nanoporous carbon, carbon dots, and graphene are reported in the field of electrochemical biosensing devices. Among them, the most extensively used carbonaceous nanomaterials utilized in electrochemical biosensors for quality determination of fruits and vegetables are carbon nanotubes and graphene.

5.2.1.1　CARBON NANOTUBES

After the first use of carbon nanotubes in the construction of sensors by Britto et al., carbon nanotubes-based electrochemical (bio)sensors developed greatly. The great mechanical flexibility, high surface-to-volume ratio, high efficiency of electron transport, the abundant presence of edge-plane-like defects, exceptional electronic properties, unique electrochemical stability, and excellent thermal conductivity, allow being used as very attractive materials for the construction of electrochemical (bio)sensors. Moreover, their high surfaces of carbon nanotubes can be readily functionalized with a variety of chemical groups for bioreceptor immobilization that it is gaining importance in the construction of electrochemical biosensors. Nanocomposites prepared by a combination of carbon nanotubes and other materials, such as metal nanoparticles, polymers, or ionic liquids are used to improve the electrocatalytic properties and better immobilization of biorecognition elements.

5.2.1.2   GRAPHENE MATERIALS

Graphene, the world's first two-dimensional nanomaterial, which is a mono-layer of carbon atoms in a hexagonal lattice displaying the highest values of mechanical strength among the known materials (fracture strength, 125 GPa), surprising charge-transfer capabilities (200 000 cm^2 V^{-1} s^{-1}), excellent electrical conductivity, ultra-large specific surface area (2630 m^2 g^{-1}), simple preparation methods, high thermal conductivity (~5000 W m^{-1} K^{-1}) and favorable biocompatibility that could possibly be used in electronic instruments such as ultrasensitive electrochemical (bio)sensors. Furthermore, graphene can be readily oxidized into graphene oxide (GO) to add solubility and dispersibility to them in aqueous media by introducing functional groups such as hydroxyl, epoxy, carbonyl, and carboxyl. These hydrophilic groups allow immobilizing biomolecules easily on the GO-modified surfaces, which are very important features in biosensing-based devices.

5.2.2   METAL NANOMATERIALS

Metal nanomaterials have found extensive use in electrochemical biosensors, especially noble metal nanoparticles, e.g., palladium nanoparticles, silver nanoparticles, platinum nanoparticles, and gold nanoparticles, due to their high specific surface area, small size (1–100 nm), high stability, conductivity, biocompatibility, excellent chemical, physical, and electronic properties (different from bulk material), plenty absorption sites to antibodies, enzymes, antigen, and DNA, and flexibility in making new and advanced biosensor devices, which subsequently led to improvements in the detection signal. Beside noble metal nanoparticles, iron, and iron-oxide magnetic nanoparticles provide a good experimental platform for the construction of electrochemical biosensor with lower cost and easier preparation. Generally, metal nanoparticles served as "conductive wires" to connect the biorecognition layer with the electrode surface and shorten the space between them, which resulted in a significant improvement in the electron transport rates. Metal oxide nanoparticles are often utilized as immobilizer agents of bioreceptors due to their biocompatibility, while semiconductor nanoparticles are often applied as labeling factors. The conductivity and electrocatalytic properties of metal nanomaterials are strictly correlated to their size, distribution, composition, shape, structure, and immobilization/dispersion, which these factors are controlled by the synthesis protocol.

5.3 ELECTROCHEMICAL DETECTION

Most researchers use electrochemical methods as a transducer of developed biosensors because equipment for electrochemical biosensor-transducer can be simpler, less expensive, portable, and miniaturized. Based on their instrumental principle, the electrochemical biosensors can employ potentiometric, amperometric, conductometric, and impedimetric transducers translating the chemical information of the analytes into a quantifiable electrochemical signal. These transducers generate a measurable charge accumulation or potential (potentiometric), a measurable current (amperometric), changes in the conductivity of the solution between two electrodes (conductometric) and changes both resistance and reactance in the biosensor (impedimetric). Glucose and lactate biosensors are two of the most commonly used electrochemical biosensors. Glucose biosensor is called glucowatch and is used to measure blood glucose in diabetic patients. Lactate bioassay is designed to measure blood lactate concentrations. This sensor operates on an amperometric method, in which the reaction of lactate oxidation is carried out by lactate dehydrogenase.

5.3.1 AMPEROMETRIC BIOSENSORS

The transducers of these sensors operate with the constant voltage setting between a working electrode and a standard electrode. By applying this potential, electron transfer reactions occur, and a special current through the oxidation and/or reduction of electroactive species, as well as biorecognition events, are produced that is proportional to the concentration of the analyte in the solution. These types of biosensors are more useful because of their high sensitivity and linear ranges. Amperometric biosensors are divided into three types of type I, type II, and type III. The type I biosensors operate on the basis of the electrochemical activity of products derived from biological reactions. The limitation of these biosensors is the need for overpotential, which is why type II biosensors were designed in which they use an intermediate for electron transfer. These intermediates are small redox molecules that by interacting with bioactive elements, electron transfer from the biological element to the electrode surface was occur. In type III biosensors, the bioreceptor acts as an electrocatalyst and causes the transfer of electrons between the analyte molecules and the electrode. This type of sensor does not need to any intermediate, and the working potential is the redox potential of the biological element, resulting in a reduction in the overpotential. That's why the selectivity of this type of biosensors is very high and, furthermore, because the biorecognition element is immobilized at the electrode surface, the sensitivity is very high.

5.3.2 POTENTIOMETRIC BIOSENSOR

This kind of electrochemical biosensor works in an inert position where no current exists. The basis of their work is the accumulation of charge density on the electrode surface, which is produced by the biocatalyst processes. The potential between the two electrodes is measured when there is no current between them. The most important potentiometric sensors are ion-selective electrodes, with a special membrane in their structure, which specifically allows only certain ions to pass. Due to the specific passage of ions, the electrical charge around the membrane will be distributed unevenly, and the potential between the two electrodes is measured.

5.3.3 CONDUCTOMETRIC BIOSENSORS

This class of transducers operates based on changing the conductivity of the medium through the interaction between the analyte and the bioreceptor. In the conductometric biosensors, the conductivity of the solution is measured by establishing a relatively high frequency alternating electric field between the two electrodes. The conductivity of a solution depends on the type and number of ions contained therein. Different ions, based on their charge and mobility, contribute to conductivity. Most enzymes catalyze reactions that cause changes in the conductivity of the environment through the production and use of ions. The conductometric method lacks the selectivity necessary for biosensors; however, in conductometric biosensors by introducing a specific enzyme as a biorecognition layer, a selective biosensor can be created.

5.3.4 IMPEDIMETRIC BIOSENSOR

This transducer works based on assessing the impedance change produced by binding of analytes to biorecognition elements immobilized onto the solid electrode surface. In general, if the analyte binds to bioreceptor, the interfacial characteristics change. Actually, electrochemical impedance spectroscopy (EIS) provides a fingerprint of the interfacial region. The impedance can be determined in the presence or absence of a redox couple, which is referred to as faradic and non-faradic impedance measurement, respectively. The faradic biosensors sense the binding of the analyte to bioreceptors on the surface of modified electrodes by assessing the variation in the faradaic current (interfacial electron transport resistance) due to steric hindrance produced by the biomolecular interaction and/or by the electrostatic repulsion between the free charges of the target molecules and the electroactive species in the supporting electrolyte. The EIS is a non-destructive (ND) technique due to small amplitude perturbation from a steady-state.

5.4   QUALITY MEASUREMENT OF FRUITS AND VEGETABLE

Recent examples of electrochemical biosensors that have been utilized for quality determination of fruits and vegetables are summarized in Table 5.2 and discussed in more detail in subsections, for each group of biorecognition elements.

5.4.1   DNA AND APTAMER-BASED BIOSENSOR

An aptasensor for *Cryptosporidium parvum* oocysts was constructed based on a simple plan (Iqbal et al., 2015). Firstly, DNA aptamers were selected for the first time against *C. parvum* oocysts using the systematic evolution of ligands by exponential enrichment (SELEX), and to use the aptamers to sense the presence of this parasite in some fruits. A screen-printed carbon electrode was modified with gold nanoparticles, and the resulting electrode used as a substrate for attachment of thiolated aptamer. Square wave voltammetry (SWV) was used as an electrochemical detection method in a solution containing ferri-ferrocyanide as redox probe. After the binding between the oocysts and immobilized aptamer, the current intensity of redox probe increased and the peak position shifted towards more cathodic potentials. The reason for these results is due to the conformational change in the aptamer after binding to the oocysts and surface reformation, which makes it possible for redox probe to penetrate more freely to the electrode surface. Both the change in peak current intensity and the shift of peak potential were used for detecting the target oocysts, with a detection limit of approximately 100 oocysts.

TABLE 5.2 Applications of Electrochemical Sensors and Biosensors for Quality Assessment of Fruits a Vegetables

Analyte	Receptor	Biosensor Configuration	Detection Technique	Working Range	LOD	Matrix	Comments	References
C. parvum oocysts	Aptamer	AuNP/SPCE	SWV	150–800 oocysts	100 oocysts	Mango and pineapple	Both the variation in peak current intensity and the shift of peak potential were used for detecting the target	Iqbal et al., 2015
Acetamiprid	Aptamer	AgNP/NG/GCE	EIS	0.1–5000 pM	33 fM	Cucumber– and tomato		Jiang et al., 2015
Acetamiprid	Aptamer	AuNP/Au	EIS	5–600 nM	1 nM	Tomato	Dissociation constant for acetamiprid-aptamer complex determined	Fan et al., 2013
Chlorpyrifos	Aptamer	Fe@MWCNTs/ OMC/GCE	CV	$1-10^5$ ng mL^{-1}	0.33 ng mL^{-1}	Pakchoi, Lettuce, Leek	Sensor selectivity was evaluated in presence of other pesticides.	Jiao et al., 2016
Aeromonas hydrophila	DNA	MWCNT/CPE and Au	SWV		0.1 pM	Vegetables	Fish sample was also analyzed. Comparison between MWCNT/ CPE and Au performed.	Ligaj et al., 2014
Malathion	Enzyme (AChE)	AChE/Pd-Cu Nanowires/GCE	Amperometry	5–1000 ppt	4.5 pM	Courgettes, Carrots, Lettuces, Orange,	Results compared with LC-MS/MS data.	Song et al., 2017

TABLE 5.2 *(Continued)*

Analyte	Receptor	Biosensor Configuration	Detection Technique	Working Range	LOD	Matrix	Comments	References
Carbaryl and methomyl	Enzyme (AChE)	GC/MWCNT/ PANI/AChE	SWV	9.9–49.6 and 4.9–29.2 μM	1.4 and 0.95 μM	Apple, Broccoli, and cabbage	Results compared with HPLC data.	Cesarino, et al., 2012
Pirimicarb	Enzyme (laccase)	Laccase/ MWCNT/GCE	SWV	0.9–10 μM	1.8×10^{-7} M	Lettuces and tomato	Sensor selectivity was evaluated in the presence of pro-vitamin A, vitamins B1 and C	Oliveira et al., 2013
Monocrot– ophos and Dichlorvos	Enzyme (AChE)	Chitosan/AChE/ ZnO/Pt	CV		0.036 and 0.012 nM	Orange	An analytical method was also constructed by establishing 24 linear regression models	Sundarmu rugasan et al., 2016
Formetanate hydro– chloride	Enzyme (laccase)	Laccase/AuNPs/ Au	SWV	0.943 –11.3 μM	9.5×10^{-8} M	Mango and grapes		Ribeiro et al., 2014

TABLE 5.2 *(Continued)*

Analyte	Receptor	Biosensor Configuration	Detection Technique	Working Range	LOD	Matrix	Comments	References
Naphthalene acetic acid	ABP1	ABP1/graphene/lipid/GCE	Potentiometry	0.05–10 μM	10 nM.	Apples, Bananas, Beans, Cabbage, Carrots, Celery, Cherries, Dill, Grapefruits, Lemons, Lettuce, Melon, Nectarines, Onions, Oranges, Parsley, Peaches, Pears (green and red), Pineapple, Potatoes, Red beets, Spinach, Strawberries, Tangerines, Tomatoes, and Watermelon	Effect of potential interferents was evaluated on the biosensor response.	Bratakou et al., 2016
Chlorpyrifos	Anti-chlorpyrifos monoclonal antibody	MWCNTs-THI-CHIT/GCE	EIS	0.1–100,000 ng/mL	0.046 ng/mL	Cabbage, Lettuce, Chinese chives	—	Sun et al., 2012
Parathion	Anti-parathion antibody	Graphene/SPCE	EIS	0.1–1000 ng L^{-1}	52 pg L^{-1}	Tomato and carrot	Results compared with HPLC data.	Mehta et al., 2016

TABLE 5.2 (Continued)

Analyte	Receptor	Biosensor Configuration	Detection Technique	Working Range	LOD	Matrix	Comments	References
Botrytis cinerea	B. cinerea-specific monoclonal antibody	SPCE-CNT	Amperometry	0–100 µg mL⁻¹	0.02 µg mL⁻¹	Apple	The electrochemical system compared with a commercial spectrophotometric system SAPS-ELISA-Kit.	Fernández-Baldo et al., 2009
Listeria monocytogenes	Mouse monoclonal IgG1 antibody to L. monocytogenes	Au	EIS	0–2000 CFU/mL	4 CFU/mL	Tomato pulp	Control experiments performed to test for both non-specific adsorption and cross-reactivity to S. enterica.	Radhakrishnan et al., 2013
Carbofuran	Anti-carbofuran monoclonal antibodies	Glutaraldehyde/cysteine/Au	EIS	1.0×10^{-7}–1.0×10^{-3} g/L	0.1 ng/mL	Cabbage, Tomato, Chinese chives	Soil sample was also analyzed.	Dai et al., 2017
Carbofuran	Carbofuran antibodies	Gelatin/Ab/GA/L-Cys/Au	EIS	1.0×10^{-1}–1.0×10^{3} ng/mL	1.0×10^{-1} ng/mL	Cabbage, Tomato, Lettuce	Soil and water samples were also analyzed.	Liu et al., 2015
Carbofuran	MIP	MIP/rGO@Au/GCE	DPV	0.05–20 µM	0.02 µM	Cabbage and cucumber		Tan et al., 2015

TABLE 5.2 (Continued)

Analyte	Receptor	Biosensor Configuration	Detection Technique	Working Range	LOD	Matrix	Comments	References
Phoxim	MIP	MIP/graphene/ GCE	DPV	0.8 – 140 μM	0.02 μM	Cucumber	Adsorption model, imprinting factor, and binding rate constant of the imprinted sensor were measured.	Tan et al., 2015
Paraoxon	MIP	MIP-CP, MIP/ PECH-GC and MIP/Graphite-PECH-GC	SWV	3.8 – 750 nM	1 nM	Cabbage	Water samples were also analyzed.	Alizadeh, 2010
Diazinon	MIP	nano-MIP-CP	SWV	2.5 – 100 nM	0.079 nM	Apple	Water samples were also analyzed.	Motaha–rian et al., 2016

Abbreviations: AuNP: gold nanoparticles; SPCE: screen-printed carbon electrode; SWV: square wave voltammetry; AgNP: silver nanoparticles; NG: nitrogen-doped graphene; GCE: glassy carbon electrode; Fc: ferrocene; MWCNTs: multi-walled carbon nanotubes; OMC: ordered mesoporous carbon; AChE: acetylcholinesterase; PANI: polyaniline; ABP1: Auxin-Binding Protein Receptor; THI: thionine; CHIT: Chitosan; Ab: antibody; GA: Glutaraldehyde; rGO: reduced graphene oxide; L-Cys: L-cysteine; PECH: poly epichlorohydrin.

For the investigation of the affinity of the proposed aptamer to oocyst, 14 aptamer clones were selected and SWV measurements were performed before and after 1-hour incubation with Cryptosporidium oocysts (3,000 oocysts in 30 µL). The value of ΔI (difference in current intensity before and after binding) used as an indicator of the affinity of the respective aptamer to the oocyst. Furthermore, the aptasensor was capable of sensing the oocysts in the mango and pineapple highlighted the efficacy of the aptasensor for food analysis applications.

An ultrasensitive and low-cost aptamer-based sensor for the quantitative detection of acetamiprid was developed based on a nanocomposite of nitrogen-doped graphene, decorated with silver nanoparticles (Jiang et al., 2015). The thiolated aptamer was immobilized on the glassy carbon electrode modified with the nanocomposite, and after soaking in droplet of the acetamiprid and formation of the target-aptamer complex, an increased impedimetric signal was captured on the electrode, the variation of EIS response being associated to the concentration of acetamiprid in the sample. The amplified signal and high sensitivity of the biosensor is due to the presence of the nanocomposite on the surface of sensor that facilitates the electron transfer and provides a large accessible surface area for loading a large amount of aptamer. The proposed aptasensor showed a low detection limit of 33 fM in spiked cucumber and tomatoes with excellent reproducibility and stability.

In another work, an aptasensor for acetamiprid sensing in tomatoes was developed based on the gold electrode modified with gold nanoparticles, which was employed as an electrode modifier for immobilization of aptamer (Fan et al., 2013). By adding acetamiprid, the binding of acetamiprid to surface-confined aptamer caused an increase of EIS signal (Figure 5.5).

A novel and an ultrasensitive aptamer-based amperometric sensor for the quantitative detection of chlorpyrifos were improved based on a simple design with double-assisted signal amplification strategy (Jiao et al., 2016). A glassy carbon electrode was modified with mesoporous carbon function-alized by chitosan, and ferrocene-branched chitosan composite dispersed multiwalled carbon nanotubes. The unique sandwich structure, aromaticity, unique oxidation-reducing aptitude, and good biological compatibility made ferrocene have a broad application prospects in electrochemical sensors. Chitosan with rich amino groups displays a tough film-forming aptitude. The porous matrix of chitosan films affords an ideal location for immobilization of the biorecognition elements and facilitates diffusion of substrate and/or inhibitors. In addition, chitosan shows the advantages of inexpensiveness, nontoxicity, susceptibility to chemical modification and biocompatibility.

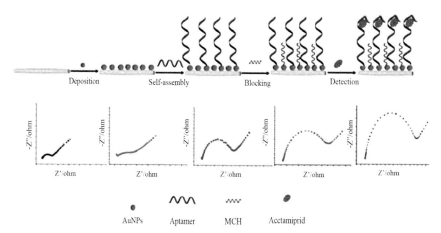

FIGURE 5.5   Aptasensor design and the procedure of acetamiprid detection presented by Fan et al. (2013) (Reprinted with permission. © 2013 Elsevier.)

Due to its desired characteristics, chitosan has extensively been utilized as an ideal matrix for immobilization of biomolecules in biosensor devices. Ferrocene-branched chitosan composite and MWCNTs for the development of electrochemical aptasensor amplified electrochemical signal, also, it also organized the effective immobilization of aptamers on the transducer. Furthermore, mesoporous carbon functionalized by chitosan has also a high specific surface area, high porosity, and ideal dispersibility which were utilized to powerfully capture larger amounts of material. The biosensor was characterized by a detection limit of 0.33 ng mL^{-1}. Although the aptasensor concept has some degree of difficulty (preparation of modified electrode), the sensor displayed good stability and reproducibility, and its application for the ultrasensitive detection of chlorpyrifos residues in vegetables and fruits demonstrated that it is adequate for real sample analysis.

Two kinds of electrochemical DNA sensor for the detection of pathogenic bacteria *Aeromonas hydrophila* are reported using DNA capture probes at the sequence 5′GTCAAGACGGTGGTGGGCTG as sensing element (Ligaj et al., 2014). The detection layer of biosensor I is a gold electrode modified with a self-assembled monolayer (SAM) of mercaptohexanol and thiolated DNA probe. In the biosensor II, DNA capture probes are covalently bound to the MWCNT-modified carbon paste (CP) electrode. Hybridization was detected by using Hoechst 33258 (biosensor I) and daunomycin (biosensor II) as electroactive hybridization indicators. After the DNA probes immobilized in the detection layers of the biosensor was hybridized with target

DNA isolated from *Aeromonas hydrophila*, square wave voltammetric peak currents of Hoechst 33258 and daunomycin were increased ranging from 75% to 135% and 34–92% for biosensor I and II, respectively. The electrochemical nucleic acid biosensors were used successfully for the detection of pathogenic strains of *Aeromonas hydrophila* in vegetable samples without performing PCR amplification prior to analysis.

5.4.2 ENZYMATIC BIOSENSORS

Bimetallic nanowires (Pd-Cu NW) have been used to develop acetylcholinesterase (AChE)-based electrochemical biosensor for the determination of organophosphate pesticides in fruit and vegetable samples (Song et al., 2017). The immobilized AChE could catalyze the hydrolysis of acetylthiocholine chloride (ATCl), and therefore an oxidation peak resulted from the oxidation of thiocholine. In the presence of increasing organophosphate pesticides, the corresponding peak current significantly decreased due to the inhibition of the activity of the AChE that could be ascribed to the interaction between pesticides and serine hydroxyl groups of AChE. The detection mechanism is based on the inhibition of AChE. The Pd-Cu NW provides a large active surface area for the immobilization of AChE, remarkable catalytic activity and conductivity and excellent electron mobility, and favorable chemical stability. The sensor showed a very low detection limit (4.5 pM) in the broad linear range of 5–1000 ppt.

An amperometric pesticide biosensor based on AChE immobilized in a core-shell structure of CNTs and polyaniline to be used to detect carbamate in fruits and vegetables by chronoamperometry (Cesarino et al., 2012). The conducting polymer of polyaniline can provide a suitable environment for the immobilization of biomolecules. The resulting nanostructures not only embrace the advantages of both CNTs and polyaniline but also exhibited enhancement of electrical conductivity as well as mechanical properties compared with pure conducting polymer or CNTs. The biosensor is an appropriate method to measure pesticides in foods, presenting reasonable stability, and not needing to exploit mediators (phthalocyanine) or cross-linking agents (glutaraldehyde (GA)). The cross-linking agents commonly bind to the active sites of enzymes, thus inhibiting their activity.

A biosensor was developed based on CP electrode modified with MWCNTs and immobilized laccase (Oliveira et al., 2013). The biosensor displayed an exceptional electron transfer rate for determination of 4-aminophenol, used as an enzyme substrate, and a stability of approximately 30 days. Pirimicarb was quantified by inhibition of laccase with a low detection limit of 0.04 mg kg^{-1} on a fresh weight vegetable basis. Among the fruits and vegetables tomato and lettuce samples were selected and analyzed by this method, and no interferences from pro-vitamin A, vitamins B$_1$, and C, and glucose were detected.

Sundarmurugasan et al. prepared an AChE-zinc oxide modified platinum electrode that was used to simultaneously detect monocrotophos and dichlorvos, categorized by the World Health Organization as two of the highly toxic organophosphate pesticides, in orange samples with a detection limit of 0.036 and 0.012 nM, respectively (Sundarmurugasan et al., 2016). The presence of ZnO nanoparticles with good biocompatibility and high isoelectric point (pI=9.4) when compared to the AChE enzyme (pI = 6.6) on the surface of biosensor improves the immobilization and electron transfer aptitude. Increased response to ATCl was observed by the introduction of ZnO nanospheres. AChE was immobilized the surface of ZnO nano-interface by a simple physical entrapment method. Because the developed biosensor can simultaneously detect monocrotophos and dichlorvos in orange samples, it is not necessary to provide a sensor for measuring MCP and DDVP separately in orange samples.

The enhanced electrochemical signal using gold electrode modified with AuNPs to design an enzymatic biosensor based on laccase for formetanate hydrochloride detection was also reported by Ribeiro et al. By impedimetric experiments, the authors assessed the enhancement of the conductivity of the electrode upon incorporation of AuNPs, by efficiently dropping the resistance to charge transfer until 350 Ω (starting from and 960 Ω in the absence of AuNPs) (Ribeiro et al., 2014). The principle of this biosensor is related to the capacity of formetanate hydrochloride to inhibit the laccase catalytic reaction that happens in the presence of aminophenol as substrate. This biosensor was effectively used to detect formetanate hydrochloride. The LOD of 9.5×10^{-8} M (0.02×10^{-4} mg/kg on a fresh fruit weight basis) was achieved. The developed biosensor was positively used for formetanate hydrochloride detection in mango and grapes.

5.4.3 IMMUNOSENSORS

A miniaturized potentiometric naphthalene acetic acid were developed with an experimental thin lipid film biosensing set-up developed on graphene nanosheets, incorporated with Auxin-binding protein 1 receptor (Bratakou et al., 2016). The incorporation of the Auxin-binding protein 1 receptor in the lipid blend, before polymerization, holds its bioactivity inside the lipid matrix. The data suggest a high affinity towards hormone and non-interference of biologically inactive 2-NAA and indole-3-acetic acid. The biosensor provides fast response times (order of min) and LOD of nanomolar for phytohormone. According to the authors, the sensor can be established into a portable device for the fast on-site monitoring of fruits and vegetables by non-skilled personnel.

An amperometric immunosensor was developed based on multi-walled carbon nanotubes/thionine/chitosan nanocomposite film for insecticide chlorpyrifos detection, which belongs to organophosphate family, in fruits and vegetables such as cabbage or lettuce (Sun et al., 2012). Chitosan has exceptional film-making and adhesion aptitude and can dissolve MWCNTs adequately without assembling into bundles. Association of chitosan with MWCNTs provides a suitable microenvironment for immobilizing molecules and enhancing the electrocatalytic activity of MWCNTs. Thionine with a large volume of hydrophilic amino functional groups and planar aromatic structure makes it possible to interact with MWCNTs through π–π stacking forces strongly and subsequently enhancing the electroactivity of MWCNTs. As stated above for enzymatic biosensors, covalent bonding is the most used strategy for immobilization of biomacromolecules on the electrode surface. The anti-chlorpyrifos monoclonal antibody was covalently immobilized onto the surface of MWCNTs-THI-CHIT/GCE using the crosslinking agent GA. The linear curve ranging from 0.1 ng/mL to 1.0×10^5 ng/mL with a detection limit of 0.046 ng/mL was observed.

In another study, graphene sheets (Gs) modified screen-printed carbon electrodes were electrochemically treated with the 2-aminobenzyl amine for non-enzymatic label-free immunosensing of parathion (Mehta et al., 2016). The biorecognition event has occurred directly without the use of any crosslinkers. Generate the active-NH_2 functional groups on the surface of the electrode upon functionalization with 2-aminobenzyl amine allows simple immobilization of anti-parathion antibodies (Figure 5.6). In the impedimetric mode, the reported immunosensor showed a very low limit of detection of 52 pg L^{-1} within the concentration range of 0.1–1000 ng L^{-1}. The applicability of the biosensor was tested via the analysis of real samples (tomato and carrot) and cross-validation against standard (HPLC) method and the fabricated immunosensor showed specificity to parathion against other pesticides.

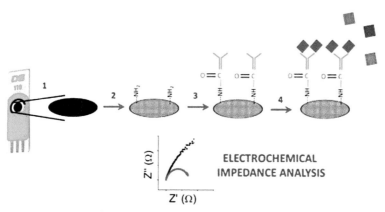

ELECTROCHEMICAL
IMPEDANCE ANALYSIS

Step 1: Drop-casting of graphene on carbon screen printed electrode
Step 2: Electro-catalyzed amine (-NH$_2$) functionalization of graphene with 2-aminobenzylamine
Step 3: Immobilization of anti-parathion antibodies on –NH$_2$ functionalized graphene SPE
Step 4: Immunosensing of parathion with above sensor

ANTIBODY   PARATHION   NON-SPECIFIC PESTICIDES

FIGURE 5.6   Schematic of the graphene-based screen-printed immunosensor for parathion (Reprinted with permission from Mehta et al., 2016. © Elsevier.)

Besides pesticides, there are a number of electrochemical immunosensors for detection of other biological elements such as fungi in fruits and vegetable samples using electrochemical immunosensors (Radhakrishnan et al., 2013). A biosensor based on screen-printed electrode modified with CNT integrated to both continuous-flow systems and microfluidic systems was developed for quantification of *Botrytis cinerea*, a plant-pathogenic fungus, producing a disease known as grey mold. The biosensor was constructed by immobilization of purified *B. cinerea* antigens on a rotating disk, and the detection of *B. cinerea* is based on a competitive immunoassay method. The immunologically reaction between *B. cinerea*-specific monoclonal antibody with the immobilized antigens is detected by a horseradish peroxidase enzyme and 4-tertbutyl catechol as an enzymatic mediator. The oxidation of 4-tertbutyl catechol to 4-tertbutyl o-benzoquinone was occurred by the catalysis activity of HRP in the presence of hydrogen peroxide. The electrochemical reduction back to 4-tertbutyl catechol is detected on SPCE-CNT at −0.15 V. The response current achieved from the product of the enzymatic reaction is proportional to the activity of the enzyme and subsequently, to the concentration of antibodies attached to the surface of the immunosensor of interest, but the response current is inversely proportional to the amount of the *B. cinerea* antigens present in the fruit sample. The response time is 30 min, and the LOD for an electrochemical method and the ELISA procedure are 0.02 and 10 µg mL^{-1}, respectively.

Rapid and sensitive biosensor methods for *Listeria monocytogenes,* an important Gram-positive rod-shaped foodborne bacterium that causes an extremely life-threatening infection, are currently mostly demonstrative (Radhakrishnan et al., 2013). One example is the impedimetric biosensor reported by Radhakrishnan et al. (2013) that used impedance spectroscopy and mouse monoclonal IgG1 antibody to *L. monocytogenes* immobilized onto the gold electrode, and achieved a calculated DL of 4 CFU/ml for *L. monocytogenes* in tomato pulp without enrichment. The impedance change arising from non-specific adsorption is measured by comparing the imped-ance change at the proposed electrode to that at a control electrode modified with mouse monoclonal IgG 1 antibody to glyceraldehyde 3-phosphate dehydrogenase (GAPDH), and found to be insignificant.

A label-free microcantilever-based immunosensor as a next-generation electromechanical technique with broad application in biological detection was reported for carbofuran determination (Dai et al., 2017). The sensor was prepared by modifying the gold-coated microcantilever by l-cysteine/ GA using crosslinking and immobilization of anti-carbofuran monoclonal antibodies as the receptor molecules on the modified microcantilever. The antigen-antibody specific binding under an experimental environment provides an increase in the tensile stress (due to antibody conformational change) which led to microcantilever deflection. The immunosensor achieved a detection limit of 0.1 ng mL^{-1}, and it showed a good linear relationship over the range from 1.0×10^{-7} to 1.0×10^{-3} g L^{-1}. Meanwhile, the immunosensor was applied for carbofuran determination in vegetable samples with accept-able results.

Liu et al. (2015) constructed an EIS-based immunosensor sensor for the determination of carbofuran by immobilizing L-cysteine and carbo-furan antibodies on a gold electrode by the applying GA as a crosslinker (Figure 5.7) (Liu et al., 2015). The analysis principle of this sensor based on the formation of antibody-antigen complexes and it could reduce the electron transfer rate between the electrode and an electrically active probe molecule, causing an extra increase in the charge-transfer resistance (Rct). The difference of Rct before and after immunization was measured to achieve the sensing of target analytes. The immunosensor can be used for the detection of carbofuran residual in agricultural and environmental samples.

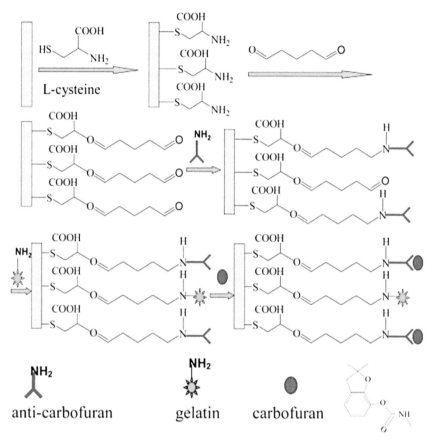

FIGURE 5.7 The schematic design of the preparation of immunosensor based on the interaction between antibody and antigen (Reprinted with permission from Liu et al., 2015. © Elsevier.)

5.4.4 MIP-BASED BIOSENSORS

A new MIP electrochemical sensor is fabricated based on reduced GO and gold nanoparticles (rGO@Au) decorated on a glassy carbon electrode for the detection of carbofuran. rGO@Au can increase imprinted cavities and recognition capacity and therefore increase the electroanalytical signal by attaching the analyte on the MIP receptor (Tan et al., 2015). The MIP was synthesized via polymerization around the template molecule in the presence of methyl acrylic acid, ethylene glycol maleic rosinate acrylate, and azobisisobutyronitrile as functional monomer, cross-linker, and initiator, respectively. This fabricated MIP has successfully detected the carbofuran pesticide in the range of 5.0×10^{-8}–2.0×10^{-5} mol L^{-1} with a detection limit of 2.0×10^{-8} mol L^{-1}. This sensor has been used efficaciously to detect carbofuran in the vegetable samples, demonstrating that the sensor will be of useful applications.

Graphene, a cross-linker (ethylene glycol-maleic rosinate acrylate) and a monomer (acrylamide) for making highly sensitive molecularly imprinted sensors are offered for phoxim (template) determination (Tan et al., 2015). The MIP film was formed on the graphene-modified GCE surface via a free radical polymerization method. The doping of graphene in the preparation of a phoxim membrane led to the enhancement of the electroanalytical signal of the MIP-based sensor regarding response and detection sensitivity. The MIP electrochemical sensor was effectively engaged for the detection of phoxim in cucumber samples.

An electrochemical sensor based on MIPs is also reported for paraoxon detection (Alizadeh, 2010). Synthesis of MIP carried out using methacrylic acid and ethylene glycol dimethacrylate as functional monomer and cross-linker agent, respectively. Three different techniques for MIP embedding to electrode surface were examined. These strategies are 1-MIP particles joining in the carbon paste (MIP-CP), 2-coupling of MIP with the GC surface by using poly epychloro hydrine (PECH) (MIP/PECH-GC) and 3-thin layer attachment of MIP and graphite mixture onto the GCE (MIP/Graphite-PECH-GC). Comparison of these strategies with each other was performed based on the sensor response to an analyte of interest, the selectivity against other structurally similar compound, and the washing effect on the electrode responses in the case of both analyte and competitor compound. MIP-CPE showed the higher response for paraoxon detection, higher paraoxon selectivity, more sensitivity, and wider linear range in comparison to other examined strategies. The results showed that the sensor retained paraoxon more firmly against washing step, demonstrating high affinity of the sensor to the target molecule. The developed MIP-CP electrode was used as a high selective sensor for paraoxon determination in vegetable samples.

An electrochemical sensor was developed by mixing MIP nanoparticles with graphite powder to prepare CP electrode, showing that MIP particles at nano-dimension showed a much better response than microparticles (due to increase in the number of recognition sites) for highly sensitive electrochemical sensing of diazinon pesticide in apple fruit samples (Motaharian et al., 2016). Cyclic voltammetry (CV) and SWV methods were used for electrochemical measurements. The sensor has long-term stability, easy preparation, and electrode regeneration providing a rapid and economical method for measuring.

5.5   CONCLUSION

So far, many efforts have been made to develop appropriate identification systems for testing the quality of fruits and vegetables. Low cost, portability, ease of use, quick detection times, and minimal side accessories are essential features of effective fruits and vegetable quality testing. It was shown that based on known principles of electrochemical biosensor, this type of analysis has all the items listed. Despite broad advances in biosensor technology and in line with previous studies in the literature, none of the current rapid sensing methods for fruits and vegetables, both commercial samples and samples that are in the development stage, do not meet all the requirements for food applications due to regulatory constraints. Therefore, further progress is needed in terms of validation parameters, analysis time, portability, and autonomy. Also, the electrochemical biosensors seem to be the most suitable and reliable technology to meet the regulatory requirements and overcome the barriers to industrial implementation, as it may be possible to achieve a lab-on-chip device, with the favorite analytical properties and compatible with specific industrial needs, at a low cost and affordable price.

KEYWORDS

- **acetylcholinesterase**
- **acetylthiocholine chloride**
- **cyclic voltammetry**
- **electrochemical impedance spectroscopy**
- **glutaraldehyde**
- **graphene oxide**

REFERENCES

Alagiri, M., Rameshkumar, P., & Pandikumar, A., (2017). Gold nanorod-based electrochemical sensing of small bio-molecules: A review. *Microchim. Acta, 184*, 3069–3092.

Alizadeh, T., (2010). Comparison of different methodologies for integration of molecularly imprinted polymer and electrochemical transducer in order to develop a paraoxon voltammetric sensor. *Thin Solid Films, 518*, 6099–6106.

Ansari, N., Yazdian-Robati, R., Shahdordizadeh, M., Wang, Z., & Ghazvini, K., (2017). Aptasensors for quantitative detection of *Salmonella Typhimurium. Anal. Biochem., 533*, 18–25.

Ansari, S., (2017). Combination of molecularly imprinted polymers and carbon nanomaterials as a versatile bio-sensing tool in sample analysis: Recent applications and challenges. *TrAC, Trends Anal. Chem., 93,* 134–151.

Attar, A., Ghica, M. E., Amine, A., & Brett, C. M. A., (2015). Comparison of cobalt hexacyanoferrate and poly(neutral red) modified carbon film electrodes for the amperometric detection of heavy metals based on glucose oxidase enzyme inhibition. *Anal. Lett., 48,* 659–671.

Bratakou, S., Nikoleli, G. P., Siontorou, C. G., Karapetis, S., Nikolelis, D. P., & Tzamtzis, N., (2016). Electrochemical biosensor for naphthalene acetic acid in fruits and vegetables based on lipid films with incorporated auxin-binding protein receptor using graphene electrodes. *Electroanalysis, 28,* 2171–2177.

Cesarino, I., Moraes, F. C., Lanza, M. R. V., & MacHado, S. A. S., (2012). Electrochemical detection of carbamate pesticides in fruit and vegetables with a biosensor based on acetyl cholinesterase immobilized on a composite of polyaniline-carbon nanotubes. *Food Chem., 135,* 873–879.

Chang, K., Deng, S., & Chen, M., (2015). Novel biosensing methodologies for improving the detection of single nucleotide polymorphism. *Biosens. Bioelectron., 66,* 297–307.

Cui, L., Lu, M., Li, Y., Tang, B., & Zhang, C. Y., (2018). A reusable ratiometric electrochemical biosensor on the basis of the binding of methylene blue to DNA with alternating AT base sequence for sensitive detection of adenosine. *Biosens. Bioelectron., 102,* 87–93.

Dai, Y., Wang, T., Hu, X., Liu, S., Zhang, M., & Wang, C., (2017). Highly sensitive microcantilever-based immunosensor for the detection of carbofuran in soil and vegetable samples. *Food Chem., 229,* 432–438.

Di Gennaro, P., Bruzzese, N., Anderlini, D., Aiossa, M., Papacchini, M., Campanella, L., & Bestetti, G., (2011). Development of microbial engineered whole-cell systems for environmental benzene determination. *Ecotoxicol. Environ. Saf., 74,* 542–549.

Ensafi, A. A., Heydari-Bafrooei, E., & Amini, M., (2012). DNA-functionalized biosensor for riboflavin based electrochemical interaction on pretreated pencil graphite electrode. *Biosens. Bioelectron., 31,* 376–381.

Ensafi, A. A., Jamei, H. R., Heydari-Bafrooei, E., & Rezaei, B., (2014). Development of a voltammetric procedure based on DNA interaction for sensitive monitoring of chrysoidine, a banned dye, in foods and textile effluents. *Sens. Actuat. B, 202,* 224–231.

Ensafi, A. A., Karbalaei, S., Heydari-Bafrooei, E., & Rezaei, B., (2016). Bio-sensing of naringin in marketed fruits and juices based on its interaction with DNA. *J. Iran Chem. Soc., 13,* 19–27.

Etienne, M., Zhang, L., Vilà, N., & Walcarius, A., (2015). Mesoporous materials-based electrochemical enzymatic biosensors. *Electroanalysis, 27,* 2028–2054.

Fan, L., Zhao, G., Shi, H., Liu, M., & Li, Z., (2013). A highly selective electrochemical impedance spectroscopy-based aptasensor for sensitive detection of acetamiprid. *Biosens. Bioelectron., 43,* 12–18.

Farka, Z., Juřík, T., Kovář, D., Trnková, L., & Skládal, P., (2017). Nanoparticle-based immunochemical biosensors and assays: Recent advances and challenges. *Chem. Rev., 117,* 9973–10042.

Felix, F. S., & Angnes, L., (2018). Electrochemical immunosensors: A powerful tool for analytical applications. *Biosens. Bioelectron., 102,* 470–478.

Fernández-Baldo, M. A., Messina, G. A., Sanz, M. I., & Raba, J., (2009). Screen-printed immunosensor modified with carbon nanotubes in a continuous-flow system for the *Botrytis cinerea* determination in apple tissues. *Talanta, 79*, 681–686.

Gui, R., Jin, H., Guo, H., & Wang, Z., (2018). Recent advances and future prospects in molecularly imprinted polymers-based electrochemical biosensors. *Biosens. Bioelectron., 100*, 56–70.

Hansen, P. D., & Unruh, E., (2017). Whole-cell biosensors and bioassays. *Compr. Anal. Chem., 77*, 35–53.

Heydari-Bafrooei, E., & Askari, S., (2017). Ultrasensitive aptasensing of lysozyme by exploiting the synergistic effect of gold nanoparticle-modified reduced graphene oxide and MWCNTs in a chitosan matrix. *Microchim. Acta, 184*, 3405–3413.

Heydari-Bafrooei, E., & Shamszadeh, N. S., (2017). Electrochemical bioassay development for ultrasensitive aptasensing of prostate specific antigen. *Biosens. Bioelectron., 91*, 284–292.

Heydari-Bafrooei, E., Amini, M., & Ardakani, M. H., (2016). An electrochemical aptasensor based on TiO_2/MWCNT and a novel synthesized Schiff base nanocomposite for the ultrasensitive detection of thrombin. *Biosens. Bioelectron., 85*, 828–836.

Huang, Y., Tan, J., Cui, L., Zhou, Z., Zhou, S., Zhang, Z., Zheng, R., et al., (2018). Graphene and Au NPs co-mediated enzymatic silver deposition for the ultrasensitive electrochemical detection of cholesterol. *Biosens. Bioelectron., 102*, 560–567.

Iqbal, A., Labib, M., Muharemagic, D., Sattar, S., Dixon, B. R., & Berezovski, M. V., (2015). Detection of *cryptosporidium parvum* oocysts on fresh produce using DNA aptamers. *PLoS One, 10*, e0137455.

Jiang, D., Du, X., Liu, Q., Zhou, L., Dai, L., Qian, J., & Wang, K., (2015). Silver nanoparticles anchored on nitrogen-doped graphene as a novel electrochemical biosensing platform with enhanced sensitivity for aptamer-based pesticide assay. *Analyst, 140*, 6404–6411.

Jiao, Y., Jia, H., Guo, Y., Zhang, H., Wang, Z., Sun, X., & Zhao, J., (2016). An ultrasensitive aptasensor for chlorpyrifos based on ordered mesoporous carbon/ferrocene hybrid multiwalled carbon nanotubes. *RSC Advances, 6*, 58541–58548.

Kang, D., Sun, S., Kurnik, M., Morales, D., Dahlquist, F. W., & Plaxco, K. W., (2017). New architecture for reagent less, protein-based electrochemical biosensors. *J. Am. Chem. Soc., 139*, 12113–12116.

Korani, A., Salimi, A., & Karimi, B., (2017). Guanine/ionic liquid derived ordered mesoporous carbon decorated with AuNPs as efficient NADH biosensor and suitable platform for enzymes immobilization and biofuel cell design. *Electroanalysis, 29*, 2646–2655.

Kurbanoglu, S., Ozkan, S. A., & Merkoçi, A., (2017). Nanomaterials-based enzyme electrochemical biosensors operating through inhibition for bio-sensing applications. *Biosens. Bioelectron., 89*, 886–898.

Ligaj, M., Tichoniuk, M., Gwiazdowska, D., & Filipiak, M., (2014). Electrochemical DNA biosensor for the detection of pathogenic bacteria *Aeromonas hydrophila*. *Electrochim. Acta, 128*, 67–74.

Liu, C., Sun, C., Tian, J., Wang, Z., Ji, H., Song, Y., Zhang, S., Zhang, Z., He, L., & Du, M., (2017). Highly stable aluminum-based metal-organic frameworks as bio-sensing platforms for assessment of food safety. *Biosens. Bioelectron., 91*, 804–810.

Liu, L., Xu, D., Hu, Y., Liu, S., Wei, H., Zheng, J., Hu, X., & Wang, C., (2015). Construction of an impedimetric immunosensor for label-free detecting carbofuran residual in agricultural and environmental samples. *Food Control, 53*, 72–80.

Malekzad, H., Jouyban, A., Hasanzadeh, M., Shadjou, N., & De La Guardia, M., (2017). Ensuring food safety using aptamer based assays: Electroanalytical approach. *TrAC, Trends Anal. Chem., 94*, 77–94.

Malvano, F., Albanese, D., Pilloton, R., Di Matteo, M., & Crescitelli, A., (2017). A new label-free impedimetric affinity sensor based on cholinesterase's for detection of organophosphorous and carbamic pesticides in food samples: Impedimetric versus amperometric detection. *Food Bioprocess Technol., 10*, 1834–1843.

Manzanares-Palenzuela, C. L., Martín-Fernández, B., Sánchez-Paniagua, L. M., & López-Ruiz, B., (2015). Electrochemical genosensors as innovative tools for detection of genetically modified organisms. *TrAC, Trends Anal. Chem., 66*, 19–31.

Manzano, M., Viezzi, S., Mazerat, S., Marks, R. S., & Vidic, J., (2018). Rapid and label-free electrochemical DNA biosensor for detecting hepatitis A virus. *Biosens. Bioelectron., 100*, 89–95.

Mehta, J., Vinayak, P., Tuteja, S. K., Chhabra, V. A., Bhardwaj, N., Paul, A. K., Kim, K. H., & Deep, A., (2016). Graphene modified screen printed immunosensor for highly sensitive detection of parathion. *Biosens. Bioelectron., 83*, 339–346.

Minaei, M. E., Saadati, M., Najafi, M., & Honari, H., (2015). DNA electrochemical nanobiosensors for the detection of biological agents. *J. Appl. Biotechnol. Rep., 2*, 175–185.

Motaharian, A., Motaharian, F., Abnous, K., Hosseini, M. R. M., & Hassanzadeh-Khayyat, M., (2016). Molecularly imprinted polymer nanoparticles-based electrochemical sensor for determination of diazinon pesticide in well water and apple fruit samples. *Anal. Bioanal. Chem., 408*, 6769–6779.

Nalini, S., Nandini, S., Reddy, M. B. M., Suresh, G. S., Savio, M. J., Neelagund, S. E., Naveenkumar, H. N., & Shanmugam, S., (2016). A novel bioassay based gold nanoribbon biosensor to aid the preclinical evaluation of anticancer properties. *RSC Advances, 6*, 60693–60703.

Oliveira, T. M. B. F., Fátima, B. M., Morais, S., De Lima-Neto, P., Correia, A. N., Oliveira, M. B. P. P., & Delerue-Matos, C., (2013). Biosensor based on multi-walled carbon nanotubes paste electrode modified with laccase for pirimicarb pesticide quantification. *Talanta, 106*, 137–143.

Ontiveros, G., Hernandez, P., Ibarra, I. S., Dominguez, J. M., & Rodriguez, J. A., (2017). Development of a biosensor modified with nanoparticles for sensitive detection of gluten by chronoamperometry. *ECS Trans., 76*, 103–107.

Patris, S., Vandeput, M., & Kauffmann, J. M., (2016). Antibodies as target for affinity biosensors. *TrAC, Trends Anal. Chem., 79*, 239–246.

Pilas, J., Yazici, Y., Selmer, T., Keusgen, M., & Schöning, M. J., (2017). Optimization of an amperometric biosensor array for simultaneous measurement of ethanol, formate, D-and L-lactate. *Electrochim. Acta, 251*, 256–262.

Puiu, M., & Bala, C., (2018). Peptide-based biosensors: From self-assembled interfaces to molecular probes in electrochemical assays. *Bioelectrochemistry, 120*, 66–75.

Radhakrishnan, R., Jahne, M., Rogers, S., & Suni, I. I., (2013). Detection of listeria monocytogenes by electrochemical impedance spectroscopy. *Electroanalysis, 25*, 2231–2237.

Rapini, R., & Marrazza, G., (2017). Electrochemical aptasensors for contaminants detection in food and environment: Recent advances. *Bioelectrochemistry, 118*, 47–61.

Ribeiro, F. W. P., Barroso, M. F., Morais, S., Oliveira, M. B. P. P., Viswanathan, S., De Lima-Neto, P., Delerue-Matos, & Correia, A. N. C., (2014). Simple laccase-based biosensor for formetanate hydrochloride quantification in fruits. *Bioelectrochemistry, 95*, 7–14.

Sadrabadi, N. R., Ensafi, A. A., Heydari-Bafrooei, E., & Fazilati, M., (2016). Screening of food samples for zearalenone toxin using an electrochemical bioassay based on DNA-zearalenone interaction. *Food Anal. Methods, 9*, 2463–2470.

Sánchez-Paniagua, L. M., Redondo-Gómez, E., & López-Ruiz, B., (2017). Electrochemical enzyme biosensors based on calcium phosphate materials for tyramine detection in food samples. *Talanta, 175*, 209–216.

Seo, H. B., & Gu, M. B., (2017). Aptamer-based sandwich-type biosensors. *J. Biol. Eng., 11*, 11.

Silva, N. F. D., Magalhães, J. M. C. S., Freire, C., & Delerue-Matos, C., (2018). Electrochemical biosensors for *Salmonella*: State of the art and challenges in food safety assessment. *Biosens. Bioelectron., 99*, 667–682.

Song, D., Li, Y., Lu, X., Sun, M., Liu, H., Yu, G., & Gao, F., (2017). Palladium-copper nanowires-based biosensor for the ultrasensitive detection of organophosphate pesticides. *Anal. Chim. Acta, 982*, 168–175.

Sun, X., Cao, Y., Gong, Z., Wang, X., Zhang, Y., & Gao, J., (2012). An amperometric immunosensor based on multi-walled carbon nanotubes-thionine-chitosan nanocomposite film for chlorpyrifos detection. *Sensors, 12*, 17247–17261.

Sundarmurugasan, R., Gumpu, M. B., Ramachandra, B. L., Nesakumar, N., Sethuraman, S., Krishnan, U. M., & Rayappan, J. B. B., (2016). Simultaneous detection of monocrotophos and dichlorvos in orange samples using acetyl cholinesterase-zinc oxide modified platinum electrode with linear regression calibration. *Sens. Actuat. B, 230*, 306–313.

Tan, X., Hu, Q., Wu, J., Li, X., Li, P., Yu, H., Li, X., & Lei, F., (2015). Electrochemical sensor based on molecularly imprinted polymer reduced graphene oxide and gold nanoparticles modified electrode for detection of carbofuran. *Sens. Actuat. B, 220*, 216–221.

Tan, X., Wu, J., Hu, Q., Li, X., Li, P., Yu, H., Li, X., & Lei, F., (2015). An electrochemical sensor for the determination of phoxim based on a graphene modified electrode and molecularly imprinted polymer. *Anal. Methods, 7*, 4786–4792.

Turco, A., Corvaglia, S., Mazzotta, E., Pompa, P. P., & Malitesta, C., (2018). Preparation and characterization of molecularly imprinted mussel inspired film as antifouling and selective layer for electrochemical detection of sulfamethoxazole. *Sens. Actuat. B, 255*, 3374–3383.

Vasilescu, A., & Marty, J. L., (2016). Electrochemical aptasensors for the assessment of food quality and safety. *TrAC, Trends Anal. Chem., 79*, 60–70.

Verma, N., & Bhardwaj, A., (2015). Biosensor technology for pesticides: A review. *Appl. Biochem. Biotechnol., 175*, 3093–3119.

Zamora-Gálvez, A., Morales-Narváez, E., Mayorga-Martinez, C. C., & Merkoçi, A., (2017). Nanomaterials connected to antibodies and molecularly imprinted polymers as bio/receptors for bio/sensor applications. *Appl. Mater. Today, 9*, 387–401.

Zappi, D., Caminiti, R., Ingo, G. M., Sadun, C., Tortolini, C., & Antonelli, M. L., (2017). Biologically friendly room temperature ionic liquids and nano materials for the development of innovative enzymatic biosensors. *Talanta, 175*, 566–572.

CHAPTER 6

ELECTRONIC NOSE (E-NOSE) APPLICATION IN THE FOOD INDUSTRY

SWARAJYA LAXMI NAYAK, BINDVI ARORA, SHRUTI SETHI, and R. R. SHARMA

Division of Food Science and Postharvest Technology, ICAR-Indian Agricultural Research Institute, New Delhi, India, E-mail: docsethi@gmail.com (S. Sethi)

ABSTRACT

Subjective evaluation of organoleptic attributes of foods is a widely used system for understanding the consumer's behavior and psychology of food acceptance but it usually entails large variability in consumer responses. In contrast, objective evaluation of the sensory attributes of foods can provide consistency during research and development as it assesses the empirical properties of foods. An electronic nose (E-nose) is a device that employs empirical analysis to mimic the human sense of smell. Various components of the e-nose are designed to function as mechanical analogs of the biological nose. The structural variation of odoriferous molecules is exploited for detection by sensors in the e-nose. These detectors can produce electronic signals that generate corresponding qualitative and quantitative data for the evaluation of odoriferous substances in the food material. The use of e-nose is expanding in the food industry, for both quality control and R&D activities. However, technological limitations such as limited sensitivity and database requirement during the analysis exists and further research in the improvement of the e-nose technology is imperative. This chapter describes the functioning of e-nose and details its numerous applications that have been identified by the researchers from academic as well as food industry sectors.

6.1 INTRODUCTION

The electronic nose or e-nose is a device that mimics the human sense of smell. An e-nose is a device that identifies the specific components of an odor or flavor in food by analyzing its chemical makeup. It is principally used in the evaluation and quantification of olfaction/olfactory responses due to different stimuli. The technology follows the principle of a growing research area called biomimetics, or biomimicry, which involves imitation of natural phenomena for the purpose of solving complex human problems. The e-nose technology is highly useful in sensory studies. By using e-nose in sensory evaluation of foods, the attribute of odor intensity can be quantified. This also ensures objective evaluation and avoids the inclusion of errors as in subjective evaluation done by humans. e-Nose detects the smell more effectively than the human nose. It uses a series of gas sensors that are overlapped selectively along with a pattern reorganization component such as in neural network. The stages of the recognition process are similar to human olfaction and are performed for identification, comparison, quantification, and other applications, including data storage and retrieval. e-Nose finds application in the fields of agriculture, environment, water, biomedical, food, cosmetics, etc. Also, the e-nose can detect the hazardous or poisonous gases which are difficult to perceive by human sniffers.

The odor is composed of diverse molecules with varying sizes and shapes. This structural variety of odor molecules combined with the corresponding receptor in human nose can generate a neurological response of the smell that is sent to the brain. e-Nose utilizes the same principal of identification by the biological model. The receptors of biological system are replaced by sensors in e-nose. The receptors can send electronic signals on contact with odor molecules to generate quantitative data on odor.

FMCG industries were the primary users of e-nose for both quality control and R&D activities. Currently, e-nose is a popular application equipment in pharmaco-medical, pollution control, and gas industries. It can detect gas leakages, pollutants, and odors specific to certain diseases. A plethora of applications is possible with the utilization of e-nose in diverse sectors of the industry.

The only limitation with e-nose is the elevated cost of equipment hardware and operations. Recent research on e-nose focuses on the development of cost-effective, small portable devices with increased sensitivity.

6.2 COMPARISON OF E-NOSE WITH HUMAN NOSE

An e-nose is a mechanical version of biological human nose. A simulation of individual organs of the human nose is combined in series to develop e-nose (Table 6.1). The inhaling action of nose is replicated by using a sampler with pumping action followed by a filter to remove impurities (if any) in the sample. The filter of e-nose acts as mucous in the bio-nose. Sensors take the place of the receptors on the olfactory epithelium. Sensors interact with the odor molecules on the principle of structural binding similar to bio-nose. The interaction results in a response that transfers the signal to circuitry neural network in a pattern similar to that of nerve impulses from bio-nose. Thus, e-nose is a simulated version of bio-nose developed with an aim to perform objective odor analysis.

6.3 PRINCIPLE

The e-nose consists majorly of three components: sample delivery system, detecting system, and computing system (Figure 6.1). The chemical sensors of e-nose are connected to a pattern-recognition system that responds to odors passing over the sensor. As different odors cause dissimilar pattern recognition responses in the sensors, identification is possible. The responses provide a signal pattern characteristic of a particular aroma by recognizing the particular molecule responsible for that aroma. The computer evaluates the signal pattern and can compare the aromas of different samples, using pattern recognition.

TABLE 6.1 Comparison of Components of Bio-Nose and e-Nose

Bio-Nose	E-Nose
Inhaling	Pumping
Mucus	Filter
Olfactory epithelium	Sensors
Binding with proteins	Interaction
Enzymatic proteins	Reaction
Cell membrane depolarized	Signal
Nerve impulses	Circuitry and neural network

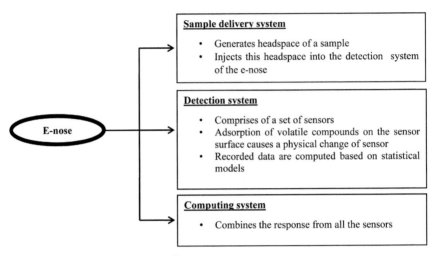

FIGURE 6.1　Schematic diagram of e-nose.

6.4　APPLICATIONS OF E-NOSE IN THE FOOD INDUSTRY

The e-nose mimics human olfaction and functions on a non-separate mechanism, i.e., the smell or flavor is perceived by bio-nose as a single overlapping effect. With the introduction of e-nose, an objective evaluation of a singular olfactory response was made possible. e-Nose has a plethora of identified applications in different commercial industries such as agriculture, environmental, food, water, biomedical, cosmetics, etc. The current chapter will focus majorly on applications of e-nose in the food industry.

6.5　APPLICATION OF E-NOSE ON FRESH AND PROCESSED VEGETABLES AND FRUITS

An array of volatile organic compounds (VOCs) is developed in fruits and vegetables that impart a characteristic aroma and unique flavor. These aroma and flavor compounds play a key role in the consumer acceptability of the fruit/vegetable (either fresh or in the processed form). Traditionally human panelists evaluate the quality of fruit/vegetable to assess salability in fresh markets. In recent times, scientists have explored the potential use of electronic-nose with specialized sensors for estimation of aroma volatiles in vegetables and fruits. Table 6.2 presents an overview of recent applications of e-nose in determination of aroma in vegetables and fruits (fresh and processed).

TABLE 6.2 Applications of e-Nose in Evaluation of Fruits and Vegetables

Product	Sensor	Purpose	Volatile Compounds Detected by e-Nose	References
Dates	Electronic nose with 32 individual polymer sensors blended with carbon black composite	Classify date-pits based on their volatile compounds generated by different degrees of heating with an accuracy of 93.3%	Guiacol, furfural, ethanol, and butanol	Rahman et al., 2018
Potato	Metal-oxide-based gas sensors	Detect soft rot in potato	Organic compounds such as hydrocarbons, amines, and aromatic compounds	Rutolo et al., 2016
Apple	Electronic nose combined with chemometrics	Detection of *Zygosaccharomyces rouxii* fungi, responsible for apple juice *spoilage*	Methane, sulfur, terpenes, etc.	Wang et al., 2016
Vegetable Soup	Oxide sensor-based electronic nose	Detect spoilage of vegetable soups caused by *Enterobacteriaceae*	Dimethyl sulfide 1-methyl-2-(3-methylpentyl) cyclopropane, N-nitroso-methyl-ethylamine, and 1-hexanol.	Gobbi et al., 2015
Onion	Metal oxide semiconductor-based E-nose	Quality evaluation of onion	Detect methlypropyl sulfide and 2-nonanone volatile compounds emitted by rotten onions	Konduru et al., 2015
Strawberry	Electronic nose containing an array of 10 different metal oxide sensors	Detection of different postharvest fungal diseases namely *Botrytis* sp. (BC), *Penicillium* sp. (PE) and *Rhizopus* sp. (RH)	Nitrogen oxides, ammonia, hydrogen, etc.	Pan et al., 2014

TABLE 6.2 *(Continued)*

Product	Sensor	Purpose	Volatile Compounds Detected by e-Nose	References
Tomato	Metal oxide nanowire (MOX-NW) sensors based e-nose	Distinguished tomato samples spoilage by *Candida milleri*	Determine volatile profile of the tomato which are produce from the metabolism of yeast	Sberveglieri et al., 2014
Potato	Electronic nose equipped with a metal oxide sensor array	Detection of potato spoilage namely brown rot and ring rot caused by *Ralstonia solanacearum*, *Clavibacter michiganensis*, respectively	Aldehydes, alcohols, ketones, acids, esters, hydrocarbons, amines, furans, and sulfur compounds emitted by rotten potatoes	Biondi et al., 2014
Orange	E-nose comprising four chemiresistive sensors	Determine fungal activity rapidly at an early stage of *Penicillium digitatum* infection in oranges	Alcohols, volatile halogenated organic compounds and carbonyl compounds in infected orange	Gruber et al., 2013
Orange	E-nose coupled with olfactive sensors	Determination of green mold infection in oranges	Terpenes, myrcene, α-pinene ethano	Pallottino et al., 2012
Blueberry	Polymer gas sensor array electronic nose	Detection of blueberry diseases such as gray mold, anthracnose, and Alternaria rot, caused by *Colletotrichum gloeosporioides*, *Botrytis cinerea*, and *Alternaria sp.* respectively	Styrene, 1-methyl-2-(1-methylethyl, benzene, eucalyptol, undecane, 5-methyl-2-(1-methylethyl)-2-cyclohexen-1-one, and thujopsene emitted from the fruit due to infection.	Li et al., 2010
Onion	E-nose coupled with gas sensor array	Distinguished between healthy and infected onion (sour skin *and Botrytis neck rot)*	1-Nonanol, heptanoic acid, 2,4-Octanedione	Li et al., 2011

TABLE 6.2 (Continued)

Product	Sensor	Purpose	Volatile Compounds Detected by e-Nose	References
Tomato	E-nose coupled with thin film metal oxide gas sensors	Detect microbial contamination in canned tomatoes	Dimethyl sulfide, 3-methyl-furan, 6-Methyl-5-hepten-2-one and others	Concina et al., 2009
Olives	Electronic nose consist of 12 metal oxide-based sensors	Quality evaluation of fermented green table olives	Esters, acids, alcohols, aldehydes, ketones, and phenols involved in olive fermentation	Panagou et al., 2008
Apple	E-nose contains 32 internal thin film carbon black polymer sensors	Distinguish between healthy and wounded apple based on the volatile compounds emitted by the apples	Detect volatile compounds emitted from wounded apple	Li et al., 2007
Potato crisps	Metal oxide semiconductor (MOS) based electronic nose	Detect rancid potato crisps	Acetone, acetic acid, pentanal, hexanal, heptanal, hexanoic acid, 3-octen-2-one, 2-octenal, 2,3-octanedione, 2,4-decadienal, etc.	Vinaixa et al., 2005
Oranges and Apples	E-nose equipped with metalloporphyrins coated sensors	Detection of skin damage and mealiness in oranges and apples	Detect different volatile compounds produce during the oxidizing process	Natale et al., 2001
Tomato	E-nose based on metalloporphyrin-coated quartz microbalance sensors	Differentiate damaged tomato from healthy tomato	Diacetyl, acetyl methyl carbinol (AMC), ethanol, and acetic acid	Sinesio et al., 2000

TABLE 6.3 Applications of e-Nose in Evaluation of Beverages

Product	Sensor	Purpose	Volatile Compounds Detected by e-Nose	References
Mandarin juice	Metal oxide semiconductor sensors based e-nose	Predict food additives in the fruit juice	Detect benzoic acid and chitosan in juice	Qiu and Wang, 2017
Beer	E-nose contained 10 metal oxide semiconductor-type chemical sensors	Determination of effect of triadimefon (fungicide) in beer	Ethyl acetate, esters, and amino acids in beer samples	Kong et al., 2016
Apple juice	E-nose contains 10 metal oxide semiconductor (MOS) chemical sensors	Detection of apple juice spoilage caused by *Alicyclobacillus acidoterrestris*	Ammonia, aromatic compounds, nitrogen oxides, hydrogen, etc. in beverages	Huang et al., 2015
Orange apple juice	Electronic nose equipped with gas sensor array	Differentiated between *Alicyclobacillus acidoterrestris* affected juice and healthy juice	Detect the foul odor emitted from the affected fruit juice	Hartyani et al., 2013
Fruit Juice (peach orange and apple)	E-nose equipped with a Metal Oxide Semiconductor sensor array	Detection of fungal contamination in fruit juice by *Alicyclobacillus acidocaldarius* and *A. acidoterrestris*	α-pinene, β-myrcene, and octanal emitted from the affected fruits	Gobbi et al., 2010
Orange Juice	Electronic nose equipped with metal oxide semiconductor sensors	Detect *Alicyclobacillus acidoterrestris* in contaminated orange juice	Guaiacol and 2,6-dibromophenolin in *A. acidoterrestris* contaminated juice	Bianchi et al., 2010
Whiskey	E-nose containing nanotube SnO_2 gas sensors	Determine the contamination in whiskeys	Methanol	Wongchoosuk et al., 2010
Soft drinks	E-nose couple with six thin film semiconductor metal oxide sensors (MOX)	Diagnosis of contamination by *Alicyclobacillus* spp. In soft drinks (metabolite: guaiacol)	Guaiacol	Concina et al., 2010

TABLE 6.3 (Continued)

Coffee	E-nose coupled with matrix of eight metal-oxide gas sensors	Detection of defects in Colombian coffee samples	Detect volatiles emitted from the coffee beans	Rodriguez et al., 2009
Alcoholic beverages	Electronic nose coupled with eighteen metal oxide semiconductor gas sensors	Detect unpleasant odor in alcohol	1-hexanol, 2,3,4-trichloroanisole (TCA) and 2,3 butanedione or diacetyl	Ragazzo et al., 2009
Wine	Metal oxide sensor-based electronic nose	Detect ochratoxin in contaminated wine	Detect volatile organic compounds produce from the fungus	Cabanes et al., 2009
Red wine	Electronic nose coupled with metal oxide-gas sensor	Detect spoiled red wine	Ethano, 4-ethylguaiacol, 3-methylbutanol associated with red wine spoilage	Berna et al., 2008
Beer	Electronic nose with micro-mechanized semiconductor sensors	Detect off-flavor in beer	Ethyl acetate and acetaldehyde	Santos and Lozano, 2008
Red wine	E-nose combined with chemometric sensors	Detect Brettanomyces yeast in spoiled red wine	4-ethylphenol (4EP)	Cynkar et al., 2007
Alcoholic beverages	E-nose coupling with metal-oxide semiconductor sensors and surface acoustic wave (SAW) sensors	Aroma characterization of alcoholic beverages	Ethyl acetate, hexanal, etc.	Ragazzo et al., 2005
Beer	Fingerprint Mass Spectrometry based e-nose	Detection of aroma in Japanese commercial beers	Diacetyl and/or dimethyl sulfide	Kojima et al., 2005
Soymilk	Electronic nose with six metal-oxide sensors	Determine the shelf life of soymilk samples	Ammonia, hydrogen sulfide, and alcohols, etc.	Ko et al., 2000

6.6 APPLICATIONS OF E-NOSE FOR BEVERAGES

The development of advanced and modern instrumentation techniques has made the detection and quantification of food odorants as well as contaminants easier. Amongst them, e-nose is one of the techniques which can correlate well with the flavor volatiles of the beverages. e-Nose is also very useful in the evaluation of external contamination or adulteration and to get a reliable result in less time than conventional methods. In the beverage industry, e-nose can be used to detect and classify beverages according to odors, identify spoilage during storage and correlate with other biochemical and human perception studies, detect off-flavors in alcoholic as well as non-alcoholic beverages, etc. Table 6.3 lists the novel and recent application of e-nose in the beverage industry.

6.7 APPLICATION OF E-NOSE ON DETERMINATION OF OIL QUALITY

As highlighted in previous sections, a potential application of e-nose can be to determine the oil quality also. The technology can be used for a multitude of applications in oil industry such as detection of off-flavors developed during rancidity, flavor reversion during oil storage, evaluation of off-odor compounds during utilization and re-utilization of oils during frying, etc. e-Nose can also help in classification of oils from different sources and grading of oil in terms of its quality. Table 6.4 lists the recent researched applications of e-nose in the oil industry.

6.8 APPLICATION OF E-NOSE ON DAIRY PRODUCTS

Sensor techniques such as e-nose offer many possibilities for the analysis of volatile compounds in dairy products. Dairy products are characterized by specific flavoring compounds such as diacetyl that is responsible for the flavor of butter, lactones attributed for ghee flavor, etc. Thus, e-nose finds multiple applications in the dairy industry such as classification of dairy products on the basis of the characteristic flavor compound, identification of endpoint of fermentation, detection of off-flavors, etc. Table 6.5 lists the recent advances in the application of e-nose in the dairy industry.

TABLE 6.4 Applications of e-Nose in Assessment of Edible Oils

Product	Sensor	Purpose	Volatile Compounds Detected by e-Nose	References
Waste cooking oil	E-nose with olfactory sensors	Characterized waste cooking oil from different sources	Identify odors produced from waste cooking oil	Siqueira et al., 2018
Peony seed oil	E-nose containing metal oxide semiconductor sensors	Rapid detection of adulterated peony seed oil	α-linolenic acid (ALA) and omega-3 fatty acid in oil	Wei et al., 2018
Rosa damascene oil	E-nose with metal oxide semiconductor (MOS) sensors	Classification of essential oil composition in *Rosa damascena*	Benzene, n-hexane, H_2S, etc.	Gorji-Chakespari et al., 2017
Edible oil	E-nose is composed of an array of chemical gas sensors	Rapid assessment of the degree of oxidation in edible oils	Detect mixture of volatiles released from samples	Xu et al., 2016
Sunflower oil, vegetable oil, olive oil	Metal oxide sensors based e-nose	Determine defects such as fusty, moldy, muddy, rancid, and winey in oils	2-isobutyl-3-methoxypyrazine, ethanol, etc.	Lerma-García et al., 2010
Virgin olive oil	E-nose coupled with metal oxide sensors	Quality evaluation of virgin olive oil	Alcohols, aldehydes, and acids mainly responsible for unpleasant sensory	García-González and Aparicio, 2002b
Virgin olive oil	Metal oxide-based e-nose	Determining level of rancid and fusty in virgin olive oils	Ethyl propionate, butyl acetate, hexanal, etc.	García-González and Aparicio, 2002a

6.9 APPLICATION OF E-NOSE ON MEAT PRODUCTS

Animal flesh products such as poultry, beef, pork, etc., also have their characteristic aroma attributed to various compounds. e-Nose can successfully identify and quantify these aroma molecules on the basis of its reaction detection array system. e-Nose can serve various applications in the meat industry such as assessing meat quality, detection of meat spoilage, classification of meat, etc. Table 6.6 depicts the recent applications of e-nose in the meat industry.

TABLE 6.5   Applications of e-Nose in Evaluation of Dairy Products

Product	Sensor	Purpose	Volatile Compounds Detected by e-Nose	References
Goat milk	E-nose systems containing metal oxide semiconductor gas sensor	Screening different antibiotics in dairy products.	Nitrogen oxides, sulfur compounds in goat milk	Ding et al., 2015
Milk	Bioelectronic nose	Detect spoiled milk	Hexanal	Park et al., 2012
Milk	E-nose coupled with metal oxide semiconductor sensor	Discriminate among different levels of sample contamination	Aflatoxin M1 (AFM1)	Barbiroli et al., 2007
Milk	E-nose based on gas-sensor array	Distinguish complex odors derived from secondary metabolites microbial contamination in milk	Secondary volatile metabolites such as 3-Hydroxy 2-butanone	Haugen et al., 2006
Ewe's milk	E-nose constituted of 12 MOS and 10 MOSFET sensors	Detection of spoiled raw ewe's milk	Aflatoxin M1	Benedetti et al., 2006
Cow milk	E-nose using gas metal oxide semiconductive sensor	Differentiate mastitic and healthy milk	Ketones, sulfides, acids, and amines	Eriksson et al., 2005
Fat-free milk	E-nose attached with 12 sensors	Quality evaluation of milk	Offensive odor from *Bacillus coagulans* and *Pseudomonas fluorescens* contaminated milk	Korel and Balaban, 2002
Skimmed milk	E-nose containing 14 conducting polymer sensors	Detect microbial spoilage in milk	Undesired off-odors	Magan et al., 2001

TABLE 6.6 Applications of e-Nose in Evaluation of Meat Products

Product	Sensor	Purpose	Volatile Compounds Detected by e-Nose	References
Friesian bulls meat	Electronic nose coupled with 10 metal oxide sensors	Assessing meat quality for detection of illegal administration of dexamethasone	Methane, Hydrocarbon, Aldehydes, and Ketones	Barbera et al., 2018
Beef	E-nose coupled with gas sensors	Classify beef samples and predict the microbial load on meat surface	Anisaldehyde, Benzaldehyde, trans-cynnamoil aldehyde	Kodogiannis, 2017
Dry-cured meat	E-nose with metal oxide sensors	Screening of meat products	Ethanol, Hexanal, Heptanal, Styrene	Lippolis et al., 2016
Oysters	E-nose with microfluidic gas sensors	Quality evaluation of seafoods	Trimethylamine	Lee et al., 2015
Catfish meat	Electronic-nose coupled with organic matrix-coated polymer sensor	Detecting off-flavor in fillets	Malodorous compounds	Wilson et al., 2013
Meat	E-nose coupled with tin dioxide semiconductor sensors	Detection of meat spoilage	Ethanol and trimethylamine (TMA)	Längkvist et al., 2013
Beef fillets	E-nose with an array of quartz crystal non-selective sensors	Organoleptic evaluation of beef fillets	Off-odors	Papadopoulou et al., 2013
Seafoods	bioelectronic nose with protein-based sensor	Determination of seafood quality	Triethylamine, Ammonia	Lim et al., 2013
Boar meat	Metal oxide gas sensor-based e-nose	Quality evaluation of boar meat	Unpleasant odor	Kirsching et al., 2012

TABLE 6.6 (Continued)

Product	Sensor	Purpose	Volatile Compounds Detected by e-Nose	References
Pork	E-nose with 18 metal oxide sensors	Estimate total viable microorganisms in pork	Off-odors	Wang et al., 2012
Ham	E-nose with metal oxide sensors	Distinguish between toxin producing and non-producing strains of *Penicillium*	Ochratoxin A	Leggieri et al., 2011
Seafoods	Electronic-nose coupled with gas sensors	Quality assessment of seafood	Formaldehyde	Zhang et al., 2009
Meat	Metal oxide-sensor-based e-nose	Useful for detecting meat spoilage	Methane, Hydrogen sulfide	Balasubramanian et al., 2009
Red meat	Metal oxide semiconductor-based e-nose	Determine microbiological quality in meat	Hydrogen sulfide	Barbri et al., 2008
Beef	Electronic nose with metal oxide gas sensor	Detect contamination in beef samples	Alcohol, Hydrogen sulfide, Carbon-di-oxide	Balasubramanian et al., 2008
Poultry	E-nose with 14 gas sensors	Measure odor released from poultry farm	Ammonia, Hydrogen-sulfide, Amine compounds	Pan and Yang, 2007
Meat	E-nose with polymer-based sensors	Distinguished between spoiled and unspoiled meat samples	Volatile chemical mixture	Panigrahi et al., 2006
Broiler chicken	E-nose with 11 MOS-sensors	Determine the degree of spoilage in broiler chicken cuts	Dimethyl sulfide	Rajamaki et al., 2006
Beef	E-nose with organic polymer sensors	Identify meat samples contaminated with *S. typhimurium*	Volatile organic compounds, nalidixic acid	Balasubramanian et al., 2005
Beef	E-nose with 10 metal oxide semiconductor sensor	Detect spoilage in vacuum packed beef	Off-odors produced from spoilage	Blixt and Borch, 1999

6.10 E-NOSE APPLICATION ON GRAINS AND BAKERY PRODUCTS

Cereal and bakery products are an integral part of the human diet across the globe. Baking generates many volatile compounds in products responsible for their characteristic aroma. e-Nose serves to detect these volatile compounds for purposes of identification, quality evaluation, classification, etc. Table 6.7 lists the recent applications of e-nose in grains and the bakery industry.

6.11 LIMITATIONS

1. High equipment and operational costs.
2. Sensors require regular calibration with odorous molecules to be tested.
3. The e-noses can only discriminate and are not capable of identifying new odorants.
4. E-noses are not capable of selective identification of the odorants responsible for the contamination of food products.
5. Selective and sensitive analysis of particular compounds is not achievable in a multi-ingredient product.

6.12 CONCLUSION

E-nose is a versatile equipment that can evaluate the aroma of food products in an objective way similar to human panels but avoids human error and ensures reproducibility. e-Nose demonstrates the pattern recognition capabilities of the neural network paradigm in sensor analysis of food. e-Noses find application in various branches of science and technology ranging from food to bio-medical to surgical sciences. With multiple advantages, e-nose is the technology of the future for sensory and qualitative analysis of foods. The only major disadvantage of e-nose is the high cost associated with the equipment and its operation. Future research should thrust on the development of cost-effective, portable e-nose with sensors capable of identifying diverse volatile flavor compounds.

TABLE 6.7   Applications of e-Nose in Evaluation of Grains and Bakery Products

Product	Sensor	Purpose	Volatile Compounds Detected by e-Nose	References
Rapeseed	Polymer-composite sensors based e-nose	Estimate fungal contamination in rapeseed during storage	Volatile organic compounds	Gancarz et al., 2017a
Rapeseed	Metal oxide gas sensor-based e-nose	Quality evaluation of rapeseed	Ergosterol content, organic compounds	Gancarz et al., 2017b
Cookies made from black pepper and cardamom	E-nose equipped with metal oxide gas sensor	Rancidity assessment in cookies	1,8-cineole, piperine	Dutta et al., 2017
Grain	Bioelectronic nose with olfactory sensors	Detect mycotoxigenic fungi contamination in grain	1-octen-3-ol	Ahn et al., 2015
Coffee beans	E-nose coupled with metal oxide sensors	Microbiological safety evaluation in coffee beans	off-flavor	Sberveglieri et al., 2014
Wheat	E-nose consist of metal oxide gas sensor	Quality evaluation of durum wheat	Deoxynivalenol	Lippolis et al., 2014
Maize	Metal oxide chemical sensors based electronic-nose	Screening of maize cultivar	Detect fumonisin content	Gobbi et al., 2011
Durum wheat	E-nose equipped with metal-oxide-semiconductor (MOS) sensors	Discriminate between contaminated and non-contaminated durum wheat	Deoxynivalenol	Campagnoli et al., 2011

TABLE 6.7 (*Continued*)

Product	Sensor	Purpose	Volatile Compounds Detected by e-Nose	References
Maize and wheat	E-nose with metal oxide sensors	Classify natural contaminated and non-contaminated maize and wheat	Deoxynivalenol and aflatoxins	Campagnoli et al., 2009
Wheat	Electronic nose with 32 gas sensor	Analysis of fungal contamination in wheat	Trichodiene	Perkowski et al., 2008
Barley	E-nose with conducting polymer sensors	Estimate ergosterol content in barley samples	2-methyl-1-propanol, 3-methyl-butanol, and 1-octen-3-ol	Balasubramanian et al., 2007
Bakery product	E-nose with gas sensors	Detect fungal spoilage in bakery products	2-methylfuran, 2-methyl-1-propanol, 2-pentanone, 2-methyl-1 butanol, 3-octen2-ol and 1-octen-3-ol	Marin et al., 2007
Durum wheat	E-nose with 10 metal-oxide sensors	Discriminate wheats of different storage age and degrees of insect damage	Aldehydes, ketones, alcohols	Zhang and Wang, 2007
Cereals	E-nose with gas sensors	Detecting mycotoxins contamination in cereals	Unpleasant odors released from fungal contamination	Presicce et al., 2006
Wheat	E-nose using chemical sensors	Discriminate between non-infected and infected wheat samples	Undesirable flavors	Paolesse et al., 2006
Maize	E-nose coupled with metal oxide sensor	Detect fungal contamination of maize	Fumonisins	Falasconi et al., 2005
Bakery products	Bloodhound BH-114 electronic nose	Evaluate physiological and microbial spoilage in bread	Pentanol, 2-pentylfuran, styrene, hexanal	Needham et al., 2005

TABLE 6.7 *(Continued)*

Product	Sensor	Purpose	Volatile Compounds Detected by e-Nose	References
Durum wheat	E-nose with 12-element metal-oxide chemosensor	Estimate ergosterol content and mycotoxins in wheat	Ergosterol, citrinin, ochratoxin A	Abramson et al., 2005
Bakery products	E-nose incorporated with gas sensor	Differentiate between spoiled and non-spoiled bakery products	3-methyl-1-butanol, 2-methyl-1-propanol, and 1-octen-3-ol	Vinaixa et al., 2004
Wheat	E-nose consist of metal oxide gas sensor	Evaluate microbial spoilage in wheat	Unpleasant odor	Lacy-Costello et al., 2003
Bread	Surface polymer sensor-based e-nose	Detect mold contaminated bread	Alcohols, ketones, 2-methylisoborneol	Keshri et al., 2002
Barley	E-nose made up of 10 metal oxide semiconductor sensor	Detect ochratoxin contaminated barley samples	Aldehydes, 2-hexanone, 3-octanone	Olsson et al., 2002
Wheat, barley, corn	Electronic nose with conducting polymer sensor	Indicates spoilage of grains by detecting the volatile compounds	3-methyl-1-butanol, 1-octen-3-ol, 8-carbon ketones and alcohols	Magan and Evans, 2000
Cereals	E-nose with 14 surface polymer sensor	Detect contaminated cereal	Off-odor produced from *Fusarium moniliforme* and *F. proliferatum*	Keshri and Magan, 2000
Barley	Electronic nose coupled with chemical gas sensor	Accurate prediction of ergosterol in barley grains	2-hexenal, benzaldehyde, nonanal, 3-octanone, methylheptanone, and trimethylbenzene	Olsson et al., 2000
Wheat	E-nose with conducting polymer sensor	Determination of wheat quality	Volatile organic compounds	Evans et al., 2000

KEYWORDS

- **acetyl methyl carbinol**
- **aflatoxin M1**
- **metal oxide semiconductor**
- **surface acoustic wave**
- **trimethylamine**
- **α-linolenic acid**

REFERENCES

Abramson, D., Hulasare, R., York, R., White, N., & Jayas, D., (2005). Mycotoxins, ergosterol, and odor volatiles in durum wheat during granary storage at 16% and 20% moisture content. *Journal of Stored Products Research, 41*, 67–76.

Ahn, J. H., Lim, J. H., Park, J., Oh, E., Son, H. M., Hong, S., & Park, T. H., (2015). Screening of target specific olfactory receptor and development of olfactory biosensor for the assessment of fungal contamination in grain. *Sensors and Actuators B: Chemical, 210*, 9–16.

Balasubramanian, S., Panigrahi, S., Kottapalli, B., & Wolf-Hall, C., (2007). Evaluation of an artificial olfactory system for grain quality discrimination. *LWT-Food Science and Technology, 40*, 1815–1825.

Balasubramanian, S., Panigrahi, S., Logue, C., Doetkott, C., Marchello, M., & Sherwood, J., (2008). Independent component analysis-processed electronic nose data for predicting *Salmonella typhimurium* populations in contaminated beef. *Food Control, 19*, 236–246.

Balasubramanian, S., Panigrahi, S., Logue, C., Gu, H., & Marchello, M., (2009). Neural networks integrated metal oxide-based artificial olfactory system for meat spoilage identification. *Journal of Food Engineering, 91*, 91–98.

Balasubramanian, S., Panigrahi, S., Logue, C., Marchello, M., & Sherwood, J., (2005). Identification of *Salmonella*-inoculated beef using a portable electronic nose system. *Journal of Rapid Methods and Automation in Microbiology, 13*, 71–95.

Barbera, S., Tarantola, M., Salac, G., & Nebbia, C., (2018). Canonical discriminant analysis and meat quality analysis as complementary tools to detect the illicit use of dexamethasone as a growth promoter in Friesian bull. *The Veterinary Journal, 235*, 54–59.

Barbiroli, A., Bonomi, F., Benedetti, S., Mannino, S., Monti, L., Cattaneo, T., & Iametti, S., (2007). Binding of aflatoxin M 1 to different protein fractions in ovine and caprine milk. *Journal of Dairy Science, 90*, 532–540.

Barbri, N. E., Llobet, E., Bari, N. E., Correig, X., & Bouchikhi, B., (2008). Electronic nose based on metal oxide semiconductor sensors as an alternative technique for the spoilage classification of red meat. *Sensors, 8*, 142–156.

Benedetti, S., Iametti, S., Bonomi, F., & Mannino, S., (2005). Head space sensor array for the detection of aflatoxin M1 in raw ewe's milk. *Journal of Food Protection®, 68*, 1089–1092.

Berna, A. Z., Trowell, S., Cynkar, W., & Cozzolino, D., (2008). Comparison of metal oxide-based electronic nose and mass spectrometry-based electronic nose for the prediction of red wine spoilage, *Journal of Agricultural and Food Chemistry, 56*, 3238–3244.

Bianchi, F., Careri, M., Mangia, A., Mattarozzi, M., Musci, M., Concina, I., & Gobbi, E., (2010). Characterization of the volatile profile of orange juice contaminated with *Alicyclobacillus Acidoterrestris*. *Food Chemistry, 123*, 653–658.

Biondi, E., Blasioli, S., Galeone, A., Spinelli, F., Cellini, A., Lucchese, C., & Braschi, I., (2014). Detection of potato brown rot and ring rot by electronic nose: From laboratory to real scale. *Talanta, 129*, 422–430.

Blixt, Y., & Borch, E., (1999). Using an electronic nose for determining the spoilage of vacuum packaged beef. *International Journal of Food Microbiology, 46*, 123–134.

Cabañes, F. J., Sahgal, N., Bragulat, M. R., & Magan, N., (2009). Early discrimination of fungal species responsible of ochratoxin A contamination of wine and other grape products using an electronic nose. *Mycotoxin. Research, 25*, 187–192.

Campagnoli, A., Cheli, F., Polidori, C., Zaninelli, M., Zecca, O., Savoini, G., Pinotti, L., & Dell'Orto, V., (2011). Use of the electronic nose as a screening tool for the recognition of durum wheat naturally contaminated by deoxynivalenol: A preliminary approach. *Sensors, 11*, 4899–4916.

Campagnoli, A., Dell'Orto, V., Savoini, G., Cheli, F., Pardo, M., & Sberveglieri, G., (2009). Screening cereals quality by electronic nose: The example of mycotoxins naturally contaminated maize and durum wheat. *AIP Conference Proceedings* (pp. 507–510). AIP.

Concina, I., Bornšek, M., Baccelliere, S., Falasconi, M., Gobbi, E., & Sberveglieri, G., (2010). *Alicyclobacillus* spp.: Detection in soft drinks by electronic nose. *Food Research International, 43*, 2108–2114.

Concina, I., Falasconi, M., Gobbi, E., Bianchi, F., Musci, M., Mattarozzi, M., Pardo, M., et al., (2009). Early detection of microbial contamination in processed tomatoes by electronic nose. *Food Control, 20*, 873–880.

Cynkar, W., Cozzolino, D., Dambergs, B., Janik, L., & Gishen, M., (2007). Feasibility study on the use of a head space mass spectrometry electronic nose (MS e-nose) to monitor red wine spoilage induced by *Brettanomyces* yeast. *Sensors and Actuators B: Chemical, 124*, 167–171.

Ding, W., Zhang, Y., Kou, L., & Jurick, W. M., (2015). Electronic nose application for the determination of *Penicillin* G in saanen goat milk with fisher discriminate and multilayer perceptron neural network analyses. *Journal of Food Processing and Preservation, 39*, 927–932.

Dutta, S., Bhattacharjee, P., & Bhattacharyya, N., (2017). Assessment of shelf life of black pepper and small cardamom cookies by metal oxide-based electronic nose using spoilage index. *Food and Bioprocess Technology*, pp. 1–11.

Eriksson, A., Waller, K. P., Svennersten-Sjaunja, K., Haugen, J., Lundby, F., & Lind, O., (2005). Detection of mastitis milk using a gas-sensor array system (electronic nose). *International Dairy Journal, 15*, 1193–1201.

Evans, P., Persaud, K. C., Mc Neish, A. S., Sneath, R. W., Hobson, N., & Magan, N., (2000). Evaluation of a radial basis function neural network for the determination of wheat quality from electronic nose data. *Sensors and Actuators B: Chemical, 69*, 348–358.

Falasconi, M., Gobbi, E., Pardo, M., Della, T. M., Bresciani, A., & Sberveglieri, G., (2005). Detection of toxigenic strains of *Fusarium verticillioides* in corn by electronic olfactory system. *Sensors and Actuators B: Chemical, 108*, 250–257.

Gancarz, M., Wawrzyniak, J., Gawrysiak-Witulska, M., Wiącek, D., Nawrocka, A., & Rusinek, R., (2017a). Electronic nose with polymer-composite sensors for monitoring fungal deterioration of stored rapeseed. *International Agrophysics, 31*, 317–325.

Gancarz, M., Wawrzyniak, J., Gawrysiak-Witulska, M., Wiącek, D., Nawrocka, A., Tadla, M., & Rusinek, R., (2017b). Application of electronic nose with MOS sensors to prediction of rapeseed quality. *Measurement, 103*, 227–234.

García-González, D. L., & Aparicio, R., (2002a). Detection of defective virgin olive oils by metaloxide sensors. *European Food Research and Technology, 215*, 118–123.

García-González, D. L., & Aparicio, R., (2002b). Detection of vinegary defect in virgin olive oils by metal oxide sensors. *Journal of Agricultural and Food Chemistry, 50*, 1809–1814.

Gobbi, E., Falasconi, M., Concina, I., Mantero, G., Bianchi, F., Mattarozzi, M., Musci, M., & Sberveglieri, G., (2010). Electronic nose and *Alicyclobacillus* spp. spoilage of fruit juices: An emerging diagnostic tool. *Food Control, 21*, 1374–1382.

Gobbi, E., Falasconi, M., Torelli, E., & Sberveglieri, G., (2011). Electronic nose predicts high and low fumonisin contamination in maize cultures. *Food Research International, 44*, 992–999.

Gobbi, E., Falasconi, M., Zambotti, G., Sberveglieri, V., Pulvirenti, A., & Sberveglieri, G., (2015). Rapid diagnosis of *Enterobacteriaceae* in vegetable soups by a metal oxide sensor-based electronic nose. *Sensors and Actuators B: Chemical, 207*, 1104–1113.

Gorji-Chakespari, A., Nikbakht, M. A., Sefidkon, F., Ghasemi-Varnamkhasti, M., & Valero, L. E., (2017). Classification of essential oil composition in *Rosa damascena* Mill. genotypes using an electronic nose. *Journal of Applied Research on Medicinal and Aromatic Plants, 4*, 27–34.

Gruber, J., Nascimento, H. M., Yamauchi, E. Y., Li, R. W., Esteves, C. H., Rehder, G. P., Gaylarde, C. C., & Shirakawa, M. A., (2013). A conductive polymer-based electronic nose for early detection of *Penicillium digitatum* in post-harvest oranges. *Materials Science and Engineering: C, 33*, 2766–2769.

Hartyáni, P., Dalmadi, I., & Knorr, D., (2013). Electronic nose investigation of *Alicyclobacillus acidoterrestris* inoculated apple and orange juice treated by high hydrostatic pressure. *Food Control, 32*, 262–269.

Haugen, J. E., Rudi, K., Langsrud, S., & Bredholt, S., (2006). Application of gas-sensor array technology for detection and monitoring of growth of spoilage bacteria in milk: A model study. *Analytica Chimica Acta, 565*, 10–16.

Huang, X. C., Yuan, Y. H., Wang, X. Y., Jiang, F. H., & Yue, T. L., (2015). Application of electronic nose in tandem with chemometric analysis for detection of *Alicyclobacillus acidoterrestris* spawned spoilage in apple juice beverage. *Food and Bioprocess Technology, 8*, 1295–1304.

Karlshøj, K., Nielsen, P. V., & Larsen, T.O., (2007). Prediction of *Penicillium expansum* spoilage and patulin concentration in apples used for apple juice production by electronic nose analysis, *Journal of Agricultural and Food Chemistry, 55*, 4289–4298.

Keshri, G., & Magan, N., (2000). Detection and differentiation between mycotoxigenic and Non-mycotoxigenic strains of two *Fusarium* spp. using volatile production profiles and hydrolytic enzymes. *Journal of Applied Microbiology, 89*, 825–833.

Keshri, G., Voysey, P., & Magan, N., (2002). Early detection of spoilage molds in bread using volatile production patterns and quantitative enzyme assays. *Journal of Applied Microbiology, 92*, 165–172.

Kirsching, A., Bázár, G., Házás, Z., & Romvári, R., (2012). Classification of meat with boar taint using an electronic nose. *Acta Agriculturae Slovenica, 100*, 99–103.

Ko, S. H., Park, E. Y., Han, K. Y., Noh, B. S., & Kim, S. S., (2000). Development of neural network analysis program to predict shelf-life of soymilk by using electronic nose. *Food Engineering Progress, 4*(3), 193–198.

Kodogiannis, V. S., (2017). Application of an electronic nose coupled with fuzzy-wavelet network for the detection of meat spoilage. *Food and Bioprocess Technology, 10*, 730–749.

Kojima, H., Araki, S., Kaneda, H., & Takashio, M., (2005). Application of a new electronic nose with fingerprint mass spectrometry to brewing. *Journal of the American Society of Brewing Chemists, 63*, 151–156.

Konduru, T., Rains, G. C., & Li, C., (2015). A customized metal oxide semiconductor-based gas sensor array for onion quality evaluation: System development and characterization. *Sensors, 15*, 1252–1273.

Kong, Z., Li, M., An, J., Chen, J., Bao, Y., Francis, F., & Dai, X., (2016). The fungicide triadimefon affects beer flavor and composition by influencing *Saccharomyces cerevisiae* metabolism. *Scientific Reports, 6*, 33552.

Korel, F., & Balaban, M., (2002). Microbial and sensory assessment of milk with an electronic nose, *Journal of Food Science, 67*, 758–764.

Lacy-Costello, B., Ewen, R., Gunson, H., Ratcliffe, N. M., Sivanand, P., & Spencer-Phillips, P. T., (2003). A prototype sensor system for the early detection of microbially linked spoilage in stored wheat grain. *Measurement Science and Technology, 14*, 397.

Längkvist, M., Coradeschi, S., Loutfi, A., & Rayappan, J. B. B., (2013). Fast classification of meat spoilage markers using nanostructured ZnO thin films and unsupervised feature learning. *Sensors, 13*, 1578–1592.

Lee, S. H., Lim, J. H., Park, J., Hong, S., & Park, T. H., (2015). Bioelectronic nose combined with a micro fluidic system for the detection of gaseous trimethylamine, *Biosensors and Bioelectronics, 71*, 179–185.

Leggieri, M. C., Pont, N. P., Battilani, P., & Magan, N., (2011). Detection and discrimination between ochratoxin producer and non-producer strains of *Penicillium nordicum* on a ham-based medium using an electronic nose. *Mycotoxin Research, 27*, 29–35.

Lerma-García, M., Cerretani, L., Cevoli, C., Simó-Alfonso, E., Bendini, A., & Toschi, T. G., (2010). Use of electronic nose to determine defect percentage in oils. Comparison with sensory panel results. *Sensors and Actuators B*: *Chemical, 147*, 283–289.

Li, C., Heinemann, P., & Sherry, R., (2007). Neural network and Bayesian network fusion models to fuse electronic nose and surface acoustic wave sensor data for apple defect detection. *Sensors and Actuators B: Chemical, 125*, 301–310.

Li, C., Krewer, G. W., Ji, P., Scherm, H., & Kays, S. J., (2010). Gas sensor array for blueberry fruit disease detection and classification. *Postharvest Biology and Technology, 55*, 144–149.

Li, C., Schmidt, N. E., & Gitaitis, R., (2011). Detection of onion postharvest diseases by analyses of headspace volatiles using a gas sensor array and GC-MS. *LWT-Food Science and Technology, 44*, 1019–1025.

Lim, J. H., Park, J., Ahn, J. H., Jin, H. J., Hong, S., & Park, T. H., (2013). A peptide receptor-based bioelectronic nose for the real-time determination of seafood quality. *Biosensors and Bioelectronics, 39*, 244–249.

Lippolis, V., Ferrara, M., Cervellieri, S., Damascelli, A., Epifani, F., Pascale, M., & Perrone, G., (2016). Rapid prediction of ochratoxin A-producing strains of *Penicillium* on dry-cured meat by MOS-based electronic nose. *International Journal of Food Microbiology, 218*, 71–77.

Lippolis, V., Pascale, M., Cervellieri, S., Damascelli, A., & Visconti, A., (2014). Screening of deoxynivalenol contamination in durum wheat by MOS-based electronic nose and identification of the relevant pattern of volatile compounds. *Food Control, 37*, 263–271.

Magan, N., & Evans, P., (2000). Volatiles as an indicator of fungal activity and differentiation between species, and the potential use of electronic nose technology for early detection of grain spoilage. *Journal of Stored Products Research, 36*, 319–340.

Magan, N., Pavlou, A., & Chrysanthakis, I., (2001). Milk-sense: A volatile sensing system recognizes spoilage bacteria and yeasts in milk. *Sensors and Actuators B: Chemical, 72*, 28–34.

Marin, S., Vinaixa, M., Brezmes, J., Llobet, E., Vilanova, X., Correig, X., Ramos, A., & Sanchis, V., (2007). Use of a MS-electronic nose for prediction of early fungal spoilage of bakery products. *International Journal of Food Microbiology, 114*, 10–16.

Natale, C. D., Macagnano, A., Martinelli, E., Paolesse, R., Proietti, E., & Amico, A. D., (2001). The evaluation of quality of post-harvest oranges and apples by means of an electronic nose, *Sensors and Actuators B: Chemical, 78*, 26–31.

Needham, R., Williams, J., Beales, N., Voysey, P., & Magan, N., (2005). Early detection and differentiation of spoilage of bakery products. *Sensors and Actuators B: Chemical, 106*, 20–23.

Olsson, J., Börjesson, T., Lundstedt, T., & Schnürer, J., (2000). Volatiles for mycological quality grading of barley grains: Determinations using gas chromatography-mass spectrometry and electronic nose. *International Journal of Food Microbiology, 59*, 167–178.

Olsson, J., Börjesson, T., Lundstedt, T., & Schnürer, J., (2002). Detection and quantification of ochratoxin A and deoxynivalenol in barley grains by GC-MS and electronic nose. *International Journal of Food Microbiology, 72*, 203–214.

Pallottino, F., Costa, C., Antonucci, F., Strano, M. C., Calandra, M., Solaini, S., & Menesatti, P., (2012). Electronic nose application for determination of *Penicillium digitatum* in Valencia oranges. *Journal of the Science of Food and Agriculture, 92*, 2008–2012.

Pan, L., & Yang, S., (2007). A new intelligent electronic nose system for measuring and analyzing livestock and poultry farm odors. *Environmental Monitoring and Assessment, 135*(1–3), 399–408.

Pan, L., Zhang, W., Zhu, N., Mao, S., & Tu, K., (2014). Early detection and classification of pathogenic fungal disease in post-harvest strawberry fruit by electronic nose and gas chromatography-mass spectrometry. *Food Research International, 62*, 162–168.

Panagou, E. Z., Sahgal, N., Magan, N., & Nychas, G. J., (2008). Table olives volatile fingerprints: Potential of an electronic nose for quality discrimination. *Sensors and Actuators B: Chemical, 134*, 902–907.

Panigrahi, S., Balasubramanian, S., Gu, H., Logue, C., & Marchello, M., (2006). Neural-network integrated electronic nose system for identification of spoiled beef. *LWT-Food Science and Technology, 39*, 135–145.

Paolesse, R., Alimelli, A., Martinelli, E., Di Natale, C., D'Amico, A., D'Egidio, M. G., Aureli, G., Ricelli, A., & Fanelli, C., (2006). Detection of fungal contamination of cereal grain samples by an electronic nose. *Sensors and Actuators B: Chemical, 119*, 425–430.

Papadopoulou, O. S., Panagou, E. Z., Mohareb, F. R., & Nychas, G. J. E., (2013). Sensory and microbiological quality assessment of beef fillets using a portable electronic nose in tandem with support vector machine analysis. *Food Research International, 50*, 241–249.

Park, J., Lim, J. H., Jin, H. J., Namgung, S., Lee, S. H., Park, T. H., & Hong, S., (2012). A bioelectronics sensor-based on canine olfactory nanovesicle-carbon nanotube hybrid structures for the fast assessment of food quality. *Analyst, 137*, 3249–3254.

Perkowski, J., Buśko, M., Chmielewski, J., Góral, T., & Tyrakowska, B., (2008). Content of trichodiene and analysis of fungal volatiles (electronic nose) in wheat and triticale grain naturally infected and inoculated with *Fusarium culmorum. International Journal of Food Microbiology, 126*, 127–134.

Presicce, D., Forleo, A., Taurino, A., Zuppa, M., Siciliano, P., Laddomada, B., Logrieco, A., & Visconti, A., (2006). Response evaluation of an E-nose towards contaminated wheat by *Fusarium poae* fungi. *Sensors and Actuators B: Chemical, 118*, 433–438.

Qiu, S., & Wang, J., (2017). The prediction of food additives in the fruit juice based on electronic nose with chemometrics. *Food Chemistry, 230*, 208–214.

Ragazzo, S. J., Chalier, P., & Ghommidh, C., (2005). Coupling gas chromatography and electronic nose for dehydration and desalcoholization of alcoholized beverages: Application to off-flavor detection in wine. *Sensors and Actuators B: Chemical, 106*, 253–257.

Ragazzo, S. J., Chalier, P., Chevalier, L. D., Calderon, S. M., & Ghommidh, C., (2009). Off-flavors detection in alcoholic beverages by electronic nose coupled to GC. *Sensors and Actuators B: Chemical, 140*, 29–34.

Rahman, M. S., Al-Farsi, K., Al-Maskari, S. S., & Al-Habsi, N. A., (2018). Stability of electronic nose (e-nose) as determined by considering date-pits heated at different temperatures. *International Journal of Food Properties, 21*, 1532–2386.

Rajamäki, T., Alakomi, H. L., Ritvanen, T., Skyttä, E., Smolander, M., & Ahvenainen, R., (2006). Application of an electronic nose for quality assessment of modified atmosphere packaged poultry meat. *Food Control, 17*, 5–13.

Rodríguez, J., Durán, C., & Reyes, A., (2009). Electronic nose for quality control of Colombian coffee through the detection of defects in cup tests. *Sensors, 10*, 36–46.

Rutolo, M. F., Iliescu, D., Clarkson, J. P., & Covington, J. A., (2016). Early identification of potato storage disease using an array of metal-oxide-based gas sensors. *Postharvest Biology and Technology, 116*, 50–58.

Santos, J. P., & Lozano, J., (2015). *Real time Detection of Beer Defects with a Hand Held Electronic nose, Electron Devices (CDE)* (pp. 1–4). 2015 10th Spanish Conference on, IEEE.

Sberveglieri, G., Zambotti, G., Falasconi, M., Gobbi, E., & Sberveglieri, V., (2014). *MOX-NW Electronic Nose for Detection of Food Microbial Contamination* (pp. 1376–1379). SENSORS, 2014 IEEE.

Sberveglieri, V., Carmona, E. N., Comini, E., Ponzoni, A., Zappa, D., Pirrotta, O., & Pulvirenti, A., (2014). A novel electronic nose as adaptable device to judge microbiological quality and safety in foodstuff. *Bio. Med. Research International.*

Sinesio, F., Natale, C. D., Quaglia, G. B., Bucarelli, F. M., Moneta, E., Macagnano, A., Paolesse, R., & Amico, A. D., (2000). Use of electronic nose and trained sensory panel in the evaluation of tomato quality. *Journal of the Science of Food and Agriculture, 80*, 63–71.

Siqueira, A. F., Melo, M. P., Giordani, D. S., Galhardi, D. R. V., Santos, B. B., Batista, P. S., & Ferreira, A. L. G., (2018). Stochastic modeling of the transient regime of an electronic nose for waste cooking oil classification. *Journal of Food Engineering, 221*, 114–123.

Vinaixa, M., Marín, S., Brezmes, J., Llobet, E., Vilanova, X., Correig, X., Ramos, A., & Sanchis, V., (2004). Early detection of fungal growth in bakery products by use of an electronic nose based on mass spectrometry. *Journal of Agricultural and Food Chemistry, 52*, 6068–6074.

Vinaixa, M., Vergara, A., Duran, C., Llobet, E., Badia, C., Brezmes, J., Vilanova, X., & Correig, X., (2005). Fast detection of rancidity in potato crisps using e-noses based on mass spectrometry or gas sensors. *Sensors and Actuators B: Chemical, 106*, 67–75.

Wang, D., Wang, X., Liu, T., & Liu, Y., (2012). Prediction of total viable counts on chilled pork using an electronic nose combined with support vector machine. *Meat Science, 90*, 373–377.

Wang, H., Hu, Z., Long, F., Guo, C., Yuan, Y., & Yue, T., (2016). Early detection of *Zygosaccharomyces Rouxii* spawned spoilage in apple juice by electronic nose combined with chemometrics., *International Journal of Food Microbiology, 217*, 68–78.

Wei, X., Shao, X., Wei, Y., Cheong, L., Pan, L., & Tu, K., (2018). Rapid detection of adulterated peony seed oil by electronic nose, *J. Food Sci. Technol., 55*(6), 2152–2159.

Wilson, A. D., Oberle, C. S., & Oberle, D. F., (2013). Detection of off-flavor in catfish using a conducting polymer electronic-nose technology. *Sensors, 13*, 15968–15984.

Wongchoosuk, C., Wisitsoraat, A., Tuantranont, A., & Kerdcharoen, T., (2010). Portable electronic nose based on carbon nanotube-SnO_2 gas sensors and its application for detection of methanol contamination in whiskeys. *Sensors and Actuators B: Chemical, 147*, 392–399.

Xu, L., Yu, X., Liu, L., & Zhang, R., (2016). A novel method for qualitative analysis of edible oil oxidation using an electronic nose, *Food Chemistry, 202*, 229–235.

Zhang, H., & Wang, J., (2007). Detection of age and insect damage incurred by wheat, with an electronic nose, *Journal of Stored Products Research, 43*, 489–495.

Zhang, S., Xie, C., Bai, Z., Hu, M., Li, H., & Zeng, D., (2009). Spoiling and formaldehyde-containing detections in octopus with an E-nose. *Food Chemistry, 113*, 1346–1350.

CHAPTER 7

IR SPECTROSCOPY FOR QUALITY OF FRUITS AND VEGETABLES

VASUDHA BANSAL and NITYA SHARMA

*Department of Food Engineering and Nutrition,
Center of Innovative and Applied Bioprocessing (CIAB),
Knowledge City, Sector-81, Mohali – 140306, Punjab, India,
E-mail: vasu22bansal@gmail.com (V. Bansal)*

ABSTRACT

Sensor-based evaluation of the quality of fresh produce (fruits and vegetables) is in great demand owing to the accurate detection and lower time consumption. Thereby, detection of quality of fruits and vegetables using non-destructive (ND) infra-red spectroscopy (IR) offers the huge potential in postharvest. The investigation of the quality of fresh produce using physical parameters (in terms of shape, color, tissue injury, and browning) plays an inevitable role in their visible acceptance by consumers. In addition, the rapid detection of chemical composition in terms of total soluble solids, phenolic compounds, and sugars add strength to the competitive acceptance in the international and national market. Therefore, the optimization of IR parameters for quality evaluation of fruits and vegetables is efficacious in preventing the wastage of fresh produce.

7.1 INTRODUCTION

Fruits and vegetables are largely consumed as the major part of the diet worldwide owing to the presence of potent bioactive compounds. These compounds are of great clinical importance owing to their properties for preventing lifestyle disorders. Therefore, fresh produce of fruits and vegetables are in demand in order to retain their maximum benefits without any processing. Thereby, reliable and quick detection of the quality of these produce is most essential. The term quality has many expressions which are required to fulfill in order to succeed in the marketing of the fruits and vegetables. The intrinsic parameters in terms of nutritional value and safety cannot be measured by the consumers visually; they depend on the visual quality of the product that exists in their mind (Butz et al., 2005). Thereby, the detection of quality parameters (such as color, shape, texture, firmness) needs a quick investigation and infrared spectroscopy (IR) has been one of the non-invasive techniques that are able to establish as the tool for analyzing the postharvest quality of fruits and vegetables.

The spectroscopic analysis uses the interaction of atoms and molecules and electromagnetic radiation to study the chemical and physical characteristics. The photons of a certain wavelength or frequency are absorbed or emitted depending on their energy levels.

7.2   EVALUATION OF PHYSICAL PARAMETERS USING IR

External quality parameters like color, texture, physical defects, and morphological characteristics, play an important role in the direct sensorial evaluation of fruits and vegetables. The external quality evaluation, therefore, affects the consumer's acceptability and buying behavior and helps in preventing losses due to the spread of infection and contamination from defective fruits and vegetables to the whole batch. Thus, a non-invasive and rapid inspection of the external quality of fruits and vegetables using IR has become a promising technique for the detection of several attributes simultaneously. Amongst the various IR spectral regions, the visible region (350–800 nm) and the near-infrared (NIR) (800–2500 nm) are used to evaluate the physical parameters and compositional analyzes of fruits and vegetables, respectively (Dufour, 2009). The postharvest quality of fruits and vegetables is characterized on the basis of size, color, firmness, surface appearance, and other important traits like dry matter content, total soluble solids, titratable acidity, and aromatic compounds. But, the use of spectroscopy alone does not provide the spatial information of fruits and vegetables, thus limiting its scope for research. To overcome this drawback, imaging spectral techniques are now being widely used, thus enabling the acquisition of fruit/vegetable images and spectral information, simultaneously. Multispectral and hyperspectral imaging techniques commonly use the waveband in the Vis/NIR region to detect almost all major structures and functional groups of organic compounds for fruits and vegetables. This section will give a comprehensive overview of the firmness and physical defects that have been commonly studied using spectroscopic and imaging spectral techniques, and the chemometric methods used to measure the external quality of fruits and vegetables. Some of the studies have been detailed in Table 7.1.

TABLE 7.1 Summary of Physical Characteristics of the Agriculture-Based Produce for Determination of Quality Using IR

Sl. No.	Sample	Method	IR Range	IR Mode/ Scan	Resolution Time/no. of Scans	Chemometric Analytical Tool	Physical Profiling	References
1.	Tommy Atkins Mangoes	Spectroscopy	950 and 1650 nm	Reflectance	50 scans, 0.50s	PCA, PLSR	Soluble solids content, dry matter, titratable acidity, pulp firmness	Marques et al., 2016
2.	Golden Delicious Apples	Spectroscopy	800–2500 nm	Reflectance	—	PLSR	Quality-soluble solids content, firmness, color, nutritional value indices-total phenolics, total flavonoids, antioxidant activity	Giovanelli et al., 2014
3.	Golden Delicious, Royal Gala, Jonagold, Braeburn, Pink Lady, Fuji Apples	Spectroscopy	380–1690 nm	Reflectance	100 scans, 30 ms	PLSR, Functional analysis of variance (FANOVA)	Fruit firmness, soluble solids content	Bobelyn et al., 2010
4.	Cogshall Mangoes	Spectroscopy	800–2300 nm	Reflectance	—	PCA, PLSR	Total soluble solid content, acidity, dry matter, flesh color	Nordey et al, 2017

TABLE 7.1 (Continued)

Sl. No.	Sample	Method	IR Range	IR Mode/Scan	Resolution Time/no. of Scans	Chemometric Analytical Tool	Physical Profiling	References
5.	Mangoes (cv. Nam Dokmai subcv. Si Thong)	Spectroscopy	700–1100 nm	Reflectance	16 scans	Discriminant analysis (DA)	Fruit firmness, TSS, TA, moisture content, ripening index	Rungpichayapichet et al., 2016
6.	Oranges (Sanguinelli, Valencia, Salustiana, Navelate)	Spectroscopy	350–2500 nm	Reflectance	100 scans, 100 ms	PLSR	Soluble solids content, acidity, maturity index, flesh firmness, juice volume, fruit weight, rind weight, fruit color index, juice color index	Cayuela and Weiland, 2010
7.	Pears (Clara Frijs)	Spectroscopy	300–2500 nm	Absorbance	64 scans	PLSR	Dry matter, soluble solids content, starch pattern index (SPI)	Travers et al., 2014
8.	Qilin watermelons	Spectroscopy	220–1020 nm	Diffuse transmittance	200 ms	MC-UVE-GA-PLSR	Weight, length, diameter, rind thickness and soluble solids content	Jie et al., 2013

TABLE 7.1 (Continued)

Sl. No.	Sample	Method	IR Range	IR Mode/ Scan	Resolution Time/no. of Scans	Chemometric Analytical Tool	Physical Profiling	References
9.	Cabbage (var. Capitata)	Spectroscopy	310–1100 nm	Reflectance	10 scans	PLSR	Morphological properties, moisture content, soluble solids content, ascorbic acid	Kramchote et al., 2014
10.	Pears	Spectroscopy	200–1100 nm	Transmittance	100 ms	PCA, Correlation, PLSR	Brown core, SSC	Sun et al., 2016
11.	Apple	Hyperspectral imaging	400–2500 nm	Reflectance, hyperspectral imaging	5 s	Supervised classification models	Surface bruising	Baranowski et al., 2013
12.	Fuji Apples	Multispectral imaging	450–1000 nm	Reflectance, hyperspectral imaging	55 ms	PCA	Surface bruising	Huang et al., 2015
13.	Fuji Apples	Hyperspectral imaging	318–1099 nm	Reflectance, hyperspectral imaging	26 ms	PLSR	Textural features, SSC	Fan et al., 2016
14.	Excursion pickling cucumbers	Hyperspectral imaging	400–1000 nm	Reflectance, transmittance	2 ms	Supervised classification models	Chilling injury	Cen et al., 2016
15.	Palmer Mangoes	Spectroscopy	310–1100 nm	Absorbance	4 scans	PLSR	SSC, dry matter	Neto et al., 2017

TABLE 7.1 *(Continued)*

Sl. No.	Sample	Method	IR Range	IR Mode/ Scan	Resolution Time/no. of Scans	Chemometric Analytical Tool	Physical Profiling	References
16.	Melons (Elizabeth, Claw Brittle, Emerald)	Hyperspectral imaging	882–1719 nm	Reflectance	3 scans	PCA, PLSR	Sugariness, hardness	Sun et al., 2017
17.	Olive fruit (cv. Canino)	Spectroscopy	1100–2300 nm	Reflectance	60 ms	ANN-Genetic algorithm (GA)	Hailstorm damage	Moscetti et al., 2016
18.	Valencia orange and Star Ruby grapefruit	Spectroscopy	450–2500 nm	Reflectance	32 scans	PCA, PLSR	TSS, TA, BrimA	Neama et al., 2017
19.	Purple passion fruit	Spectroscopy	310–1100 nm	Interactance, transmittance	14 scans	PLSR	SSC, TA, ascorbic acid content, ethanol concentration, peel firmness, pulp percentage	Maniwara et al., 2014
20.	Pinggu peaches	Hyperspectral imaging	325–1100 nm	Reflectance	22 ms	PCA	Skin defects (skin injury, scarring, insect damage, puncture injury, decay, disease spots, dehiscent scarring, and anthracnose)	Li et al., 2016

TABLE 7.1 (*Continued*)

Sl. No.	Sample	Method	IR Range	IR Mode/ Scan	Resolution Time/no. of Scans	Chemometric Analytical Tool	Physical Profiling	References
21.	Peaches (Prunus persica cv. Xiahui 5)	Hyperspectral imaging	400–1000 nm	Reflectance	2.5 ms	ANN	Cold injury (firmness, extractable juice, SSC, TA, chlorophyll content)	Pan et al., 2016
22.	Honey peaches	Hyperspectral imaging	400–1000 nm	Reflectance	2.5 ms	PLS-DA, ANN, support vector machines (SVM)	Chilling injury (internal browning, taste, texture, flavor)	Sun et al., 2017
23.	Shingo pears	Hyperspectral imaging	950–1650 nm	Reflectance	—	ANOVA	Physical damages-bruise classification	Lee et al., 2014
24.	European pear fruit (Pyrus communis L. cv. Abate, Cascade, Conference, Red Comice, Wujiuxiang)	Spectroscopy	500–1010 nm	Absorbance	200 ms	PLSR, multiple linear regression (MLR)	Firmness, SSC	Wang et al., 2017

TABLE 7.1 *(Continued)*

Sl. No.	Sample	Method	IR Range	IR Mode/ Scan	Resolution Time/no. of Scans	Chemometric Analytical Tool	Physical Profiling	References
25.	Persimmon (Diospyros kaki cv. Rojo Brillante)	Spectroscopy	450–1020 nm	Reflectance	—	Linear and quadratic discriminant analysis (LDA and QDA), SVM, PCA, PLSR	Ripeness (firmness), astringency (soluble tannin content)	Munera et al., 2017
26.	Strawberries	Hyperspectral imaging	380–1734 nm	Reflectance	0.05 s, 5 ms	PCA, SVM	Ripeness (texture features)	Zhang et al., 2016
27.	Chelan and Bing cherry (Prunus avium) fruit	Spectroscopy	310–1100 nm	Interactance	2 scans	PLSR	Dry matter, SSC	Escribano et al., 2017
28.	Nanfeng mandarin fruit	Spectroscopy	350–1040 nm	Reflectance	—	PLSR	SSC, TA	Liu et al., 2010

TABLE 7.1 (*Continued*)

Sl. No.	Sample	Method	IR Range	IR Mode/ Scan	Resolution Time/no. of Scans	Chemometric Analytical Tool	Physical Profiling	References
29.	Golden Delicious, Red Delicious, Ambri, and two unknown varieties of Apple	Spectroscopy	900–1700 nm	Transmittance	50 scans, 100 ms	PLSR, MLR	TSS, TA, TSS/TA ratio	Jha and Garg, 2010
30.	Hayward kiwifruit	Spectroscopy	400–1000 nm	Transmittance	—	PCA, PLSR	SSC, acidity (pH)	Moghimi et al., 2010
31.	Banana	Hyperspectral imaging	400–1000 nm	Reflectance	200 ms	PCA, PLSR	TSS, moisture content, firmness	Rajkumar et al., 2012
32.	Mulberry fruit (*Morus australis* Poir.)	Spectroscopy	325–1075 nm	Reflectance	30 scans	PLSR, successive projections algorithm (SPA)	TSS, pH	Huang et al., 2011

TABLE 7.1 (Continued)

Sl. No.	Sample	Method	IR Range	IR Mode/ Scan	Resolution Time/no. of Scans	Chemometric Analytical Tool	Physical Profiling	References
33.	Plums cv. Black Diamond, Fortune, Friar, Golden Globe, Golden Japan, and Santa Rosa	Spectroscopy	1600–2400 nm	Reflectance	2 scans, 1–2 s	PLS-DA	SSC, firmness	Pérez-Marín et al., 2010
34.	Unripe banana banana combs cv. Grand naine	Spectroscopy	299–1100 nm	Transmittance	6 scans	PLS, MLR	TSS, pH, dry matter, acid-brix ratio (ABR)	Jaiswal et al., 2012
35.	Green bananas cv. Musa	Hyperspectral imaging	380–1023 nm	Reflectance	0.13 s	PCA, PLS-DA	Skin color, firmness	Xie et al., 2018
36.	Mangoes (cv. Chausa, Langra, Dashehri, Kesar, Maldah, Mallika, Neelam)	Spectroscopy	1200–2200 nm	Reflectance	50 scans	PLSR, MLR	TSS, pH	Jha et al., 2012

TABLE 7.1 (Continued)

Sl. No.	Sample	Method	IR Range	IR Mode/ Scan	Resolution Time/no. of Scans	Chemometric Analytical Tool	Physical Profiling	References
37.	Tomatoes cv. Momotaro	Spectroscopy	1100–2500 nm	Interactance	288 scans	PCA, PLSR	Texture, Maturity (SSC)	Sirisomboon et al., 2012
38.	Tomatoes cv. Raf	Spectroscopy	400–2500 nm; 400–1700 nm	Reflectance	1.8 scans/s; 600 scans/s	PCA, modified partial least squares regression (MPLSR)	Skin color, dry matter, SSC, fructose, glucose, TA, citric acid, malic acid	Torres et al., 2015
39.	Highbush blueberries	Hyperspectral imaging	360–2000 nm	Reflectance	200 ms	PLSR	Firmness index, SSC	Leiva-Valenzuela et al., 2013
40.	Longan fruits	Spectroscopy	400–700 nm	Reflectance	3 scans	PCA, Soft independent modeling of class analogy (SIMCA), PLSR	Bruises	Pholpho et al., 2011

7.2.1 FIRMNESS

Firmness is a commonly used parameter to predict the maturity and wholesomeness of fruits and vegetables. Considerable research has been reported on the development of firmness evaluation techniques for either sorting or grading of fruits and vegetables. NIR spectroscopy has been found to be a useful technique in monitoring the levels of maturity and aging at harvest and during storage for fruits and vegetables like apples (Giovanelli et al., 2014; Fan et al., 2016) and mangoes (Rungpichayapichet et al., 2016; Jha et al., 2012; Padda et al., 2011; Sivakumar et al., 2011). Various types of spectra are used to develop calibration models for the firmness using regression analysis, which is then evaluated using correlation coefficient and root mean square error between the predicted value and measured value. For example, internal quality (including firmness) of mango has been evaluated by developing robust NIRS calibration models using diffuse reflectance spectra with a broadband peak observed around 900–1050 nm (Rungpichayapichet et al., 2016). Though, in the case of apricots, diffuse reflectance measurements couldn't accurately predict quality trait-like firmness (Bureau et al., 2009). On the other hand, firmness of apples was online examined using Vis/NIR calibration equations developed by transmittance spectra in the range of 850–920 nm (Fan et al., 2016). Some studies have also used two types of spectra to establish a competency comparison in predicting the firmness of fruits and vegetables. Maniwara et al. (2014) studied the feasibility and comparison in interactance and transmission modes of Vis/short wave NIRS on the evaluation of firmness and other postharvest physico-chemical quality traits in purple passion fruit. The results revealed that spectra at wavelengths of 940–970 nm from interactance modes had more homogeneity than those found in transmission measurements.

Low cost and smaller NIR spectrometers that require no moving parts are now being widely used outside of the laboratory for fruit and vegetable quality and maturity analysis in the field or in storage. Some studies have also reported that handheld and portable NIR spectrometers have similar performance to the benchtop NIR spectrometers for determination of quality parameters like firmness in mangoes (Marques et al., 2016; Jha et al., 2014; Cayuela and Weiland, 2010); pears (Wang et al., 2017). According to Pérez-Marín et al. (2010) both portable diode-array spectrophotometer and handheld MEMS (micro-electro-mechanical system) based spectrophotometer proved to be a promising instrument for supermarket evaluation of quality parameters, variety, and shelf-life, thus affording an additional advantage of convenience and cost-effective.

Recently, a growing interest has been reported on an emerging technique of hyperspectral imaging to determine the quality of fruits and vegetables using both spatial and spectral information. It utilizes the advantages of conventional RGB (expand the term), NIR spectroscopy and multispectral imaging, thus enhancing the capability to identify and detect various chemical constituents of the fruit or vegetable along with their spatial distributions. For example, banana fruit quality and maturity stages (including firmness and other traits) have been successfully identified using MLR models developed based on the optimal wavelengths of 440, 525, 633, 672, 709, 760, 925, and 984 (Rajkumar et al., 2012). In a recent study, Xie et al. (2018) extracted the spectral reflectance information from the hyperspectral images to determine the firmness of ripe and unripe bananas. They used a novel two-wavelength combination method for selecting the wavelengths, which gave better results as compared to full wavelengths in predicting the ripening stages of bananas. The selected wavelengths to predict banana fruit firmness was 638, 675, 681, and 696 nm. Similarly, hyperspectral imaging has also been applied to determine the firmness of various other fruits and vegetables like melons using PLSR, principal component analysis (PCA), support vector machine (SVM) and artificial neural network (ANN), thus, concluding that PLSR produced the most accurate results and predicted the fruit hardness using five selected wavelengths for each variety (Sun et al., 2017). Wei et al. (2014) predicted firmness of persimmons using PLSR analysis of spectra and texture features of images at specified wavelengths of 518, 711, and 980 nm. Amongst other fruits and vegetables, firmness using hyperspectral imaging at wavelength 790 nm has been evaluated with maximum correlation for apple, peach, pear, kiwifruit, plum, cucumber, zucchini squash and tomato (three ripeness stages) combined with inverse algorithms based on diffusion theory model; blueberries using prediction models developed by PLSR in the wavelength range of 500–1000 nm (Leiva-Valenzuela et al., 2013); strawberries using classification models built by SVM using datasets from hyperspectral images at 441.1–1013.97 nm (Zhang et al., 2016); and many more. Apart from this, some past studies have also quantified light backscattering profiles from fruits and vegetables using multispectral imaging to predict their firmness. Lu (2004) concluded that the backpropagation neural network with inputs of ratios of scattering profile combinations with wavelengths at 680, 880, 905, and 940 nm gave the best predictions for apple fruit firmness.

7.2.2   EXTERNAL DEFECTS

External defects on fruits and vegetables, like bruises, chilling injuries, brown spots, etc, are important as they affect the consumer's satisfaction and lead to spoilage and loss of nutritional value. While handling fruits and vegetables, there is a greater possibility of developing physical defects due to impact, compression, vibration, or abrasion. Although, there are various automatic sorting systems that help in removing bruised fruits and vegetables from the lot, but because of their undesirable accuracy a manual sorting method has to be used. Moreover, common bruises like browning and softening of tissue are difficult to identify during the first few days as they develop gradually depending upon various external factors. Hyperspectral imaging in VNIR and SWIR spectral ranges has a good application in detecting the number of days after bruising. Bruises on apples were best detected on hyperspectral images at three effective wavelengths, i.e., 780, 850, and 960 nm, using a 325–1100 nm and segmented PCA (Huang et al., 2014). While, Baranowski et al. (2013) stated that the use of supervised classification methods had a good applicability of hyperspectral imaging in spectral ranges of 400–1000 nm, as well as 1000–2500 nm, to distinguish time after bruising in apples. In the case of pears, selected optimal NIR reflectance image ratio (1074 and 1016 nm: R1074/R1016), determined using simple F-value statistics, could efficiently detect bruised areas (Lee et al., 2014). Apart from NIR, visible reflectance spectroscopy along with PLS-DA could successfully classify bruised and non-bruised longan fruits (Pholpho et al., 2011).

Amongst the various external defects, chilling injury is a common physiological disorder in fruits and vegetables that results in the damage of cell membranes caused due to improper storage temperatures. The symptoms for chilling injury in fruits and vegetables include surface bruising, peel, and pulp discoloration, tissue shrinkage, pitting, and off-flavors. Hyperspectral imaging techniques has demonstrated the potential for the detection of chilling injuries in fruits and vegetables. Cen et al. (2016) found that the sequential forward selection method along with the SVM classifier from the spectral transmittance images in the short-near infrared region (675–1000 nm) resulted in the best online detection of chilling injuries in cucumbers. Similarly, hyperspectral reflectance imaging and ANN prediction model was used to detect chill damages in peaches by successfully determining eight optimal wavelengths (487, 514, 629, 656, 774, 802, 920 and 948 nm) in one study (Pan et al., 2016) and six optimal wavelengths (580, 599, 650, 675, 710 and 970 nm) in another (Sun et al., 2017).

Advancement in computing speed and data transfer rates have undoubtedly increased the speed and efficiency of spectroscopic devices, but data handling still poses a challenge to practical applications. In the case of hyperspectral imaging, the acquisition of both spectral and spatial information is time-consuming and requires standardized calibration and model transfer procedures. In a recent study, Vis-NIR hyperspectral imaging in association with two-wavelength (781 nm/848 nm) pixel-by-pixel ratio of reflectance has also been used for detecting different types of skin defects on bi-colored peaches (Li et al., 2016). Here, instead of using a hyperspectral system with a large number of spectral channels, only two selected optimal wavelengths could provide higher speeds. In a different study, the similar problem for speed was resolved by using an accurate and automated single-ray NIR-based detection system was used, instead of hyperspectral system, to detect hailstorm damaged olive fruits with optimal wavelengths around 1320, 1400–1550, 1650 and 2000–2220 nm (Moscetti et al., 2016).

7.3 EVALUATION OF CHEMICAL PARAMETERS USING IR

Artificial vision machines are becoming the powerful tools for the visual inspection of the quality of fruits and vegetables. These abilities are now going beyond the visible electromagnetic spectrum. Therefore, the emergence of hyperspectral imaging has come in to play, where at any particular wavelength any external or internal defect can be inspected (Cubero et al., 2011). The non-destructive (ND) measurements and analysis are preferable owing to their very less wastage and advantage of reproducibility. Among the ND techniques, NIR has been the most advanced in terms of spectra capturing, analysis, and interpretation in order to examine the internal quality of fruits and vegetables (Magwazaet al., 2012). Evaluation of internal quality includes the concentration of organic acids (citric acid, malic acid, tartaric acid, fumaric acid, and succinic acid) (Beullens et al., 2006), presence of pigments as carotenoids, lycopene), (Baranska et al., 2006) phenolic compounds, flavonoids (Sinelli et al., 2008) and vitamin C as mentioned in Table 7.2. The quality pattern of these characteristics decides the acceptance of fruits and vegetables internally and their intact presence will keep the physical parameters intact. In order to determine the chemical constituents in fruits and vegetables, Vis-NIR has been an established technique for the evaluation of their postharvest quality. Wavelengths related to visible part of the electromagnetic spectrum can also show the absorption bands for chlorophyll as well as anthocyanin (Carlini et al., 2000).

The peaks of IR are narrow and diagnostic whereas, peaks of NIR are broad. The broad bands of NIR result from overlapping of absorptions owing to the combinations of vibration modes involving the chemical bonds of C-H, O-H, N-H, and S-H (Golic et al., 2003). Similarly, this method also allowed us the possibility of highlighting the most important wave numbers for each type of polysaccharide: 1,740, 1,610, and 1,240 cm denoting pectins or 1,370 and 1,317 cm denoting hemicelluloses and cellulose, respectively (Szymanska-Chargot, 2013). The molecular vibrations in the NIR region are the strong absorption bands, particularly in the biological material. That is why; the band of O-H dominates in the hydrated material. Moreover, fruits and vegetables are majorly composed of water, and thereby, their bands are more visible in the NIR region. Hydrated samples are then characterized in terms of their interactions of hydrogen bonding among water, sugar, acids, and proteins (Guthrie et al., 2005).

7.3.1   ORGANIC ACIDS AND SUGARS

New paths have been followed by IR spectroscopy other than the chromatography-based analytical tools for the determination of the quality of fruits and vegetables in terms of organic acids. ATR-FTIR has used as the rapid detection tool for sugars and acids in the tomatoes (Beullens et al., 2006). With the usage of advanced optics, the measurement of signal-noise ratio, the presence of organic acids can be checked. The samples of tomato (in the form of powder) were analyzed by taking the 64 scans with a resolution of 4 cm^{-1} in the range of 3250–800 cm^{-1}. A PCA was performed on the data received by ATR-FTIR. The region between 1200 to 900 cm^{-1} showed the major regions for sugars (sucrose, glucose, fructose) and organic acids (malic, tartaric, citric, succinic, and fumaric acid) owing to the stronger stretching vibration of C-O bonds (Buellens et. al., 2006). Organic acids typically constitute to about 10% in the fruits. There are typically three modes which are used for the measurements and configurations of the fruits and vegetables (Cayuela, 2008). These modes are reflectance, transmittance, and interactance.

TABLE 7.2 Summary of Chemical Characteristics of the Agriculture-Based Produce for Determination of Quality Using IR

Sl. No.	Fruit/Product	Chemical Profiling	IR Range	IR Mode/ Scans	Resolution/ Time	Analytical Tool	References
1.	Apple	Vitamin C and phenolics	400–2500 nm	Reflectance		LS-SVM	Pissard et al., 2013
2.	Olives	Phenolics in the ripening stage	1100–2300 nm	Reflectance	10 spectra	PLS, PCA	Bellincontro et al., 2012
3.	Red wine	Phenolic compounds	190–2500 nm	Reflectance	Scanning in duplicate	PLS, PCA	Martelo-Vidal and Vázquez, 2014
4.	Grapes	Flavonoids	1100–2200 nm	Reflectance	32	PLS, PCA	Gallego et al., 2011
5.	Bell peppers	Vitamin C, carotenoids	477–950	Reflectance	10	PLS regression models	Ignat et al., 2012
6.	Mandarin	Vitamin C	400–1040 nm		0.2 sampling interval	Root mean square	Xudong et al., 2009
7.	Tomato	Sugars (Glucose, Fructose, Sugar) Organic Acids (citric acid, malic acid, tartaric acid, fumaric acid, succinic acid)	3250–800 cm^{-1}	ATR-FTIR	4 cm^1	e-tongue	Beullens et al., 2006

TABLE 7.2 *(Continued)*

Sl. No.	Fruit/Product	Chemical Profiling	IR Range	IR Mode/ Scans	Resolution/ Time	Analytical Tool	References
8.	Kiwi	TSS	750–1050 nm 800–1000 nm 500–1050 nm		1.25 s	PCA, CDA	Clark et al., 2004
9.	Mango		700–1100 nm	50	25 s	MLR, PLS, NSAS	Saranwong et al., 2004
10.	Blueberry	TSS, total flavonoids, total anthocyanins, Vitamin C	12000–3600 cm^{-1}	Reflectance/64		PLS, RMSE	Sinelli et al., 2008
11.	Apple	TSS	810–999 nm	10	30 s	Regression	Ventura et al., 1998
12.	Olive	Oleic acid and linoleic acid	400 to 1700 nm		5 nm	PCA, PLS	León et al., 2004
13.	Tomato	Lycopene, β carotene	650–4000 cm^{-1} 4000–15000 cm^{-1}	ATR/30	4 cm^{1}	PLS	Baranska et al., 2006
14.	Tomato, potato, pumpkin, carrot, and celery root	Polysaccharides	1,800–1,200 cm 1,200–800				Szymanska-Chargot et al., 2013
15.	Apple		380–950 nm 950–1690 nm	100	100 ms	PLS	Bobelyn et al., 2010

TABLE 7.2 (Continued)

Sl. No.	Fruit/Product	Chemical Profiling	IR Range	IR Mode/ Scans	Resolution/ Time	Analytical Tool	References
16.	Apricot	Total acidity, TSS,	650–1200 nm			FDA (Factorial discriminant analysis)	Camps et al., 2009
17.	Cherry, Apricot	TSS	600 to 1100 nm 400–2500 nm			PLS	Carlini et al., 2000
18.	Oranges	Total acidity, TSS,	570–1048 nm 1100–1850 nm		1s	PLS, PCA	Cayuela, 2008
19.	Oranges	Total acidity, TSS,	350–2500 nm 1100–2300 nm			PLS	Cayuela and Weiland, 2010
20.	Orange juice	TSS, pH	325–1075 nm		3.5 nm	PLS, PCA	Cen et al., 2006
21.	Orange juice	Citric acid, tartaric acid	325–1075 nm		3.5 nm	PLS, PCA	Cen et al., 2007
22.	Mandarin	Total acidity, TSS,	808 nm laser light		6500 to 4 ms	Monte Carlo simulation	Fraser et al., 2003
23.	Stone fruit (Peaches, nectarine, plum)	TSS	306–1150 nm		30 ms	PLS1 regressions	Golic and Walsh, 2006

TABLE 7.2 *(Continued)*

Sl. No.	Fruit/Product	Chemical Profiling	IR Range	IR Mode/ Scans	Resolution/ Time	Analytical Tool	References
24.	Mandarin	Total acidity, TSS, firmness	350–2500 nm			PCA, PCR, PLS	Gomez et al., 2006
25.	Mandarin	TSS and dry matter	720–950-nm			PLS	Guthrie et al., 2005
26.	Melon	TSS	695–1045 nm	4	200 ms	MPLS	Guthrie et al., 2006
27.	Orange	TSS	350–1800 nm	Reflectance/30		Multiplicative scatter correction (MSC), PLSR, PCA	Liu et al., 2010
28.	Mandarin	Total acidity, TSS,	600–980 nm			iPLS	Liu et al., 2010
29.	Plum	Total acidity, TSS, sugar to acid ratio	800–2700 nm	16		PLS	Louw et al., 2010
30.	Citrus fruit Gannan	TSS	12500–4000 cm−1			PLS, PCR	Huishan et al., 2005
31.	Valencia orange	TSS, vitamin C	800–2500 nm	32		PLS	Magwaza et al., 2011
32.	Apricot	Sugars and acids	3700–2800 nm			PLS	Bureau et al., 2009

Good calibrations of the fruit are usually obtained in the transmittance mode. Whereas, in order to view the surface of the fruits and vegetables, reflectance mode constitutes the good results owing to the direct illumination of the source by the detector. Correspondingly, the presence of acids is peculiar characteristics for the fruits as well as for their derived products i.e., juices. NIR spectra have the capability for determining the complex attributes of their complex structure (Cen at al., 2007). The sugar and acids in fruits like apple and tomato were investigated by Tsuchikawa and Hamada, (2004) and Pedro and Ferreira, (2005). The application of MIR spectroscopy was explored for analysis of sugars and organic acids in apricot fruit in the range of 1500–900 cm⁻¹. The investigation was done in order to study the ripening stages of different apricots (Bureau et al., 2009). The authors divided the spectra in three regions 3700–2800 cm⁻¹, 1800–1470 cm⁻¹, 1500–900 cm⁻¹. The variations of different cultivators of apricot were shown under 1500–900 cm⁻¹ in terms of sugars and organic acids (i.e., glucose, fructose, citric acid, malic acid). Specifically, the absorption of citric acid was taken place between 1150 and 1300 cm⁻¹. The calibration model of PLS was applied between the spectral region of 1500–900 cm⁻¹.

7.3.2 BIOACTIVE COMPOUNDS

Similarly, the evaluation of the bioactive compounds (anthocyanins, flavonoids, and vitamin C) was reported by Sinelli et al. (2008). The potential of the MIR and NIR was investigated on the nutritional content of blueberries. The reflectance was done in the reflectance region in the range of 12000–3000 cm⁻¹ with an average of 64 scans in the case of FT-NIR. However, FT-MIR was done in the range of 4000–700 cm⁻¹ with 16 scans at the resolution of 4 cm⁻¹. It was reported that the peaks at 10344 cm⁻¹ was related the stretch of -OH in the NIR spectra and peaks at 8458 and 7054 cm⁻¹ related to the -CH stretch. The sugar absorptions were visible at 1220–968 cm⁻¹ (fructose, glucose, and sucrose). Leon et al. (2004) also reported the evaluation of oleic acid and linoleic acid were detected by NIRS under the spectral range of 400–1700 nm. The diffused reflectance mode of NIR was explored for detecting the quality of Valencia oranges (Magwaza et al., 2011) under the wavelength range of 780–2500 nm with PbS detector on the multi-purpose analyzer spectrometer, whereas, NIR spectra was taken under 800–2500 nm.

NIR depends on the low absorptivity and scattering which favors to analyze food and agricultural products with minimal prior sampling of samples.

Determination of phenolics was reported by Martelo-Vidal and Vázquez (2014) in red wines under the reference range of 190–2500 nm and verified with the chemometrics using PCA and PLS models of calibration. Aromatics and sugars in red wines were detected by O-H stretch in 990 nm. Similarly, determination of phenolics was reported by Gallego et al. (2011), under the grape skin using NIR spectroscopy. The scanning was done under reflectance mode in the range of 1100–2200 nm. The spectral regions under 1140 and 1320 nm was found with the stretching of -OH functional group. MPLS and PLS were used as calibration models which showed the best validation for determination of flavonoids. The detection of phenolic compounds in order to evaluate the ripening stage of olives was done using NIR in reflectance mode with PCA and PLS as calibration model (Bellincontro et al., 2012). The range was selected between 1100–2300 nm. The bands were observed at 1150 for stretching of OH bonds, 1200 for CH stretching, and water absorption bands were visible 1920–1950 nm.

In terms of pigments, carotenoids, and lycopene were invested with NIR (Baranska et al., 2006) under the range of 4000–15000 cm^{-1} by employing the PLS calibration tool. The intense bands of tomato were observed at 960 cm^{-1} due to the vibration of RH-C=C-RH showing for β carotene and lycopene. It was found that both fresh and dried samples of tomato were feasibly visible under IR spectroscopy.

7.3.3   VITAMIN C

The quality in terms of vitamin C was best detected under the wavelength range of 1639–1836 nm. Along with the presence of phenolics, vitamin C also plays a lead role in elevating the acceptance of fruits and vegetables. NIR spectroscopy can be used accurately to measure its content. Pissard et al. (2013) evaluated the quality of apples by measuring the parameters of vitamin C and phenolics using NIR analysis with the range of 400–2500 in reflectance mode. NIR spectrum provided the relevant information based on hydroxyl bond (O-H), amido bond (N-H), and vibration absorption of C-H bond. Good calibrations were achieved on comparing the reference and predicted values of vitamin C and phenolics. Furthermore, portable NIR analysis was done on analysis of vitamin C in mandarin fruit (Xudong et al., 2009). The evaluation was done using root mean square error of calibration with the range of 400–1040 nm. The presence of strong water bands were observed at 680 nm along with the water absorption bands at 710 and 960 nm owing to the stretching vibration of OH bond. Correspondingly, Ignat et al. (2012) reported the content of ascorbic acid in bell peppers via taking the NIR in reflectance mode under 477–950 nm. The highest concentration of ascorbic acid was found in green variety of bell pepper. PLS regression models were used to evaluate the calibration. The range of 670–695 nm were found for carotenoid content and 1350–1800 nm was associated with stretching of C-H and O-H bonds. These bonds are usually found in the chemical structure of ascorbic acid, water, and sugar.

7.4 CHEMOMETRICS-BASED MODELING

In the case of fruits and vegetables, water dominates the NIR spectrum because of its ability to highly absorb NIR radiation. In addition to this, the NIR spectrum comprises of a large set of overtones and combination bands, complex chemical composition, wavelength-dependent scattering effects, tissue heterogeneities, instrumental noise, ambient effects, and various other sources of variability. Thus, this convolutes the NIR spectrum, making the identification of absorption bands difficult. Therefore, multivariate statistical techniques or chemometrics is used to extract information about quality attributes from the NIR spectrum using model calibration. These model calibration methods are developed using chemometric tools like partial least squares regression (PLSR), multivariate linear regression (MLR), PCA, ANN, and many more. Amongst all these tools, PLSR has been reported in numerous studies in the development of prediction models for fruits and vegetables quality parameters. Magwaza et al. (2012) stated that PLSR is better than MLR as well as PCA in both developments of calibration models and validation, as PLSR analysis produces models that exclude latent variables which does not play any role in describing the variance of the quality parameter of a fruit or vegetable. The authors have also stated that PLSR is based on linear models while ANN is used to model non-linearity. ANN involves flexible learning algorithms, diverse network topology, fast learning algorithm, and higher error tolerance. Table 7.1 shows the chemometric tools used in various studies of fruits and vegetables. Different researchers have explored and compared different regression techniques on different fruits and vegetables. Apart from the aforementioned chemometric tools, various other modified tools have also been reported in the literature, which has been used to analyze the variance in spectroscopic data like functional analysis of variance (FANOVA) (Bobelyn et al., 2010), Monte-Carlo uninformative variable elimination and Genetic algorithm PLS (MC-UVE-GA-PLS) (Jie et al., 2013), supervised classification models (Cen et al., 2016; Baranowski et al., 2013), successive projections algorithm (Huang et al., 2011), modified partial least squares regression (MPLSR) (Torres et al., 2015), soft independent modeling of class analogy (SIMCA) (Pholpho et al., 2011).

7.5 CONCLUSION

The non-invasive technique of IR has been proven advantageous and time-saving for the rapid analysis of physical defects as well as chemical composition in the category of agricultural produce. The role of the calibration model cannot be neglected as they are the major tool for finding the accuracy in prediction of results. The IR technique can be easily adapted for a variety of fruits and vegetables by considering their peculiar characterizes viz. of firmness, soluble acids, sugars, and vitamins. However, furthermore, clarity is needed in establishing the models for calibration for a particular trait of agricultural products. Overall, the measurements of IR can be strongly correlated in order to predict the presence of phenolics, flavonoids, and ascorbic acid. Therefore, IR spectroscopy presents the advantage and produces promising results for measuring the physical and chemical features accumulated in the fruits and vegetables.

KEYWORDS

- artificial neural network
- factorial discriminant analysis
- IR spectroscopy
- micro-electro-mechanical system
- multivariate linear regression
- principal component analysis

REFERENCES

Baranowski, P., Mazurek, W., & Pastuszka-Woźniak, J., (2013). Supervised classification of bruised apples with respect to the time after bruising on the basis of hyperspectral imaging data. *Postharvest Biology and Technology, 86,* 249–258.

Baranska, M., Schütze, W., & Schulz, H., (2006). Determination of lycopene and β-carotene content in tomato fruits and related products: Comparison of FT-Raman, ATR-IR, and NIR spectroscopy. *Analytical Chemistry, 78*(24), 8456–8461.

Bellincontro, A., Taticchi, A., Servili, M., Esposto, S., Farinelli, D., & Mencarelli, F., (2012). Feasible application of a portable NIR-AOTF tool for on-field prediction of phenolic compounds during the ripening of olives for oil production. *Journal of Agricultural and Food Chemistry, 60*(10), 2665–2673.

Beullens, K., Kirsanov, D., Irudayaraj, J., Rudnitskaya, A., Legin, A., Nicolaï, B. M., & Lammertyn, J., (2006). The electronic tongue and ATR-FTIR for rapid detection of sugars and acids in tomatoes. *Sensors and Actuators B: Chemical, 116*(1–2), 107–115.

Bobelyn, E., Serban, A. S., Nicu, M., Lammertyn, J., Nicolai, B. M., & Saeys, W., (2010). Postharvest quality of apple predicted by NIR-spectroscopy: Study of the effect of biological variability on spectra and model performance. *Postharvest Biology and Technology, 55*(3), 133–143.

Bureau, S., Ruiz, D., Reich, M., Gouble, B., Bertrand, D., Audergon, J. M., & Renard, C. M., (2009). Application of ATR-FTIR for a rapid and simultaneous determination of sugars and organic acids in apricot fruit. *Food Chemistry, 115*(3), 1133–1140.

Butz, P., Hofmann, C., & Tauscher, B., (2005). Recent developments in noninvasive techniques for fresh fruit and vegetable internal quality analysis. *Journal of Food Science, 70*(9).

Camps, C., & Christen, D., (2009). Non-destructive assessment of apricot fruit quality by portable visible-near infrared spectroscopy. *LWT-Food Science and Technology, 42*(6), 1125–1131.

Carlini, P., Massantini, R., & Mencarelli, F., (2000). Vis-NIR measurement of soluble solids in cherry and apricot by PLS regression and wavelength selection. *Journal of Agricultural and Food Chemistry, 48*(11), 5236–5242.

Cayuela, J. A., & Weiland, C., (2010). Intact orange quality prediction with two portable NIR spectrometers. *Postharvest Biology and Technology, 58*(2), 113–120.

Cayuela, J. A., (2008). Vis/NIR soluble solids prediction in intact oranges (*Citrus sinensis* L.) cv. Valencia Late by reflectance. *Postharvest Biology and Technology, 47*(1), 75–80.

Cen, H., Bao, Y., He, Y., & Sun, D. W., (2007). Visible and near infrared spectroscopy for rapid detection of citric and tartaric acids in orange juice. *Journal of Food Engineering, 82*(2), 253–260.

Cen, H., Lu, R., Zhu, Q., & Mendoza, F., (2016). Non-destructive detection of chilling injury in cucumber fruit using hyper spectral imaging with feature selection and supervised classification. *Postharvest Biology and Technology, 111*, 352–361.

Clark, C. J., McGlone, V. A., De Silva, H. N., Manning, M. A., Burdon, J., & Mowat, A. D., (2004). Prediction of storage disorders of kiwifruit (*Actinidia chinensis*) based on visible-NIR spectral characteristics at harvest. *Postharvest Biology and Technology, 32*(2), 147–158.

Cubero, S., Aleixos, N., Moltó, E., Gómez-Sanchis, J., & Blasco, J., (2011). Advances in machine vision applications for automatic inspection and quality evaluation of fruits and vegetables. *Food and Bioprocess Technology, 4*(4), 487–504.

Dos, S. N, J. P., De Assis, M. W. D., Casagrande, I. P., Júnior, L. C. C., & De Almeida, T. G. H., (2017). Determination of 'palmer' mango maturity indices using portable near infrared (VIS-NIR) spectrometer. *Postharvest Biology and Technology, 130*, 75–80.

Dufour, E., (2009). Principles of infrared spectroscopy. In: Sun, D. W., (ed.), *Infrared Spectroscopy for Food Quality Analysis and Control* (pp. 3–27). Academic Press, Amsterdam, London.

Escribano, S., Biasi, W. V., Lerud, R., Slaughter, D. C., & Mitcham, E. J., (2017). Non-destructive prediction of soluble solids and dry matter content using NIR spectroscopy and its relationship with sensory quality in sweet cherries. *Postharvest Biology and Technology, 128,* 112–120.

Fan, S., Zhang, B., Li, J., Liu, C., Huang, W., & Tian, X., (2016). Prediction of soluble solids content of apple using the combination of spectra and textural features of hyper spectral reflectance imaging data. *Postharvest Biology and Technology, 121,* 51–61.

Ferrer-Gallego, R., Hernández-Hierro, J. M., Rivas-Gonzalo, J. C., & Escribano-Bailón, M. T., (2011). Determination of phenolic compounds of grape skins during ripening by NIR spectroscopy. *LWT-Food Science and Technology, 44*(4), 847–853.

Fraser, D. G., Jordan, R. B., Künnemeyer, R., & McGlone, V. A., (2003). Light distribution inside mandarin fruit during internal quality assessment by NIR spectroscopy. *Postharvest Biology and Technology, 27*(2), 185–196.

Giovanelli, G., Sinelli, N., Beghi, R., Guidetti, R., & Casiraghi, E., (2014). NIR spectroscopy for the optimization of postharvest apple management. *Postharvest Biology and Technology, 87,* 13–20.

Golic, M., & Walsh, K. B., (2006). Robustness of calibration models based on near infrared spectroscopy for the in-line grading of stone fruit for total soluble solids content. *Analytica. Chimica. Acta, 555*(2), 286–291.

Guthrie, J. A., Walsh, K. B., Reid, D. J., & Liebenberg, C. J., (2005). Assessment of internal quality attributes of mandarin fruit. 1. NIR calibration model development. *Australian Journal of Agricultural Research, 56*(4), 405–416.

Huang, H., Liu, L., & Ngadi, M. O., (2014). Recent developments in hyper spectral imaging for assessment of food quality and safety. *Sensors, 14*(4), 7248–7276.

Huishan, L., Yibin, Y., Huanyu, J., Yande, L., Xiaping, F., & Wang, J., (2005). Application Fourier transform near infrared spectrometer in rapid estimation of soluble solids content of intact citrus fruits. In: *2005 ASAE Annual Meeting* (p. 1). American Society of Agricultural and Biological Engineers.

Ignat, T., Schmilovitch, Z., Fefoldi, J., Steiner, B., & Alkalai-Tuvia, S., (2012). Non-destructive measurement of ascorbic acid content in bell peppers by VIS-NIR and SWIR spectrometry. *Postharvest Biology and Technology, 74,* 91–99.

Jaiswal, P., Jha, S. N., & Bharadwaj, R., (2012). Non-destructive prediction of quality of intact banana using spectroscopy. *Scientia Horticulturae, 135,* 14–22.

Jha, S. N., & Garg, R., (2010). Non-destructive prediction of quality of intact apple using near infrared spectroscopy. *Journal of Food Science and Technology, 47*(2), 207–213.

Jha, S. N., Jaiswal, P., Narsaiah, K., Gupta, M., Bhardwaj, R., & Singh, A. K., (2012). Non-destructive prediction of sweetness of intact mango using near infrared spectroscopy. *Scientia Horticulturae, 138,* 171–175.

Jha, S. N., Narsaiah, K., Jaiswal, P., Bhardwaj, R., Gupta, M., Kumar, R., & Sharma, R., (2014). Nondestructive prediction of maturity of mango using near infrared spectroscopy. *Journal of Food Engineering, 124,* 152–157.

Jie, D., Xie, L., Fu, X., Rao, X., & Ying, Y., (2013). Variable selection for partial least squares analysis of soluble solids content in watermelon using near-infrared diffuse transmission technique. *Journal of Food Engineering, 118*(4), 387–392.

Kramchote, S., Nakano, K., Kanlayanarat, S., Ohashi, S., Takizawa, K., & Bai, G., (2014). Rapid determination of cabbage quality using visible and near-infrared spectroscopy. *LWT-Food Science and Technology, 59*(2), 695–700.

Lee, W. H., Kim, M. S., Lee, H., Delwiche, S. R., Bae, H., Kim, D. Y., & Cho, B. K., (2014). Hyper spectral near-infrared imaging for the detection of physical damages of pear. *Journal of Food Engineering, 130*, 1–7.

Leiva-Valenzuela, G. A., Lu, R., & Aguilera, J. M., (2013). Prediction of firmness and soluble solids content of blueberries using hyperspectral reflectance imaging. *Journal of Food Engineering, 115*(1), 91–98.

León, L., Garrido-Varo, A., & Downey, G., (2004). Parent and harvest year effects on near-infrared reflectance spectroscopic analysis of olive (*Olea europaea* L.) fruit traits. *Journal of Agricultural and Food Chemistry, 52*(16), 4957–4962.

Li, J., Chen, L., Huang, W., Wang, Q., Zhang, B., Tian, X., Fan, S., & Li, B., (2016). Multispectral detection of skin defects of bi-colored peaches based on VIS-NIR hyper spectral imaging. *Postharvest Biology and Technology, 112*, 121–133.

Louw, E. D., & Theron, K. I., (2010). Robust prediction models for quality parameters in Japanese plums (*Prunus salicina* L.) using NIR spectroscopy. *Postharvest Biology and Technology, 58*(3), 176–184.

Lu, R., (2004). Multispectral imaging for predicting firmness and soluble solids content of apple fruit. *Postharvest Biology and Technology, 31*(2), 147–157.

Magwaza, L. S., Opara, U. L., Nieuwoudt, H., & Cronje, P., (2011). Non-destructive quality assessment of 'Valencia' orange using FT-NIR spectroscopy. In: *Proceedings of the NIR 2011, 15th International Conference on Near-Infrared Spectroscopy.*

Magwaza, L. S., Opara, U. L., Nieuwoudt, H., Cronje, P. J., Saeys, W., & Nicolaï, B., (2012). NIR spectroscopy applications for internal and external quality analysis of citrus fruit: A review. *Food and Bioprocess Technology, 5*(2), 425–444.

Maniwara, P., Nakano, K., Boonyakiat, D., Ohashi, S., Hiroi, M., & Tohyama, T., (2014). The use of visible and near infrared spectroscopy for evaluating passion fruit postharvest quality. *Journal of Food Engineering, 143*, 33–43.

Marques, E. J. N., De Freitas, S. T., Pimentel, M. F., & Pasquini, C., (2016). Rapid and non-destructive determination of quality parameters in the 'Tommy Atkins' mango using a novel handheld near infrared spectrometer. *Food Chemistry, 197*, 1207–1214.

Martelo-Vidal, M. J., & Vázquez, M., (2014). Determination of polyphenolic compounds of red wines by UV-VIS-NIR spectroscopy and chemometrics tools. *Food Chemistry, 158*, 28–34.

Moghimi, A., Aghkhani, M. H., Sazgarnia, A., & Sarmad, M., (2010). Vis/NIR spectroscopy and chemometrics for the prediction of soluble solids content and acidity (pH) of kiwifruit. *Biosystems Engineering, 106*(3), 295–302.

Moscetti, R., Haff, R. P., Monarca, D., Cecchini, M., & Massantini, R., (2016). Near-infrared spectroscopy for detection of hailstorm damage on olive fruit. *Postharvest Biology and Technology, 120*, 204–212.

Munera, S., Besada, C., Blasco, J., Cubero, S., Salvador, A., Talens, P., & Aleixos, N., (2017). Astringency assessment of persimmon by hyper spectral imaging. *Postharvest Biology and Technology, 125*, 35–41.

Ncama, K., Opara, U. L., Tesfay, S. Z., Fawole, O. A., & Magwaza, L. S., (2017). Application of Vis/NIR spectroscopy for predicting sweetness and flavor parameters of 'Valencia' orange (*Citrus sinensis*) and 'Star Ruby' grapefruit (Citrus x paradise Mac fad). *Journal of Food Engineering, 193*, 86–94.

Nordey, T., Joas, J., Davrieux, F., Chillet, M., & Léchaudel, M., (2017). Robust NIRS models for non-destructive prediction of mango internal quality. *Scientia Horticulturae, 216*, 51–57.

Padda, M. S., do Amarante, C. V., Garcia, R. M., Slaughter, D. C., & Mitcham, E. J., (2011). Methods to analyze physico-chemical changes during mango ripening: A multivariate approach. *Postharvest Biology and Technology, 62*(3), 267–274.

Pan, L., Zhang, Q., Zhang, W., Sun, Y., Hu, P., & Tu, K., (2016). Detection of cold injury in peaches by hyper spectral reflectance imaging and artificial neural network. *Food Chemistry, 192*, 134–141.

Pedro, A. M., & Ferreira, M. M., (2005). Nondestructive determination of solids and carotenoids in tomato products by near-infrared spectroscopy and multivariate calibration. *Analytical Chemistry, 77*(8), 2505–2511.

Pérez-Marín, D., Paz, P., Guerrero, J. E., Garrido-Varo, A., & Sánchez, M. T., (2010). Miniature handheld NIR sensor for the on-site non-destructive assessment of post-harvest quality and refrigerated storage behavior in plums. *Journal of Food Engineering, 99*(3), 294–302.

Pholpho, T., Pathaveerat, S., & Sirisomboon, P., (2011). Classification of longan fruit bruising using visible spectroscopy. *Journal of Food Engineering, 104*(1), 169–172.

Pissard, A., Fernández, P. J. A., Baeten, V., Sinnaeve, G., Lognay, G., Mouteau, A., & Lateur, M., (2013). Non-destructive measurement of vitamin C, total polyphenol and sugar content in apples using near-infrared spectroscopy. *Journal of the Science of Food and Agriculture, 93*(2), 238–244.

Rajkumar, P., Wang, N., EImasry, G., Raghavan, G. S. V., & Gariepy, Y., (2012). Studies on banana fruit quality and maturity stages using hyper spectral imaging. *Journal of Food Engineering, 108*(1), 194–200.

Rungpichayapichet, P., Mahayothee, B., Nagle, M., Khuwijitjaru, P., & Müller, J., (2016). Robust NIRS models for non-destructive prediction of postharvest fruit ripeness and quality in mango. *Postharvest Biology and Technology, 111*, 31–40.

Saranwong, S., Sornsrivichai, J., & Kawano, S., (2004). Prediction of ripe-stage eating quality of mango fruit from its harvest quality measured nondestructively by near infrared spectroscopy. *Postharvest Biology and Technology, 31*(2), 137–145.

Sinelli, N., Spinardi, A., Di Egidio, V., Mignani, I., & Casiraghi, E., (2008). Evaluation of quality and nutraceutical content of blueberries (*Vaccinium corymbosum* L.) by near and mid-infrared spectroscopy. *Postharvest Biology and Technology, 50*(1), 31–36.

Sirisomboon, P., Tanaka, M., Kojima, T., & Williams, P., (2012). Nondestructive estimation of maturity and textural properties on tomato 'Momotaro' by near infrared spectroscopy. *Journal of Food Engineering, 112*(3), 218–226.

Sivakumar, D., Jiang, Y., & Yahia, E. M., (2011). Maintaining mango (*Mangifera indica* L.) fruit quality during the export chain. *Food Research International, 44*(5), 1254–1263.

Sun, M., Zhang, D., Liu, L., & Wang, Z., (2017). How to predict the sugariness and hardness of melons: A near-infrared hyperspectral imaging method. *Food Chemistry, 218*, 413–421.

Szymanska-Chargot, M., & Zdunek, A., (2013). Use of FT-IR spectra and PCA to the bulk characterization of cell wall residues of fruits and vegetables along a fraction process. *Food Biophysics, 8*(1), 29–42.

Torres, I., Pérez-Marín, D., De La Haba, M. J., & Sánchez, M. T., (2015). Fast and accurate quality assessment of Raf tomatoes using NIRS technology. *Postharvest Biology and Technology, 107,* 9–15.

Travers, S., Bertelsen, M. G., Petersen, K. K., & Kucheryavskiy, S. V., (2014). Predicting pear (cv. Clara Frijs) dry matter and soluble solids content with near infrared spectroscopy. *LWT-Food Science and Technology, 59*(2), 1107–1113.

Tsuchikawa, S., & Hamada, T., (2004). Application of time-of-flight near infrared spectroscopy for detecting sugar and acid contents in apples. *Journal of Agricultural and Food Chemistry, 52*(9), 2434–2439.

Ventura, M., de Jager, A., De Putter, H., & Roelofs, F. P., (1998). Non-destructive determination of soluble solids in apple fruit by near infrared spectroscopy (NIRS). *Postharvest Biology and Technology, 14*(1), 21–27.

Wang, J., Wang, J., Chen, Z., & Han, D., (2017). Development of multi-cultivar models for predicting the soluble solid content and firmness of European pear (*Pyrus communis* L.) using portable vis-NIR spectroscopy. *Postharvest Biology and Technology, 129,* 143–151.

Wei, X., Liu, F., Qiu, Z., Shao, Y., & He, Y., (2014). Ripeness classification of astringent persimmon using hyper spectral imaging technique. *Food and Bioprocess Technology, 7*(5), 1371–1380.

Xie, C., Chu, B., & He, Y., (2018). Prediction of banana color and firmness using a novel wavelengths selection method of hyperspectral imaging. *Food Chemistry, 245,* 132–140.

Xudong, S., Hailiang, Z., & Yande, L., (2009). Nondestructive assessment of quality of Nanfeng mandarin fruit by a portable near infrared spectroscopy. *International Journal of Agricultural and Biological Engineering, 2*(1), 65–71.

Zhang, C., Guo, C., Liu, F., Kong, W., He, Y., & Lou, B., (2016). Hyperspectral imaging analysis for ripeness evaluation of strawberry with support vector machine. *Journal of Food Engineering, 179,* 11–18.

CHAPTER 8

NUCLEAR MAGNETIC RESONANCE (NMR) SPECTROSCOPY FOR QUALITY DETERMINATION OF FRUITS AND VEGETABLES

AAMIR HUSSAIN DAR,[1] HILAL AHMAD MAKROO,[1] SHAFAQ SHAH,[2] and SHAFAT KHAN[1]

[1]*Department of Food Technology, Islamic University of Science and Technology, Awantipora, Pulwama, Jammu and Kashmir, India*

[2]*Department of Food Technology Jamia Hamdard University, New Delhi, India*

ABSTRACT

To investigate chemical reactions and molecular structures, biochemists and chemists are using nuclear magnetic resonance spectroscopy, because of its versatile analytical techniques and being a powerful spectroscope. Nuclear magnetic resonance is used to study materials of liquid and solid nature and has become one of the most favored technology to be applied in the field of food technology and science for qualitative and quantitative analysis of various food products, fruit products, vegetables, meat and meat products, milk and milk products, and several alcoholic and non-alcoholic beverages. In the early 20th century, nuclear magnetic resonance was only limited to analysis of moisture content in food products because of its low resolution. In several foods matrix high resolution studies were conducted using nuclear magnetic resonance. Another type of nuclear magnetic resonance technology is magnetic resonance imaging also called MRI, which is being extensively utilized to study food products. Several techniques have been developed by the scientist to use MRI and NMR spectroscopy in the field of food Science and technology. Process optimization for online process monitoring was difficult by employing traditional analytical techniques which were time consuming as well as destroy the sample. X-ray transmission, sonication, vibration and light microscope are non-destructive techniques that can be employed for the quality analysis of food products. However, these techniques were unable to give data about complex physical properties of food products. As an alternative nuclear magnetic resonance was used for the online monitoring of a process. To analyze bio process in biological science nuclear magnetic resonance technique was successfully developed by scientists. Nuclear magnetic resonance technology has successfully been employed online food monitoring process. Food industries and control agencies have extensively utilizing nuclear magnetic resonance spectroscopy to warrant the food safety and quality of food products to meet consumer needs and expectations.

8.1 INTRODUCTION

Nuclear magnetic resonance (NMR) is a spectroscopic technique in which electromagnetic radiations are exhibited by the atomic nuclei under the influence of a magnetic field. This energy is near a precise resonance frequency which can be determined by the influence of the magnetic properties of the isotope of the atoms and the electromagnetic field; in practical submissions, the regularity is comparable to VHF and UHF television broadcast (60–1000 MHz). NMR authorizes the surveillance of specific quantum mechanical properties of the atomic nucleus. The NMR concept has many misapplications attributed to many scientific practices, done to study crystals, non-crystalline materials, and molecular physics through NMR spectroscopy. NMR has diverse and consistent applications in advanced medical imaging technologies, like that of magnetic resonance imaging (MRI). A notable characteristic of NMR is that the value of resonating frequency pertaining to a precise material is directly related to the strength of the functional magnetic field. The same peculiarity is repressed in imaging techniques, i.e., when a sample is positioned under the influence of the magnetic field which is non-uniform in nature, the field positioning of the sample's nuclei determines the resonating frequency. To generate amplified field strength with the help of superconductors, studies are being carried out. Using multidimensional (one, two, and three dimensional) frequency techniques and hyperpolarization, the efficiency of NMR can also be enhanced to a great extent.

8.2   HISTORY

The first NMR model was designed and simulated in a molecular beam by Isidor Rabi in 1938, by out-spreading the Stern Gerlach experiment. To appreciate and acknowledge his work, Rabi was given the noble prize in the year 1944. Following some years in 1946, the technique was further developed by Felix Bloch and Edward Mill Purcell the technique in case of solids and liquids that fetched them the Nobel Prize in Physics in the year 1952. Yevgeny Zavoisky had earlier discovered the NMR in 1941 but could not find it reproducible, as was done by Edward Mills Purcell and Felix Bloch, but sacked the results as not reproducible. Russell H. Varian filed the "Method and means for correlating nuclear properties of atoms and magnetic fields," U.S. Patent 2,561,490 on July 24, 1951. Varian Acquaintances developed the first NMR unit called NMR HR-30 in 1952. Purcell had functioned on the progress of radar during World War II at the Massachusetts Institute of Technology Radiation Laboratory. His work throughout that project on the construction and discovery of radio frequency (RF) power and on the captivation of such RF power by stuff laid the basis for Rabi's discovery of NMR. Rabi, Puncell, and Bloch detected the magnetic nuclei like ^{1}H and ^{31}P could engross RF energy when positioned in a magnetic field and when the RF was of an incidence specific to the uniqueness of the nuclei. When this captivation happened, the nucleus was labeled as being in resonance. Altered atomic nuclei within a molecule were observed to resonate at different (radio) frequencies for the identical magnetic field strength. The comment of such magnetic resonance frequencies of the nuclei existing in a molecule permits any proficient user to notice vital structural and chemical information about the molecule.

8.3   EVALUATION OF FOOD QUALITY

Sensory and quality assessment of food is utilized to adjust the discrimination of vegetables and fruits (O'Mahony, 1991). The food qualities are often used as a laboratory and profitable pointers of ripeness for many horticultural produces (Genizi and Cohen, 1988; Kader, 1999, 2008a, b; Fawole and Opara, 2013a). In mangoes, due to expansion the absorption of flesh acid is greater in younger fruits and fall offs for example due to the presence of sugars, which collect quickly as the fruit mellows (Ito et al., 1997). With the increase in the sweetness of the fruit, the SSC content increases while TA declines as a result of the proportion of the two surges. Similar subsiding drift in TA and surge in TSS: TA content was determined in the pomegranate fruit by Fawole and Opara (2013b). The configuration of sugars in foodstuffs can be viably examined only if a reliable, sensitive, and fast logical technique is accessible (Shanmugavelan et al., 2013). There are many logical systems (destructive as well as non-destructive (ND)) that are used to obtain chemical conformation of the model and, consequently, are advantageous to define the sensory outline of fruit and vegetables. The various instruments used for chemical characterization include high-pressure liquid chromatography, colorimeter, gas chromatography (GC), and other instrumental systems such as refractometer, hydrometer, and electronic tongue (Kader et al., 2003; Saftner et al., 2008; Shanmugavelan et al., 2013). Expecting the varying content of sugar and sweetness in the foodstuff, it is necessary to every now and then to check the rapid sugar content. In the context of the following, analytical approaches for sugar purpose may be characterized as either destructive or ND:

1. Destructive measurements methods:
   - Hydrometer;
   - Sensory evaluation;
   - Refract meter;
   - Electronic tongue.
2. Non-destructive measurements methods:
   - Visible to near-infrared (NIR) spectroscopy;
   - Hyperspectral and multispectral imaging.

8.4 NMR TECHNOLOGY AND PRINCIPLE

The Institute for Food Research in the UK has industrialized a new employed example of a novel type of NMR system, which permits numerous, examines to be accepted out on the same model concurrently. The plan has also encompassed the expansion of a low-cost MRI scanner and also an online MRI sensor for usage on conveyors in the food and pharmaceutical diligences. During food, processing many samples can undergo rapid changes which are then irreversible. NMR would habitually model to screen such fluctuations but the conformist technique has a constant count on very strong magnetic fields, which might only be formed by a large superconducting magnet by means of liquid helium-cooled coils. Though, the IFR has now established an open-access Halbach NMR spectrometer (via smaller powerful shaped magnets), which incomes that superior bulky trials can be surveyed. The

Halbach organization is economical and informal to build and is moveable. It allows NMR/MRI technology to be positively ported to the construction line and used to deliver constant exploration.

As NMR is the most influential method in defining the structure of organic compounds. NMR is used effectively in various food submissions for quality research and quality control. NMR spectroscopy is used to regulate the structure of proteins, organic acid, lipid fractions, carotenoids, amino acid profile, and the mobility of water in food. NMR spectroscopy is used to identify and quantify metabolites in food. Also, vegetables cheese, oil, fish oil, fruit juice, meat green tea, and food such as beer, wine are amid the last NMR submissions. NMR spectroscopy is applied for foodomics which is a novel correction that fetches food science and nutritional research collectively. NMR methods used for food confirmations are one and two dimensional NMR systems, N15, and P-31 NMR techniques high resolutions liquid state, 1H and 13C MR techniques, 1H HR/MAS (high resolutions angle spinning) NMR techniques. NMR is a diagnostic chemistry practice used in value regulator and examination for defining the gratified and transparency of a taster as well its molecular structure. NMR is a more commanding system used to find organizational material, and consequently, it can help to comprehend the assembly of workings in food multifaceted organization. ^{1}H and ^{31}C NMR spectroscopies are cast-off to examine all organic compounds repeatedly. NMR is a multiparametric, multinuclear, ND, and often nonsensitive technique, effectively engaged in plant and mammalian biochemistry. NMR imaging can be employed to ground on the radiation and captivate energy in the RF assortment of the electromagnetic spectrum origin. All nucleotides holding a single proton and neutron can be detected by NMR. The high eminence of spectra empowers the documentation and acknowledgment of most of the water-soluble constituents. The foundation of NMR spectroscopy is grounded on the magnetic possessions of the nucleus. The magnetism of the atomic nucleolus merely can be clarified as; Atomic nucleolus can be restrained as a spherical body and spinning around the axis of the center. Proton is found in the nucleus, so the nucleus is positively exciting. Since the nucleus comes back around, the positive charge will move in spherical orbits positioned around the axis. The undertaking of this charge on a specific course is electric current. Each electric current generates a magnetic field about its environment. The nucleus of an atom which is spinning around the axis, due to be connected, the magnetic field arises around. Consequently, the atomic nucleus works like a magnet. The magnetic moment of the magnetic field engendered by the nucleus.

NMR spectrometer principally comprises of four main segments:

- Magnet comprising highly standardized magnetic field in of pole ends;
- A very constant radio frequency transmitter;
- Radiofrequency handset; and
- Recorder.

Nuclear magnetic field spectroscopy is being cast off effectively at a variety of food systems. NMR is the most influential practice used to lighten the structure. Therefore, it to comprehend to structures of mechanism in compound food systems. The ^{1}H NMR spectroscopy can also be employed as a fingerprinting procedure. The source of chemical inconsistency inside the biomaterial can be accurately food can be quickly determined using these devices, and the intrinsic capability of NMR to bring evidence around the chemical conformation of a taster which will permit the foundation of disparity to be acknowledged. NMR methods are applied in defining the variations in heat-treated foods. In this practice, modification of water circulation and flexibility in foods could be resolute without obliteration to food. Likewise, it is one of the systems used to regulate the level of melamine in foods. In count, NMR spectroscopy is a moral process to regulate the capacity to hold water of the material, total moisture content of meat, and intramuscular fat. The high-resolution application of NMR can be used to analyze the organic mixtures of vinegar and help in the quantification of the vital components of vinegar (carbohydrates, volatile compounds of alcohols, organic acids, and amino acids). Effectively added another field of NMR spectroscopy is milk and dairy products. Studies were directed on the water fragments of casein systems, milk powder, and serum proteins by the NMR system. Besides, NMR submissions are cast-off to govern the features of the cheese. NMR is castoff to display the ripening and gage the adjustments of water and amino content in cheeses. NMR investigation of the water-soluble element of cheese organizes a binding style for model depiction. It is an enormously informal system which does not entail the maintenance or origin of the sample grounding.^{1}H NMR is cast-off as relax metric depiction of fat and water states in hard and soft cheese. NMR relaxometry has potent applications in delivering material on hydrodynamic behavior (which is the number and affinity of connections with proteins) and on the proportion of solid to liquid content for anhydrous milk fat in case of cheese. The dynamic and critical characteristics of NMR describe the molecular performance of the sample. Slackening states to the singularity of nuclei recurring to their thermodynamically steady conditions after being thrilled to higher energy stages. The un-restriction of energy is fascinated when the conversion from a lower energy level to a high energy level takes place after" the confliction ensues. For instance, the exhibition of protein water interactions and the authentication of the cumulative water movement achieved by chemical reactions and diffusive altercation are given by the Relaxation time.

Relaxation time is termed as the time engaged for relaxation. There are two sorts of relaxation.

1.   **Spin-Lattice Relaxation T_1:** Discharge of energy by enthusiastic nuclei to their overall environment.
2.   **Spin-Spin Relaxation T2:** Release energy is relocated to an adjoining nucleus by the nucleus.

On the behalf of low and high molecular flexibility, the classification of water molecules in pasta filata Mozzarella cheese as carried out mainly into two compartments namely T_{21} and T_{22} on the basis of spin relaxation spells adhering to proton powers A_1 and A_2, respectively. An increase in the amount of A_1 (and decrease in A_2) indicates that there was the migration of hydro molecules from higher to lower mobile section (which is from T_{22} to T_{21} portion) after the initial 10 days of storage. Variations in the movement of water molecules in variety pasta filata and that of the non-pasta filata type of Mozzarella cheeses are attributed to NMR technique. The outpouring of the quantity of water molecules in the less movable segment illustrates an increase in the water-holding volume of pasta filata Mozzarella. The information on the water movement and its localization can be obtained from the sloping reduction times occurring in the water protons.

NMR is an efficient method to determine water accessibility in biomaterials. Low-resolution NMR has been used to detect diverse forms of water, in case of cheese, and to obtain information on the localization of water molecules inside the biomaterial while high-resolution NMR is used to analyze the amino acid content present in the cheese.

The spectral lines of ^1H and ^{13}C-NMR in the uninvolved lipids were completely examined in the case of Pecorino Sardo Endangered Description of Origin (PS PDO) cheese. Various fatty acids (FA) like long-chain saturated ones, oleic, linoleic, butyric, capric, caproic, caprylic, trans vacinic, conjugated linoleic acid (CLA) (cis9, trans 11–18:2), and caaproleic

(9–10:1) were analyzed. The position of the isomers of acyl groups in the triacylglycerols (TAG) was ascertained. As a lipolytic procedure on cheese NMR signals were used to identify to 1,3-diacylglycerols (DAG), and free fatty acids (FFA).

^{1}H-NMR resonances of saturated aldehydes and hydroperoxides were noticed. Studies have exposed that a description of the topographical source could be completed by means of NMR spectroscopy. NMR method was castoff for the depiction of the topographical origin of buffalo milk as well as mozzarella cheese by revenues of logical and spectroscopic resolves. Much emphasis was carried out on isotopic ratios ($^{13}C/^{12}C$ and $^{15}N/^{14}N$) and other variables by the precise area of the derivation of milk tasters. While data regarding NMR which comprises of isotopic ratios turned favorable for the perception of mozzarella illustrations. NMR is a submission approach used to regulate the assembly and subtleties of proteins. The qualitatively and quantitative assessment of amino acids present in the cheese was conducted by means of high-resolution NMR experimentations.

The high-resolution spectral lines of Parmigiano-Reggiano aqueous cutting was highly influenced by the presence of free amino acids in combination with a few FA and organic acids. The precise efficacy of the spectral lines enables the recording and encrypting of major water-soluble structures. The application of NMR has proved to be quite an efficient, economical, and accurate method to analyze changes in the ripening of cheese. Shintuve-Caldarelli (2005) used the high-resolution MAS NMR for the characterization of the ripening of Parmigiano-Reggiano cheese. Characterization of the ripening of Parmigiano-Reggiano cheese revealed the presence of selected free amino acids and other low molecular weight metabolites among the most relevant compounds. A speedy and measurable ^{1}H NMR technique was advanced to observe the presence of histamine in cheese. The submission of NMR spectroscopy in the identification of CLA content in some lipids of dairy products is in simulation and a substantive alternative to outmoded instruments such as GC. The ^{1}H NMR uses latent submission as an identification technique for the screening of CLA meditations in large variants of cheese models and in the screening of CLA added in other dairy foodstuffs. In addition, green tea from diverse countries was composed and investigated by ^{1}H NMR approach. It was projected to create if the teas could be differentiated owing to the country of source or with admiration to superiority. After a wide-ranging obligation of spectra, NMR spectroscopy has been publicized to deliver much detailed evidence about the foremost considered metabolites of the teas. The structural orientation of tea was single-minded for the judgment of teas. Solicitation of ^{1}H NMR was used superiority regulator and genuineness of instant coffee by Charlton et al. The occurrence of characteristic alterations amid coffees fashioned by diverse producers, and even amongst those formed by the same producer, by recognizing 5-(hydroxymethyl)-2-furaldehyde as an indicator complex by means of the mechanical physiognomies were gritty by using NMR. A different study, ^{31}P NMR was used to regulate the quantity of mono- and diglycerides in virgin olive oils, and results initiated the measurement of other ingredients of olive oils which demeanor functional groups with imbalanced protons that could be prolonged by reckonable ^{31}P NMR spectroscopy. NMR can be castoff for foodomics since of affluence of quantification and documentation, low costs, short time desired for investigation and high figure of metabolites that can be restrained concluded a single-pass. Because of the highest understanding of NMR, attention on hydrogen is favored for foodomics trainings. NMR is a robust logical technique that can deliver material about the quantity of tasters above and outside the molecular structure, purity, and gratified of samples. It is used for willpower of the superiority physiognomies of the cheese and following mellowing effectively. NMR is tremendously consistent method that can get outcomes in a short time as well as the comfort of sample groundwork. The consequences gained from the trainings are verifying that the NMR system can be used positively in nourishments for resolving of possessions of the arrangement, nursing of amino acids and FA, monitoring of water mobility, description of topographical origin, and growth time. Food scientists have also discovered the potent applications of both NMR and MRI techniques with great scope for food examination and food indulgence. NMR helps to know about the purpose of conformation and preparation of packaging materials, recognition of food confirmation, optimization of food dispensation parameters, chemical compositional examination, and structural documentation of practical mechanisms in foods and examination of the microbiological, chemical, and physical superiority of foods. NMR applications aid in the exploration of characteristic foods such as vegetables, meat, fish, wine, cheese, fruits, beverages (i.e., tomato juice and pulp, green tea, coffee) and edible oils. Food scientists have ended the cumulative use of this skill in recent years and the drift is feasible to endure with the taster of devices for solids and multiphase systems, isotope ratio dimensions, and MRI.

NMR spectroscopy delivers a course to the sympathetic of lectin-carbo-hydrate connections. NMR is not as influential as X-ray crystallography in resolving the structures of large, multimeric proteins. Though, small monomeric proteins are appropriate and the pronounced uninvolved NMR devices deal an inscription of the ligand-binding site. The perfect condition is when a 3D construction of a lectin is accessible and a grouping of transferred NMR and exhibiting offer a construction of the compound and such projects often deliver a good in materials and time associated to the crystal meeting of the compound. NMR spectroscopy has numerous ideal features obligatory for the no embattled examination of liquid samples for endogenous metabolites. There is thus no necessity for the pre-selection of the exploratory conditions in the case of biofluid tasters such as urine, plasma, and bile, as these can be inspected deprived of the necessity for any procedure of taster pretreatment (other than adding $c.$ 10% by volume of D_2O to act as a field incidence lock for the spectrometer and buffering to diminish chemical shift disparity. NMR spectroscopy is the method that most chemists, particularly organic chemists, practice first and habitually in the structural examination. In the organic mixtures, this ND spectroscopic examination can disclose the amount of carbon and proton atoms and their connectivities, the conformations of the molecules, as thriving as virtual and entire stereo chemistries. The multiparametric and multi-layered NMR in grouping with multivariate data and observations seem to be a straight source of material for the molecular assumed of the multifaceted marvels stirring in food during thermal processing, counting denaturation, phase transitions, mass transport phenomena and the development and statement of flavor and toxic mixtures. Prognostic mathematical models in combination with spectroscopic sensors have a huge latent in food research and manufacturing to govern and display the excellence of raw supplies, food dispensation, and final foodstuffs. A mounting attention is consequently predicted in unconventional fingerprinting methods, comprising high-resolution NMR sensors. Finest ripeness at harvest is an identical central to assure the final superiority of the fruits.

Mowing of fruits at appropriate stage retains many qualitative features such as size, shape, taste, nutrients, and enables longer shelf life. Fruits chosen at early or late stage of seasons are prone to different physical distortions. Hence, an optimal mowing time is the most vital factor for many fruit cultivators to raise the productivity and minimize losses. In recent times various quick and nondestructive technologies have been used having multivariate, chemometric algorithms data fusion characteristics. Various ND scanning techniques, its trials, and limitations have been analyzed for the fruit quality evaluation. In the present global market, numerous produces are available which are reviewed stranded for their sensing system (application, estimation competence, price, convenience, and dependability).

In the present scenario, there is a major contribution of India in the production of fruits and vegetables. Mowing time is one of the most vital and influential factors that determine fruit and vegetable quality. Fruits selected at ideal harvesting time are observed to show improved and enhanced eminence limitations compared to fruits selected too early or too late. Fruits can be classified into two main components such as climacteric and non-climacteric fruits. Many fruit cultivars can be broadly categorized into two altered classes such as climacteric fruits and non-climacteric fruits. Non-climacteric fruits cannot endure their seasoning cycle once they are picked from the tree and also do not respond to ethylene strength ripening technique. On the other side, the endurance capacity of climacteric fruits is higher after gathering and after ethylene ripening technique. Power ripening techniques are not feasible as they are not fit for human consumption and more so over harm fruit quality in terms of nutritional parameters, taste, etc. Whereas in case of non-climacteric fruit models, ideal harvesting period is one of the maximum significant restrictions at the stretch of gathering. On-field difficulties have an approach associated to them, i.e., the initial threat of diverse disease spasms and proposition for right corrective action. Studies reveal ND sensing systems that have been explored by different investigators to magistrate superiority restrictions of fruit tasters. Some of them have been castoff for on-tree fruit eminence examination course, whereas others are fine matched for lab level submissions. Fruit grower was reliant on his own intellect grounded on the alteration in different parameters like shape, color, size, and weight. Chemical and automated methods are also castoff in diverse fruit foodstuffs industries to admittances the fruit superiority. Freshly, there is a mounting attention in dissimilar electronic devices to entrées the fruit excellence limitations. Although countless electronic destructive type devices (pressure based measurement, impedance measurement) have been evaluated for fruit value calculation, still many of these take a lot of time, cause a lot of changes while measurement and hence more suitable for lab-scale submissions. As the progress of science falls underway novel electronic devices which are non destructive in nature have come up to analyze the qualitative parameters. Some of the methods used for detecting qualitative parameters without causing any harm to the fruits are hyperspectral imaging, ultrasonic based stiffness method, electric nose (e-nose), etc. Several researchers succeeded in reporting some ND approaches for fruit quality screenings but most of the methods were limited to lab level explanations and haven't been commercialize for their field snags and their explanations. Progressions in aroma sensors knowledge, chemical engineering electronics and instrumentation, and artificial intellect made it imaginable to mature diversity of instruments to notice, amount, and illustrate diverse volatile organic compounds (VOCs) in numerous submissions. The approach of electronic noses (e-noses) have been intended and industrialized specifically for the automatic acquisition, finding, documentation, and organization of compound combinations of odors, vapors, and gases. The different prototypes of artificial e-nose intend their use to distinguish compound odor blend consist of dissimilar VOCs. The use of different kinds of devices such as a metal oxide (MOS), semiconductive polymer, optical, electrochemical, and surface acoustic wave (SAW) sensors have been devoted to this area of e-nose design and progress. The retort of discrete sensors are cooperatively united to yield a unique digital response design and further transformed into the form of a unique aroma signature. Composed signatures characterize as the distinctiveness of a simple or compound mixture, which could be further applied for testing, training, and organization. Fruit aroma is known to comprise of a complex combination of different VOCs whose configuration resembles detailed plant and fruit cultivar. Very less fruits share numerous aromatic features while as each fruit have their own characteristic aroma that is contingent upon the grouping of volatile organic mixes.

Kiani et al., observed the application of moveable e-nose system for analysis of various saffron stigma of *Crocas sativus* L. (Iridaceae), to determine various VOCs. E-nose possesses metal oxide assisted gas sensors (having 10 types of gas sensors) based on humidity, temperature, sensors along with straight space specimen connected with a microcontroller for real-time data recording.

NMR helps to give succeeding evidence:

- The quantity of peaks specifies altered types of nucleus.
- The location of the peak specifies the category of nucleus and chemical situation.
- The comparative zones of the peaks give the comparative numeral of each type of nucleus.
- Disturbance in the peak, specifies that precious nucleuses from each other.

8.5   NMR APPLICATIONS

8.5.1   *FRUIT QUALITY INSPECTION USING UV-VIS-NIR SPECTROSCOPY*

In the current scenario, India remains one of the largest producers of fruits and vegetables however losses are also one of the highest. Optimal Mowing time remains the major factor that determines the fruits and vegetable quality. Fruits selected at the optimum harvesting stage depict better eminence limitations when compared to that of the fruits selected early or too late.

With the quick expansion in science, knowledge, and calculation algorithms, new electronic ND sensing approaches have been discovered to extract diverse quality limitations of on-tree fruits and also to overwhelm all prevailing trials and bottlenecks with destructive methods. Little of these approaches such as e-nose, hyperspectral imaging, ultrasonic based stiffness quantity, etc. are also cast-off in non-contact detecting mode to elude any kind of injury to the on-tree fruit crop. Several investigators subsidized their examination concerning about ND approaches for fruit superiority examination and confirmation but most of them were absorbed on lab level explanations and haven't conferred the on-field snags and their explanations.

Fruits have associated diverse fascination to them and replication belongings in the light of diverse wavelength bands alter their internal parameters for example acidity, SCC, sugar, etc. Electromagnetic spectrum concealments a large range of photon energies which may act as rift in diverse bands such as radio waves, infrared, near-infrared, microwave, visible, ultraviolet, X-ray, etc. The reaction of the fruits to the visible light band (400–700 nm) and nearby infrared (NIR) band (700–2500 nm) is clearly specified by optical possessions. When a light beam cataracts on an entity, part of the incident beam gets imitated by the surface of the article, which is named as regular reflectance. Further, the transmission of residual radiation through the surface into cellular structure of the object is observed where it gets scattered by minor interfaces in the materials or fascinated by cellular parts. Also, an engrossed serving of radiation could be transmuted to diverse portions of energy such as luminescence, chemical changes, heat, etc. As a result, the general portion of the transmitted energy gets engrossed (absorption) and some portion is reflected back to the shallow and outstanding part transmitted finished object. Visible and NIR spectroscopy have continued for utilization of fruit quality dimension for many years. NIR spectroscopy has numerous compensations over the visible spectroscopy such as NIR radiation is extremely penetrating and thus it can be functional unswervingly to the taster without any sample training. Other prosperities are that they might deliver quantitative material on the chief organic workings of food foodstuffs. Numerous styles have been discovered for a diverse class of fruits like transmission mode is measured to entree sugar, acidity of thick-skinned fruits, and SSC. There is logging of scattered reflected radiation, lactation of source and detector in the same track and then illumination of fruit by the font in reflectance process by the detector. Normally different lamps are castoff as a font and diode array as an indicator while as few of exertions also have been booked in the track of LED array.

Cen et al. conducted a study to check the viability of NIR reflectance spectroscopy to monitor and manage the efficiency of various food models. Many researchers studied models and structures related to NIR reflectance spectroscopy and carried out a series of detailed trials associated with spectroscopy system design. Preprocessing and chemometric systems are used to analyze both quantitative as well as qualitative parameters applied in standardization model expansion. Jha et al. carried out an extensive study to explore various ND technologies used by various researchers for quality evaluation of intact fruits and vegetables. The utilization of various assorted practices comprising of acoustic built, color detection using machine vision, firmness detection, etc., have been tried with considerable success in recent times.

Many researchers worked on areas of wavelength band selection, NIR spectroscopy, spectral pre-processing, positioning concerning the data acquisition, and various other chemometric models especially used for many citrus cultivars.

Choi et al. studied many biomaterials under NIR spectroscopy for fruit superiority evaluation to a wider range valid for half and full program or under a reflectance mode.

8.5.2 QUALITY ASSESSMENT OF SOME HORTICULTURE PRODUCE USING NMR

8.5.2.1 BANANA

The variations in CPMG-T_2 and ostensible water self-diffusion constants, D_w, during the seasoning of banana have been explored Raffo and co-workers succeeding in the organization of earlier work on apple and potato tissue the CPMG echo deteriorations for banana were deconvoluted into three workings conforming to cytoplasmic cellular and vacuolar water. It was then initiated that the cytoplasmic and vacuolar water T_2 presented significant increases from 90 to 190 and 320 to 610 ms, correspondingly over 7 days stowage. The variations were clarified as the liberal enzymic hydrolysis of starch granules during ripening. Dimensions were accepted on a Brukerminispec functioning at 0.47 T (20 MHz), which recommends that comparable connections should be beneficial in on-line circumstances. There do not appear to be other intelligence of the properties of physiological disorders or bruising on NMR features for bananas.

8.5.2.2   BERRIES

It is contentious whether on-line NMR sensor technology is proper for on-line cataloging of small fruit such as strawberries and grapes. There is no doubt that NMR microscopy has been providing invaluable understanding into the expansion and composition of such fruit, but these are improbable to be scrutinized one-by-one in a single file over a sensor magnet. It is believable that whole gatherings or pallets of the fruit might be imaged in immobile mode on conveyors, but whether this would ever be commercially viable relics to be inspected. So consequently comprising this class of fruit with the arrangement that the off-line studies may never be oppressed commercially in an on-line sense.

8.5.2.3   DURIAN

Durian is one of the most commercially vital fresh fruits in S.E. Asia; yet cataloging immature from mature fruit by exterior actions is very challenging, so it is a prime applicant for non-invasive devices. Yantarasri and co-workers have hunted associations between soluble solids and sensory approximations of mellowness with X-ray CT and NIRS dimensions with reasonable accomplishment but more newly, they detected that MRI spin-echo image difference at 0.5 T varied with the grade of ripeness. Unfortunately, no effort was made to enumerate reduction time variations or separate oil-water peaks. However, it was recommended that the distinct alterations specified expressively lower oil content in young durian equated with the ripe and overripe fruit.

8.5.2.4   KIWI MATURITY

A measurable imaging study on kiwifruit fruit attentive on the variations in relaxation time and dispersion maps in kiwifruit as they are ready. Capacities were made at T (86 MHz) and T_1, T_2 (spin-echo), and T_2 (gradient-echo) maps were described over a 30-day maturing dated. It was exposed that all reduction times show insignificant increases during seasoning, particularly in the flesh and locule districts. The surge in reduction times during maturing seems to reverse the practical surge in free and total sugars in the tissue, which amplified between 100 and 300% and would have been probable to decrease the reduction times. Here again, the most credible description is the liberal hydrolysis of the large sum of starch granules in the unripe tissue, though this premise relics to be verified.

8.5.2.5   MANDARIN

Clark and co-workers have stated quantitative MRI studies at 4.7 T (200 MHz) of Satsuma mandarins as they ripened during a 15-week period beginning 10 weeks after anthesis. T_1, T_2, and proton density maps were conveyed, but, enquiringly, displayed no spatial ramps in the more settled fruit during the retro of cell amplification (10–25 weeks after anthesis). Fluctuations in the reduction times with mellowing were initiated but showed no obvious correlation with independent measurements of pH, Brix, and sugar focuses in articulated juice. It was deduced that the relaxation variations were being subjugated by microstructural influences associated to cell structure rather than the chemical arrangement of the cell fluids.

8.5.2.6   PAPAYA

Introduced papayas are exposed to quarantine vapor heat behavior at 47.2°C for 30 min for decontamination of fruit fly eggs and larvae. Inappropriately this handling can also cause heat impairment, which inspired a measurable MRI study of the impairment by Suzuki and co-workers. Proton density, T_1 and T_2 maps of unripe and ripe papaya, an enfolded heat-treated fruit and a thermally injured fruit were associated. Slightly amazingly, the immature and ripe papaya models had virtually equal proton masses, equal, and uniform T_1 and T_2. However, heat impairment caused misdeeds to seem in each map. Black areas seemed in the proton density maps, conforming to low proton density regions. Likewise, T_1 maps displayed regions of long and short T_1, the short zones conforming to lower polygalacturonase enzyme activity, which is accountable for the relaxing of the seasoning tissue. Regions of longer T_2 seemed to resemble to crumbled fragments of overripe tissue.

NMR spectroscopy system and imaging have been functional widely to most parts of plant organizations as discovered from the many literature and importance on issues such as plant expansion and physiology, a significant number also elaborate correlations between NMR parameters and internal superiority influences that could have thoughtful commercial standing in the fruit and vegetable sectors of the food manufacturing. The size and position of this market should not be undervalued. In the United States, the yearly income of fruits and vegetable industry is over $6 billion in income yearly and there is an endless demand for new instruments of fruit and vegetable superiority. Form this motive numeral research groups are looking for to adventure NMR-quality associations on a commercial source by evolving low-cost NMR sensors for cataloging fruit and vegetables on manufacturing conveyor belts. The trails of the transesterification of vegetable oil by applying NMR was carried out for the documentation of intermediates in the transesterification reaction. Outcomes displayed that the important methanolysis produce was SN-1, 3-diglycerides in diglycerides, and SN-2-monoglycerides was not found. These logical consequences recommend that the methanolysis reaction may occur effortlessly at the SN-2-position for both SN-tri- and SN-1, 2-diglycerides.

8.5.2.7 POTATO

Most NMR studies have investigated that potato have intensive on empathetic properties of procedures such as frying, freezing, boiling, and drying. Allowances are the NMR microimaging of vigorous and diseased potato tubers stated by Goodman and co-workers, and relaxometry replies in healthy tubers subsequent induced sharp damage. In a recent research conducted in 2019, NMR relaxometry on crude potato has been cast-off to figure the sensory touch after boiling. It is astonishing, therefore, in spite of the position of potato and its treated foodstuffs and the fact that it is disposed to internal flaws that disturb both dispensation and customer reception; some studies have shown that NMR/MRI has been cast-off to notice internal superiority faults in tubers. These include brown core and echoing heart. Expending multi-slice spin-echo imaging, Wang et al. have confirmed that both brown core and hollow heart can be perceived by MR systems. Brown core, a physiological disorder of potato categorized by medulla tissues in the center of the tuber spinning from light to dark brown in color, is identifiable by the occurrence of high concentration signal in scans with long echo times. Electron microscopy demonstrates that this disorder is going with the loss of membrane structures, organelles, cytoplasm, and a thickening of cell walls that grades in the pretentious tissue having a longer T_2 relaxation time. Cells that are present in the pit zone of the tuber may eventually split independently to form star or lens-shaped hollow holes near the center of the tuber, which is the so-called 'hollow-heart' disorder. The nonappearance of water means in the hollow heart is different by the lack of signal in images or one-dimensional outlines. Another excellence fault which is negatively contacting the marketing of potato foodstuffs has been entitled the hardening condition. Amid cooking, there are internal lumps in the potato. To overcome this problem, various biochemical and physiological inquiries are progressing to try and appreciate the roots of this condition it relics to be seen whether NMR studies on the raw potato can recognize areas probably to form internal lumps.

8.5.2.8   SQUASH

Squash such as courgette (zucchini) and cucumber is a chill-sensitive product. Revelation to chilling temperatures foundations loss of membrane integrity and water rearrangement from the cell to the intercellular planetary. Visible injury indications such as surface opposing and dark watery patches ultimately mature. Early uncovering of chill injury is vital in order to discover the requirement and correct timing of handlings such as heat shock that can contrary to the disorder. However, delaying treatment to the point where visible symptoms appear is too late for reversal. More recently, Naruke and co-workers 95 have calculated chilling injury in cucumber with very comparable fallouts. Cucumbers were stored either at 15°C or at 2°C for 9 days. Visible pitting seemed after 5 days stowage in the chilled models, but T_1 measured at 0.58 T (25 MHz) exhibited an abrupt growth after just 3 days stowage.

8.5.2.9   TOMATO RIPENESS

Tomato maturity can be well-defined on the basis of exterior color as, 1-1/4 mature green (MG), 2-1/4 breaker, 3-1/4 turning, 4-1/4 pink, 5-1/4 light red, 6-1/4 red-ripe. Four added sub-stages of the chief mature-green (MG) stage have been familiar on the base of the inside color and tissue steadiness. These substages are MG_1, MG_2, MG_3, MG_4. MG_1 are categorized as hard green locular tissue and soft seeds that cut easily. MG_2, are soft green locule tissue and hard seeds that do not censored when the tissue is sliced with a sharp blade; MG_3, have some gel-like substance in the locule, but no internal red color; and MG_4, where locular tissue is principally gel-like having red arrives. Perhaps the most significant step from an MRI standpoint is the liquefaction of the locular tissue during phases MG_1 to MG_4 since this proclaimers the onset of the later ripening steps. If this modification could be sensed rapidly and non-destructively, it would significantly diminish the inconsistency in the ripeness of tomatoes promoted for global export. Such tomatoes are selected at the MG stage and elated in boxes gassed with ethylene. But on entrance, the tomatoes display the whole choice of ripeness up to 6 (red-ripe) and entail a second costly (and damaging) hand-sorting phase. If the tomatoes, when picked, could be confidential on the MG_1 to MG_4 scale, then seasoning would be more undeviating thereby eradicating the succeeding sorting phase. This was the object for an early MRI study by Saltveit (1991) which displayed that as locular tissue ripened from stage MG_1 to MG_3, its water content augmented. This was seen as an augmented strength in the spin density map verified in the nonattendance of important relaxation attenuation. Later phases of ripening were related with a cumulative graininess in the pericarp tissue as small air pockets established in the tissue. Whether these explanations can be established into an on-line NMR sensor for tomato ordering relics to be explored.

8.5.2.10   BULBS

Though fruit and vegetable eminence has been the main lunge of this appraisal it should not be elapsed that the eminence of flower bulbs are also of chief position, particularly in countries such as the Netherlands where disseminate souks in tulips are commercially substantial. Bud abortion is a complaint of tulips related with an arid flower after establishing, while introduction often takes place during stowage. In an effort to use MRI as a conceivable indicator of tulip bulbs likely to exhibition this criticism, van Kilsdonk et al. (2002) assumed spin-echo images of tulip bulbs on a 4.7 T gadget over a course of 35 weeks stowing under surroundings helpful to bud abortion. An initiate that the mean proton density, T_1 and T_2 for the scale, shoot, and stamens all displayed reformist decrease during the storage period. This was interpreted as a progressive loss of internal quality, such that the bulb ultimately developed too weak for the fast root and shoot elongation obligatory throughout the planting phase, subsequent in bud abortion. There is therefore the opportunity of using low field NMR to display bulb practicality and water status through stowing.

The cellular origins of relaxation dissimilarity in plant tissue have been mentioned to numerous times in the preceding section. Empathetic the molecular and microstructural cause of water proton sloping and longitudinal relaxation times in plant tissue is crucial if NMR acquisition arrangements and field strengths are to be improved for observing degrees of maturity and for sensing defects in plant tissue. In an effort to vindicate relaxation time differences the early literature trusted seriously on the very loose notion of free and bound water in cellular tissue; but it is now approved by most academics that such language is ambiguous and of little extrapolative value. In statistic, there is no such thing as bound water, except one is prepared to call water hydrogen bonding biopolymers with the extended exchange lifetimes of the order of microseconds bound. In a succession of papers, Hills and co-workers displayed that in most plant tissue, in the lack of paramagnetic ions, there are three leading relaxation paths for water proton relaxation.

8.6   ADVANTAGES

Though NMR is a comparatively insensitive technique in these development submissions, it is investigating bulk belongings and this does not matter. The practice can be used to inspect procedures and consistencies deprived of the use of any indicator complexes or indeed any physical intrusion at all beyond a magnetic field.

NMR data can be used to intensify empathetic of procedures such as enzymatic and conventional chemical reactions during a manufacturing process and also to complement superiority and development control during fabrication.

8.7   INFERENCES AND FUTURE WORK

A large number of studies have been carried out and many processes have been developed to evaluate the quality of fruit and provide a detailed analysis but they have been mainly laborious, time exhuming, largely applicable to lab-scale and requiring specialized skill to operate and analyze. The qualities of a quality evaluation device should be based on a non-contact detection practices to identify the specific compound from a large variety of different fruits augmented with smart practices such as instant data assortment. NIR spectroscopy is one such technique having all the benefits mentioned above, it has proved very successful to predict and model ripeness stage, firmness, sugar content, and acidity in case of fruits. It has been found that NIR spectroscopy is a relatively expensive technique due to expensive NIr filters, detectors, and other interface electrical systems but exhibits a higher repeatability and reliability. Apart from spectroscopy, NMR is one of the efficient and cost economic techniques to carry real-time on-tree fruit superiority analysis. Lack of flexibility and spatial difficulty in the case of the on-tree fruits can be easily overcome in the case of NMR using modest linear or non-linear numerical demonstration. NMR technologies available in the commercial industry for on-tree screening are based on interaction type systems and uses a single sensing system tailed by single-level calculation modeling. Most of the prevailing explanation discussions include the harvesting time decision issue, shelf-life calculation, and early advertisement for disease spell still need a cost operative, precise, and fast explanation.

KEYWORDS

- conjugated linoleic acid
- diacylglycerols
- fatty acids
- gas chromatography
- magnetic resonance imaging
- nuclear magnetic resonance

REFERENCES

Abdullah, M. Z., (2008). *Quality Evaluation of Computer Vision Technology for Food Quality Evaluation* (p. 481). Academic Press. Inc. (London), Ltd., London.

Almeida, D. P., & Huber, D. J., (2001). Transient increase in locular pressure and occlusion of endocarpic apertures in ripening tomato fruit. *Journal of Plant Physiology, 158*(2), 199–203.

Antonucci, F., Pallottino, F., Paglia, G., Palma, A., D'Aquino, S., & Menesatti, P., (2011). Non-destructive estimation of mandarin maturity status through portable VIS-NIR spectrophotometer. *Food and Bioprocess Technology, 4*(5), 809–813.

Awad, T. S., Moharram, H. A., Shaltout, O. E., Asker, D., & Youssef, M. M., (2012). Applications of ultrasound in analysis, processing, and quality control of food: A review. *Food Research International, 48*(2), 410–427.

Balci, M., (2000). Nucleur Magnetic Resonance spectroscopy. *Nükleer Manyetik Rezonans Spektroskopisi.*

Barreiro, P., Ortiz, C., Ruiz-Altisent, M., Ruiz-Cabello, J., Fernández-Valle, M. E., Recasens, I., & Asensio, M., (2000). Mealiness assessment in apples and peaches using MRI techniques. *Magnetic Resonance Imaging, 18*(9), 1175–1181.

Barreiro, P., Zheng, C., Sun, D. W., Hernández-Sánchez, N., Perez-Sanchez, J. M., & Ruiz-Cabello, J., (2008). Non-destructive seed detection in mandarins: Comparison of automatic threshold methods in FLASH and COMSPIRA MRIs. *Postharvest Biology and Technology, 47*(2), 189–198.

Belloque, J., & Ramos, M., (1999). Application of NMR spectroscopy to milk and dairy products. *Trends in Food Science and Technology, 10*(10), 313–320.

Beullens, K., Kirsanov, D., Irudayaraj, J., Rudnitskaya, A., Legin, A., Nicolaï, B. M., & Lammertyn, J., (2006). The electronic tongue and ATR–FTIR for rapid detection of sugars and acids in tomatoes. *Sensors and Actuators B: Chemical, 116*(1–2), 107–115.

Beullens, K., Mészáros, P., Vermeir, S., Kirsanov, D., Legin, A., Buysens, S., Cap, N., Nicolaï, B. M., & Lammertyn, J., (2008). Analysis of tomato taste using two types of electronic tongues. *Sensors and Actuators B: Chemical, 131*(1), 10–17.

Bidlack, W. R., & Wang, W., (2000). *Designing Functional Foods to Phytochemicals as Bioactive Agents* (p. 241).

Blasco, J., Alamar, M. C., & Molto, E., (2003). Detection no destructiva de semillas en mandarinas mediante resonancia magnetica [Nondestructive detection of seeds in mandarins by using magnetic resonance.]. In: *Resiimenesdel II Congreso Nacional de Agroingenieria* (pp. 439–440). Cordoba, Spain.

Bloch, F., (1946). Nuclear induction. *Physical Review, 70*(7–8), 460.

Bock, C., Sartoris, F. J., & Pörtner, H. O., (2002). *In vivo* MR spectroscopy and MR imaging on non-anaesthetized marine fish: Techniques and first results. *Magnetic Resonance Imaging, 20*(2), 165–172.

Borompichaichartkul, C., Moran, G., Srzednicki, G., & Price, W. S., (2005). Nuclear magnetic resonance (NMR) and magnetic resonance imaging (MRI) studies of corn at subzero temperatures. *Journal of Food Engineering, 69*(2), 199–205.

Brescia, M. A., Monfreda, M., Buccolieri, A., & Carrino, C., (2005). Characterization of the geographical origin of buffalo milk and mozzarella cheese by means of analytical and spectroscopic determinations. *Food Chemistry, 89*(1), 139–147.

Bureau, S., Ruiz, D., Reich, M., Gouble, B., Bertrand, D., Audergon, J. M., & Renard, C. M., (2009). Application of ATR-FTIR for a rapid and simultaneous determination of sugars and organic acids in apricot fruit. *Food Chemistry, 115*(3), 1133–1140.

Butz, P., Hofmann, C., & Tauscher, B., (2005). Recent developments in noninvasive techniques for fresh fruit and vegetable internal quality analysis. *Journal of Food Science, 70*(9).

Cabrer, P. R., Van, D. J. P., Van, D. G., & Nicolay, K., (2005). MRI: Assessment of water transport in food. *Ciencia En La Frontera*, p. 59.

Caligiani, A., Acquotti, D., Palla, G., & Bocchi, V., (2007). Identification and quantification of the main organic components of vinegars by high resolution 1H NMR spectroscopy. *Analytica. Chimica. Acta, 585*(1), 110–119.

Callaghan, P. T., Clark, C. J., & Forde, L. C., (1994). Use of static and dynamic NMR microscopy to investigate the origins of contrast in images of biological tissues. *Biophysical Chemistry, 50*(1–2), 225–235.

Cavanagh, J., Fairbrother, W. J., & Palmer, III. A. G., (2007). In: Rance, III, M., & Skelton, N. J., (eds.), *Protein NMR Spectroscopy.*

Cayuela, J. A., & Weiland, C., (2010). Intact orange quality prediction with two portable NIR spectrometers. *Postharvest Biology and Technology, 58*(2), 113–120.

Cayuela, J. A., (2008). Vis/NIR soluble solids prediction in intact oranges (*Citrus sinensis* L.) cv. Valencia Late by reflectance. *Postharvest Biology and Technology, 47*(1), 75–80.

Cerdá, B., Espín, J. C., Parra, S., Martínez, P., & Tomás-Barberán, F. A., (2004). The potent in vitro antioxidant ellagitannins from pomegranate juice are metabolized into bioavailable but poor antioxidant hydroxy-6H-dibenzopyran-6-one derivatives by the colonic micro flora of healthy humans. *European Journal of Nutrition, 43*(4), 205–220.

Chaland, B., Mariette, F., Marchal, P., & De Certaines, J., (2000). ^{1}H nuclear magnetic resonance relaxometric characterization of fat and water states in soft and hard cheese. *Journal of Dairy Research, 67*(4), 609–618.

Chandrapala, J., Oliver, C., Kentish, S., & Ashokkumar, M., (2012). Ultrasonic's in food processing. *Ultrasonics Sonochemistry, 19*(5), 975–983.

Charlton, A. J., Farrington, W. H., & Brereton, P., (2002). Application of 1H NMR and multivariate statistics for screening complex mixtures: Quality control and authenticity of instant coffee. *Journal of Agricultural and Food Chemistry, 50*(11), 3098–3103.

Chen, L., & Opara, U. L., (2013). Approaches to analysis and modeling texture in fresh and processed foods: A review. *Journal of Food Engineering, 119*(3), 497–507.

Chen, P., McCarthy, M. J., & Kauten, R., (1989). NMR for internal quality evaluation of fruits and vegetables. *Trans. ASAE, 32*(5), 1747–1753.

Choi, K. H., Lee, K. J., & Kim, G., (2006). Nondestructive quality evaluation technology for fruits and vegetables using near-infrared spectroscopy. In: *Proceedings of the International Seminar on Enhancing Export Competitiveness of Asian Fruits* (pp. 18–19), Bangkok, Thailand.

Ciosek, P., & Wróblewski, W., (2007). Sensor arrays for liquid sensing-electronic tongue systems. *Analyst, 132*(10), pp.963–978.

Clark, C. J., Drummond, L. N., & MacFall, J. S., (1998). Quantitative magnetic resonance imaging of kiwifruit during growth and ripening. *J. Sci. Food Agric., 78*, 349–358.

Clark, C. J., MacFall, J. S., & Bieleski, R. L., (1998). Loss of water core from *Fuji'* apple observed by magnetic resonance imaging. *Scientia Horticulturae, 73*(4), 213–227.

Clark, C. J., Richardson, A. C., & Marsh, K. B., (1999). Quantitative magnetic resonance imaging of Satsuma mandarin fruit during growth. *Hort. Science, 34*(6), 1071–1075.

Cohen, E., (1988). The chemical composition and sensory flavor quality of '*Mineola*' tangerines. I. Effects of fruit size and within-tree position. *Journal of Horticultural Science, 63*(1), 175–178.

Consonni, R., & Cagliani, L. R., (2008). Ripening and geographical characterization of Parmigiano Reggiano cheese by 1H NMR spectroscopy. *Talanta., 76*(1), 200–205.

Cools, K., Chope, G. A., Hammond, J. P., Thompson, A. J., & Terry, L. A., (2011). Ethylene and 1-methylcyclopropene differentially regulate gene expression during onion sprout suppression. *Plant Physiology, 156*(3), 1639–1652.

Cox, B., & Parry, C., (1968). Principle. *Smith Junior High School, Mesa AZ, 480*, p. 4650.

Crowther, T., Collin, H. A., Smith, B., Tomsett, A. B., O'Connor, D., & Jones, M. G., (2005). Assessment of the flavor of fresh uncooked onions by taste-panels and analysis of flavor precursors, pyruvate and sugars. *Journal of the Science of Food and Agriculture, 85*(1), 112–120.

Cubero, S., Aleixos, N., Moltó, E., Gómez-Sanchis, J., & Blasco, J., (2011). Advances in machine vision applications for automatic inspection and quality evaluation of fruits and vegetables. *Food and Bioprocess Technology, 4*(4), 487–504.

Dale, B. M., Brown, M. A., & Semelka, R. C., (2015). *MRI: Basic Principles and Applications.* John Wiley & Sons.

Damadian, R., (1971). Tumor detection by nuclear magnetic resonance. *Science, 171*(3976), 1151–1153.

De Angelis, C. S., Curini, R., Delfini, M., Brosio, E., D'Ascenzo, F., & Bocca, B., (2000). Amino acid profile in the ripening of grana padano cheese: A NMR study. *Food Chemistry, 71*(4), 495–502.

Dicko, M. H., (2005). *Endogenous Phenolics and Starch Modifying Enzymes as Determinants of Sorghum for Food use in Burkina Faso.* Wageningen University.

Dongare, M. L., Buchade, P. B., Awatade, M. N., & Shaligram, A. D., (2014). Mathematical modeling and simulation of refractive index based brix measurement system. *Optik-International Journal for Light and Electron Optics, 125*(3), 946–949.

Downes, K., Chope, G. A., & Terry, L. A., (2009). Effect of curing at different temperatures on biochemical composition of onion (*Allium cepa* L.) skin from three freshly cured and cold stored UK-grown onion cultivars. *Postharvest Biology and Technology, 54*(2), 80–86.

Du, C. J., & Sun, D. W., (2004). Recent developments in the applications of image processing techniques for food quality evaluation. *Trends in Food Science and Technology, 15*(5), 230–249.

Duce, S. L., & Hall, L. D., (1995). Visualization of the hydration of food by nuclear magnetic resonance imaging. *Journal of Food Engineering, 26*(2), 251–257.

El Hadi, M. A. M., Zhang, F. J., Wu, F. F., Zhou, C. H., & Tao, J., (2013). Advances in fruit aroma volatile research. *Molecules, 18*(7), 8200–8229.

Erdik, E., (1998). *Spectroscopic Methods in Organic Chemistry* [*Organik Kimyada Spektroskopik Yöntemler*]. GaziBüroKitabevi.

Fawole, O. A., & Opara, U. L., (2013). Harvest discrimination of pomegranate fruit: Postharvest quality changes and relationships between instrumental and sensory attributes during shelf life. *Journal of Food Science, 78*(8).

Filler, A. G., (2010). The history, development, and impact of computed imaging in neurological diagnosis and neurosurgery: CT, MRI, and DTI. *Internet Journal of Neurosurgery, 7*(1), 5–35.

Foucat, L., Taylor, R. G., Labas, R., & Renou, J. P., (2004). Soft flesh problem in freshwater rainbow trout investigated by magnetic resonance imaging and histology. *Journal of Food Science, 69*(4).

Fourel, I., Guillement, J. P., & Le Botlan, D., (1995). Determination of water droplet size distributions by low resolution PFG-NMR: II. Solid emulsions. *Journal of Colloid and Interface Science, 169*(1), 119–124.

Gao, H., Zhu, F., & Cai, J., (2009). A review of non-destructive detection for fruit quality. In: *International Conference on Computer and Computing Technologies in Agriculture* (pp. 133–140). Springer, Berlin, Heidelberg.

Gil-Izquierdo, A., Gil, M. I., Tomás-Barberán, F. A., & Ferreres, F., (2003). Influence of industrial processing on orange juice flavanone solubility and transformation to chalcones under gastrointestinal conditions. *Journal of Agricultural and Food Chemistry, 51*(10), 3024–3028.

Glover, G. H., Flax, S. W., & Shimakawa, A., (1993). General Electric Co. *Reduction of NMR Artifacts Caused by Time Varying Linear Geometric Distortion.* U.S. Patent 5,200,700.

Gobet, M., Buchin, S., Rondeau-Mouro, C., Mietton, B., Guichard, E., Moreau, C., & Le Quéré, J. L., (2013). Solid-state 31 P NMR, a relevant method to evaluate the distribution of phosphates in semi-hard cheeses. *Food Analytical Methods, 6*(6), 1544–1550.

Golic, M., & Walsh, K. B., (2006). Robustness of calibration models based on near infrared spectroscopy for the in-line grading of stone fruit for total soluble solids content. *Analytica. Chimica, Acta, 555*(2), 286–291.

Golic, M., Walsh, K., & Lawson, P., (2003). Short-wavelength near-infrared spectra of sucrose, glucose, and fructose with respect to sugar concentration and temperature. *Applied Spectroscopy, 57*(2), 139–145.

Gómez-Sanchis, J., Gómez-Chova, L., Aleixos, N., Camps-Valls, G., Montesinos-Herrero, C., Moltó, E., & Blasco, J., (2008). Hyper spectral system for early detection of rottenness caused by *Penicillium digitatum* in mandarins. *Journal of Food Engineering, 89*(1), 80–86.

Goodman, B. A., Williamson, B., Simpson, E. J., Chudek, J. A., Hunter, G., & Prior, D. A. M., (1996). High field NMR microscopic imaging of cultivated strawberry fruit. *Magnetic Resonance Imaging, 14*(2), 187–196.

Gorinstein, S., Martin-Belloso, O., Lojek, A., Číž, M., Soliva-Fortuny, R., Park, Y. S., Caspi, A., Libman, I., & Trakhtenberg, S., (2002). Comparative content of some phytochemicals in Spanish apples, peaches, and pears. *Journal of the Science of Food and Agriculture, 82*(10), 1166–1170.

Goudappel, G. J. W., Van, D. J. P. M., & Mooren, M. M. W., (2001). Measurement of oil droplet size distributions in food oil/water emulsions by time domain pulsed field gradient NMR. *Journal of Colloid and Interface Science, 239*(2), 535–542.

Gowen, A. A., O'Donnell, C., Cullen, P. J., Downey, G., & Frias, J. M., (2007). Hyper spectral imaging: An emerging process analytical tool for food quality and safety control. *Trends in Food Science and Technology, 18*(12), 590–598.

Groves, P., Canales, A., Chávez, M. I., Palczewska, M., Díaz, D., & Jiménez-Barbero, J., (2007). NMR investigations of lectin: Carbohydrate interactions. In: *Lectins* (pp. 51–73).

Gustavsson, J., Cederberg, C., Sonesson, U., Van, O. R., & Meybeck, A., (2011). *Global Food Losses and Food Waste* (pp. 1–38). Rome: FAO.

Ha, K. L., Kanai, H., Chubachi, N., & Kamimura, K., (1991). A basic study on nondestructive evaluation of potatoes using ultrasound. *Japanese Journal of Applied Physics, Part 1, 30,* 80–82.

Hernández-Sánchez, N., Barreiro, P., & Ruiz-Cabello, J., (2006). On-line identification of seeds in mandarins with magnetic resonance imaging. *Biosystems Engineering, 95*(4), 529–536.

Hills, B. P., & Clark, C. J., (2003). Quality assessment of horticultural products by NMR. *Annual Reports on NMR Spectroscopy, 50,* 76–121.

Hills, B. P., & Le Floc'h, G., (1994). NMR studies of non-freezing water in cellular plant tissue. *Food Chemistry, 51*(3), 331–336.

Hills, B. P., (2008). NMR relaxation and diffusion studies of horticultural products. In: *Modern Magnetic Resonance* (pp. 1721–1727). Springer, Dordrecht.

Hills, B., (1995). Food processing: An MRI perspective. *Trends in Food Science and Technology, 6*(4), 111–117.

Hinrichs, R., Götz, J., & Weisser, H., (2003). Water-holding capacity and structure of hydrocolloid-gels, WPC-gels, and yogurts characterized by means of NMR. *Food Chemistry, 82*(1), 155–160.

Hong, X., Wang, J., & Qiu, S., (2014). Authenticating cherry tomato juices: Discussion of different data standardization and fusion approaches based on electronic nose and tongue. *Food Research International, 60*, 173–179.

Howell, N., Shavila, Y., Grootveld, M., & Williams, S., (1996). High-resolution NMR and magnetic resonance imaging (MRI) studies on fresh and frozen cod (Gadusmorhua) and haddock (*Melanogrammus aeglefinus*). *Journal of the Science of Food and Agriculture, 72*(1), 49–56.

Huishan, L., Yibin, Y., Huanyu, J., Yande, L., Xiaping, F., & Wang, J., (2005). Application Fourier transform near infrared spectrometer in rapid estimation of soluble solids content of intact citrus fruits. In: *2005 ASAE Annual Meeting* (p. 1). American Society of Agricultural and Biological Engineers.

Hulme, A. C., (1970). *The Biochemistry of Fruits and Their Products* (Vol. 1, 2).

Ishida, N., Kobayashi, T., Koizumi, M., & Kano, H., (1989). 1H-NMR imaging of tomato fruits. *Agricultural and Biological Chemistry, 53*(9), 2363–2367.

Ishida, N., Koizumi, M., & Kano, H., (1994). Ontogenetic changes in water in cherry tomato fruits measured by nuclear magnetic resonance imaging. *Scientia Horticulturae, 57*(4), 335–346.

Ito, T., Sasaki, K., & Yoshida, Y., (1997). Changes in respiration rate, saccharide and organic acid content during the development and ripening of mango fruit (*Mangifera indica* L. 'Irwin') cultured in a plastic house. *Journal of the Japanese Society for Horticultural Science, 66*(3–4), 629–635.

Jamshidi, B., Minaei, S., Mohajerani, E., & Ghassemian, H., (2012). Reflectance Vis/NIR spectroscopy for nondestructive taste characterization of Valencia oranges. *Computers and Electronics in Agriculture, 85*, 64–69.

Jha, S. N., & Matsuoka, T., (2000). Non-destructive techniques for quality evaluation of intact fruits and vegetables. *Food Science and Technology Research, 6*(4), 248–251.

Jie, D., Xie, L., Rao, X., & Ying, Y., (2014). Using visible and near infrared diffuse transmittance technique to predict soluble solids content of watermelon in an on-line detection system. *Postharvest Biology and Technology, 90*, 1–6.

Jin, F., Kawasaki, K., Kishida, H., Tohji, K., Moriya, T., & Enomoto, H., (2007). NMR spectroscopic study on methanolysis reaction of vegetable oil. *Fuel, 86*(7–8), 1201–1207.

Johnson, Jr. C. S., (1999). Diffusion ordered nuclear magnetic resonance spectroscopy: Principles and applications. *Progress in Nuclear Magnetic Resonance Spectroscopy, 34*(3–4), 203–256.

Jones, F. E., (1995). A new reference method for testing hydrometers. *Measurement, 16*(4), 231–237.

Kader, A. A., (1997). Fruit maturity, ripening, and quality relationships. In: *International Symposium Effect of Pre- and Postharvest Factors in Fruit Storage* (Vol. 485, pp. 203–208).

Kader, A. A., (2008). Flavor quality of fruits and vegetables. *Journal of the Science of Food and Agriculture, 88*(11), 1863–1868.

Kader, A., Hess-Pierce, B., & Almenar, E., (2003). Relative contribution of fruit constituents to soluble solids content measured by refractometer. *Hort. Science, 38*, 833.

Kaffarnik, S., Ehlers, I., Gröbner, G., Schleucher, J., & Vetter, W., (2013). Two-dimensional 31P, 1H NMR spectroscopic profiling of phospholipids in cheese and fish. *Journal of Agricultural and Food Chemistry, 61*(29), 7061–7069.

Kauffman, G., (2014). Nobel prize for MRI imaging denied to Raymond V. Damadian a decade ago. *Journal of Computing, 19*, 73–90.

Kauten, R., & McCarthy, M., (1995). Applications of NMR imaging in processing of foods. In: *Food Processing* (pp. 1–22).

Kay, L. E., (2005). NMR studies of protein structure and dynamics. *Journal of Magnetic Resonance, 173*(2), 193–207.

Keeler, J., (2011). *Understanding NMR Spectroscopy*. John Wiley & Sons.

Khairi, M. T. M., Ibrahim, S., Yunus, M. A. M., & Faramarzi, M., (2015). Contact and non-contact ultrasonic measurement in the food industry: A review. *Measurement Science and Technology, 27*(1), p012001.

Kim, S. M., Chen, P., McCarthy, M. J., & Zion, B., (1999). Fruit internal quality evaluation using on-line nuclear magnetic resonance sensors. *Journal of Agricultural Engineering Research, 74*(3), 293–301.

Kim, S. M., McCarthy, M. J., & Chen, P., (1994). *ASAE Conference Paper No. 94-6519*. Atlanta, GA.

Kitinoja, L., AlHassan, H. A., Saran, S., & Roy, S. K., (2010). *Identification of Appropriate Postharvest Technologies for Improving Market Access and Incomes for Small Horticultural Farmers in SUB-Saharan Africa and South Asia.*

Kumar, D., Singh, B. P., & Kumar, P., (2004). An overview of the factors affecting sugar content of potatoes. *Annals of Applied Biology, 145*(3), 247–256.

Kuo, M. I., Gunasekaran, S., Johnson, M., & Chen, C., (2001). Nuclear magnetic resonance study of water mobility in pasta filata and non-pasta filata mozzarella. *Journal of Dairy Science, 84*(9), 1950–1958.

Laghi, L., Picone, G., & Capozzi, F., (2014). Nuclear magnetic resonance for foodomics beyond food analysis. *TrAC Trends in Analytical Chemistry, 59*, 93–102.

Lagi, M., Bertrand, K., & Bar-Yam, Y., (2011). *The Food Crises and Political Instability in North Africa and the Middle East.*

Larrigaudiere, C., Lentheric, I., Puy, J., & Pinto, E., (2004). Biochemical characterization of core browning and brown heart disorders in pear by multivariate analysis. *Postharvest Biology and Technology, 31*(1), 29–39.

Le Gall, G., Colquhoun, I. J., & Defernez, M., (2004). Metabolite profiling using 1H NMR spectroscopy for quality assessment of green tea, *Camellia sinensis* (L.). *Journal of Agricultural and Food Chemistry, 52*(4), 692–700.

Le Gall, G., Colquhoun, I. J., Davis, A. L., Collins, G. J., & Verhoeyen, M. E., (2003). Metabolite profiling of tomato (*Lycopersicon esculentum*) using 1H NMR spectroscopy as a tool to detect potential unintended effects following a genetic modification. *Journal of Agricultural and Food Chemistry, 51*(9), 2447–2456.

Lee, W. D., Drazen, J., Sharp, P. A., & Langer, R. S., (2013). *From X-Rays to DNA: How Engineering Drives Biology*. MIT Press.

Loutfi, A., Coradeschi, S., Mani, G. K., Shankar, P., & Rayappan, J. B. B., (2015). Electronic noses for food quality: A review. *Journal of Food Engineering, 144*, 103–111.

Magwaza, L. S., Opara, U. L., Nieuwoudt, H., Cronje, P. J., Saeys, W., & Nicolaï, B., (2012). NIR spectroscopy applications for internal and external quality analysis of citrus fruit: A review. *Food and Bioprocess Technology, 5*(2), 425–444.

Magwaza, L. S., Opara, U. L., Terry, L. A., Landahl, S., Cronje, P. J., Nieuwoudt, H. H., Hanssens, A., Saeys, W., & Nicolaï, B. M., (2013). Evaluation of Fourier transform-NIR spectroscopy for integrated external and internal quality assessment of Valencia oranges. *Journal of Food Composition and Analysis, 31*(1), 144–154.

Mansfield, P., & Maudsley, A. A., (1977). Medical imaging by NMR. *The British Journal of Radiology, 50*(591), 188–194.

Marcone, M. F., Wang, S., Albabish, W., Nie, S., Somnarain, D., & Hill, A., (2013). Diverse food-based applications of nuclear magnetic resonance (NMR) technology. *Food Research International, 51*(2), 729–747.

Margulis, A. R., & Fisher, M. R., (1985). Present clinical status of magnetic resonance imaging. *Magnetic Resonance in Medicine, 2*(4), 309–327.

Martens, H., Thybo, A. K., Andersen, H. J., Karlsson, A. H., Dønstrup, S., Stødkilde-Jørgensen, H., & Martens, M., (2002). Sensory analysis for magnetic resonance-image analysis: Using human perception and cognition to segment and assess the interior of potatoes. *LWT-Food Science and Technology, 35*(1), 70–79.

McCarthy, M. J., (2012). *Magnetic Resonance Imaging in Foods*. Springer Science & Business Media.

McCarthy, M. J., Zion, B., Chen, P., Ablett, S., Darke, A. H., & Lillford, P. J., (1995). Diamagnetic susceptibility changes in apple tissue after bruising. *Journal of the Science of Food and Agriculture, 67*(1), 13–20.

McClure, C. K., (2010). *Structural Chemistry Using NMR Spectroscopy, Organic Molecules.*

Meeten, G. H., & North, A. N., (1995). Refractive index measurement of absorbing and turbid fluids by reflection near the critical angle. *Measurement Science and Technology, 6*(2), 14.

Mendoza, F., Lu, R., & Cen, H., (2014). Grading of apples based on firmness and soluble solids content using Vis/SWNIR spectroscopy and spectral scattering techniques. *Journal of Food Engineering, 125*, 59–68.

Meyer, D. A., (1996). From quantum cellular automata to quantum lattice gases. *Journal of Statistical Physics, 85*(5–6), 551–574.

Mohorič, A., Vergeldt, F., Gerkema, E., De Jager, A., Van, D. J., Van, D. G., & Van, A. H., (2004). Magnetic resonance imaging of single rice kernels during cooking. *Journal of Magnetic Resonance, 171*(1), 157–162.

Ncama, K., Opara, U. L., Tesfay, S. Z., Fawole, O. A., & Magwaza, L. S., (2017). Application of Vis/NIR spectroscopy for predicting sweetness and flavor parameters of '*Valencia*' orange (*Citrus sinensis*) and 'Star Ruby' grapefruit (Citrus x paradise Macfad). *Journal of Food Engineering, 193*, 86–94.

Neufeld, R., & Stalke, D., (2015). Accurate molecular weight determination of small molecules via DOSY-NMR by using external calibration curves with normalized diffusion coefficients. *Chemical Science, 6*(6), 3354–3364.

Newbold, J. W., Hunt, A., & Brereton, J., (2015). *Chemical Spectral Analysis through Sonification.* Georgia Institute of Technology.

Nicolai, B. M., Beullens, K., Bobelyn, E., Peirs, A., Saeys, W., Theron, K. I., & Lammertyn, J., (2007). Nondestructive measurement of fruit and vegetable quality by means of NIR spectroscopy: A review. *Postharvest Biology and Technology, 46*(2), 99–118.

Nor, F. M., Ismail, A. K., Clarkson, M., & Othman, H., (2014). An improved ring method for calibration of hydrometers. *Measurement, 48*, 1–5.

Nott, K. P., Evans, S. D., & Hall, L. D., (1999). The effect of freeze-thawing on the magnetic resonance imaging parameters of cod and mackerel. *LWT-Food Science and Technology, 32*(5), 261–268.

O'Mahony, M. I. C. H. A. E. L., (1991). Taste perception, food quality and consumer acceptance. *Journal of Food Quality, 14*(1), 9–31.

Opara, U. L., & Pathare, P. B., (2014). Bruise damage measurement and analysis of fresh horticultural produce: A review. *Postharvest Biology and Technology, 91*, 9–24.

Patel, K. K., Khan, M. A., & Kar, A., (2015). Recent developments in applications of MRI techniques for foods and agricultural produce: An overview. *Journal of Food Science and Technology, 52*(1), 1–26.

Peng, Y., & Lu, R., (2005). Modeling multispectral scattering profiles for prediction of apple fruit firmness. *Transactions of the ASAE, 48*(1), 235–242.

Peng, Y., & Lu, R., (2008). Analysis of spatially resolved hyper spectral scattering images for assessing apple fruit firmness and soluble solids content. *Postharvest Biology and Technology, 48*(1), 52–62.

Pereira, F. M. V., De Souza, C. A., Cabeça, L. F., & Colnago, L. A., (2013). Classification of intact fresh plums according to sweetness using time-domain nuclear magnetic resonance and chemometrics. *Microchemical Journal, 108*, 14–17.

Peris, M., & Escuder-Gilabert, L., (2009). A 21st century technique for food control: Electronic noses. *Analytica. Chimica. Acta, 638*(1), 1–15.

Piras, C., Marincola, F. C., Savorani, F., Engelsen, S. B., Cosentino, S., Viale, S., & Pisano, M. B., (2013). A NMR metabolomics study of the ripening process of the Fiore sardo cheese produced with autochthonous adjunct cultures. *Food Chemistry, 141*(3), 2137–2147.

Povlsen, V. T., Rinnan, Å., Van, D. B. F., Andersen, H. J., & Thybo, A. K., (2003). Direct decomposition of NMR relaxation profiles and prediction of sensory attributes of potato samples. *LWT-Food Science and Technology, 36*(4), 423–432.

Prema, D., Pilfold, J. L., Krauchi, J., Church, J. S., Donkor, K. K., & Cinel, B., (2013). Rapid determination of total conjugated linoleic acid content in select Canadian cheeses by 1H NMR spectroscopy. *Journal of Agricultural and Food Chemistry, 61*(41), 9915–9921.

Purcell, E. M., Torrey, H. C., & Pound, R. V., (1946). Resonance absorption by nuclear magnetic moments in a solid. *Physical Review, 69*(1–2), 37.

Rawel, H. M., Kroll, J., & Hohl, U. C., (2001). Model studies on reactions of plant phenols with whey proteins. *Food/Nahrung, 45*(2), 72–81.

Ritota, M., Gianferri, R., Bucci, R., & Brosio, E., (2008). Proton NMR relaxation study of swelling and gelatinization process in rice starch-water samples. *Food Chemistry, 110*(1), 14–22.

Robards, K., & Antolovich, M., (1997). Analytical chemistry of fruit bioflavonoids: A review. *Analyst, 122*(2), 11R–34R.

Robards, K., (2003). Strategies for the determination of bioactive phenols in plants, fruit, and vegetables. *Journal of Chromatography A, 1000*(1–2), 657–691.

Robards, K., Prenzler, P. D., Tucker, G., Swatsitang, P., & Glover, W., (1999). Phenolic compounds and their role in oxidative processes in fruits. *Food Chemistry, 66*(4), 401–436.

Ruan, R., & Litchfield, J. B., (1992). Determination of water distribution and mobility inside maize kernels during steeping using magnetic resonance imaging. *Cereal Chemistry*.

Ruan, R., Schmidt, S. J., Schmidt, A. R., & Litchfield, J. B., (1991). Nondestructive measurement of transient moisture profiles and the moisture diffusion coefficient in a potato during drying and absorption by NMR imaging. *Journal of Food Process Engineering, 14*(4), 297–313.

Rudnitskaya, A., Kirsanov, D., Legin, A., Beullens, K., Lammertyn, J., Nicolaï, B. M., & Irudayaraj, J., (2006). Analysis of apples varieties-comparison of electronic tongue with different analytical techniques. *Sensors and Actuators B: Chemical, 116*(1–2), 23–28.

Ryan, D., & Robards, K., (1998). Critical review. Phenolic compounds in olives. *Analyst, 123*(5), 31R–44R.

Ryan, D., Antolovich, M., Prenzler, P., Robards, K., & Lavee, S., (2002). Biotransformations of phenolic compounds in *Oleaeuropaea* L. *Scientia Horticulturae, 92*(2), 147–176.

Saftner, R., Polashock, J., Ehlenfeldt, M., & Vinyard, B., (2008). Instrumental and sensory quality characteristics of blueberry fruit from twelve cultivars. *Postharvest Biology and Technology, 49*(1), 19–26.

Saleem, A., Ahotupa, M., & Pihlaja, K., (2001). Total phenolics concentration and antioxidant potential of extracts of medicinal plants of Pakistan. *Zeitschrift. Für Naturforschung C, 56*(11–12), 973–978.

Saltveit, Jr. M. E., (1991). Determining tomato fruit maturity with nondestructive *in vivo* nuclear magnetic resonance imaging. *Postharvest Biology and Technology, 1*(2), 153–159.

Scano, P., Anedda, R., Melis, M. P., Dessi, M. A., Lai, A., & Roggio, T., (2011). 1H-and 13C-NMR characterization of the molecular components of the lipid fraction of pecorino Sardo cheese. *Journal of the American Oil Chemists' Society, 88*(9), 1305–1316.

Schievano, E., Guardini, K., & Mammi, S., (2009). Fast determination of histamine in cheese by nuclear magnetic resonance (NMR). *Journal of Agricultural and Food Chemistry, 57*(7), 2647–2652.

Shaarani, S. M., Amin, M. G., Soon, N. G., & Hall, L. D., (2010). Monitoring development and ripeness of oil palm fruit (Elaeisguneensis) by MRI and bulk NMR. *International Journal of Agriculture and Biology, 12*(1), 101–105.

Shah, N., Sattar, A., Benanti, M., Hollander, S., & Cheuck, L., (2006). Magnetic resonance spectroscopy as an imaging tool for cancer: A review of the literature. *The Journal of the American Osteopathic Association, 106*(1), 23–27.

Shanmugavelan, P., Kim, S. Y., Kim, J. B., Kim, H. W., Cho, S. M., Kim, S. N., Kim, S. Y., Cho, Y. S., & Kim, H. R., (2013). Evaluation of sugar content and composition in commonly consumed Korean vegetables, fruits, cereals, seed plants, and leaves by HPLC-ELSD. *Carbohydrate Research, 380*, 112–117.

Shintu, L., & Caldarelli, S., (2005). High-resolution MAS NMR and chemometrics: Characterization of the ripening of Parmigiano-Reggiano cheese. *Journal of Agricultural and Food Chemistry, 53*(10), 4026–4031.

Sittig, D. F., Ash, J. S., & Ledley, R. S., (2006). The story behind the development of the first whole-body computerized tomography scanner as told by Robert S. Ledley. *Journal of the American Medical Informatics Association, 13*(5), 465–469.

Snijder, A. J., Wastie, R. L., Glidewell, S. M., & Goodman, B. A., (1996). *Free Radicals and other Paramagnetic Ions in Interactions between Fungal Pathogens and Potato Tubers.*

Söbeli, C., & Kayaardı, S., (2014). New techniques in determining meat quality (Et kalitesini belirlemede yeni teknikler)/*The Journal of Food, 39*(4).

Song, H. P., Delwiche, S. R., & Line, M. J., (1998). Moisture distribution in a mature soft wheat grain by three-dimensional magnetic resonance imaging. *Journal of Cereal Science, 27*(2), 191–197.

Spyros, A., & Dais, P., (2000). Application of 31P NMR spectroscopy in food analysis. 1. Quantitative determination of the mono-and diglyceride composition of olive oils. *Journal of Agricultural and Food Chemistry, 48*(3), 802–805.

Stapley, A. G., Hyde, T. M., Gladden, L. F., & Fryer, P. J., (1997). NMR imaging of the wheat grain cooking process. *International Journal of Food Science and Technology, 32*(5), 355–375.

Sugiyama, J., & Tsuta, M., (2010). Visualization of sugar distribution of melons by hyperspectral technique. In: *Hyper spectral Imaging for Food Quality Analysis and Control* (pp. 349–368).

Tepel, M., Van, D. G. M., Statz, M., Jankowski, J., & Zidek, W., (2003). The antioxidant acetylcysteine reduces cardiovascular events in patients with end-stage renal failure: A randomized, controlled trial. *Circulation, 107*(7), 992–995.

Thybo, A. K., Andersen, H. J., Karlsson, A. H., Dønstrup, S., & Stødkilde-Jørgensen, H., (2003). Low-field NMR relaxation and NMR-imaging as tools in differentiation between potato sample and determination of dry matter content in potatoes. *LWT-Food Science and Technology, 36*(3), 315–322.

Thybo, A. K., Bechmann, I. E., Martens, M., & Engelsen, S. B., (2000). Prediction of sensory texture of cooked potatoes using uniaxial compression, near infrared spectroscopy and low field1H NMR spectroscopy. *LWT-Food Science and Technology, 33*(2), 103–111.

Thygesen, L. G., Thybo, A. K., & Engelsen, S. B., (2001). Prediction of sensory texture quality of boiled potatoes from low-field1H NMR of raw potatoes. The role of chemical constituents. *LWT-Food Science and Technology, 34*(7), 469–477.

Todt, H., Burk, W., Guthausen, G., Guthausen, A., Kamlowski, A., & Schmalbein, D., (2001). Quality control with time-domain NMR. *European Journal of Lipid Science and Technology, 103*(12), 835–840.

Tscheuschner, H. D., & Mohsenin, N. N., (1987). Physical properties of plant and animal materials. Structure, physical characteristics, and mechanical properties. 2. Auf. 891 Seiten, zahlr. Abb. und Tab, vol. 31, Gordon and Breach Science Publishers, New York, NY, USA.

Van, D. G., Van, D. J. P. M., Blonk, H., Mohoric, A., Ramos, P. C., & Van, D. R., (2005). Multi-dimensional imaging of foods using Magnetic Resonance Imaging. *Imaging Microscopy, 7*(4.5), 2–4.

Vandersypen, L. M., Steffen, M., Breyta, G., Yannoni, C. S., Sherwood, M. H., & Chuang, I. L., (2001). Experimental realization of Shor's quantum factoring algorithm using nuclear magnetic resonance. *Nature, 414*(6866), 883.

Veliyulin, E., & Aursand, I. G., (2007). 1H and 23Na MRI studies of Atlantic salmon (Salmosalar) and Atlantic cod (Gadusmorhua) fillet pieces salted in different brine concentrations. *Journal of the Science of Food and Agriculture, 87*(14), 2676–2683.

Viereck, N., Dyrby, M., & Engelsen, S. B., (2005). Monitoring thermal processes by NMR technology. In: *Emerging Technologies for Food Processing* (pp. 553–575).

Walsh, K. B., (2005). Commercial adoption of technologies for fruit grading, with emphasis on NIRS. In: *FRUTIC, 5*.

Wang, Z. M., Lee, J. S., Park, J. Y., Wu, C. Z., & Yuan, Z. H., (2008). Optimization of biodiesel production from trap grease via acid catalysis. *Korean Journal of Chemical Engineering, 25*(4), 670–674.

Wei, Z., & Wang, J., (2013). The evaluation of sugar content and firmness of non-climacteric pears based on voltammetric electronic tongue. *Journal of Food Engineering, 117*(1), 158–164.

Williamson, B., Goodman, B. A., & Chudek, J. A., (1992). Nuclear magnetic resonance (NMR) micro-imaging of ripening red raspberry fruits. *New Phytologist, 120*(1), 21–28.

Woertz, K., Tissen, C., Kleinebudde, P., & Breitkreutz, J., (2011). Taste sensing systems (electronic tongues) for pharmaceutical applications. *International Journal of Pharmaceutics, 417*(1–2), 256–271.

Wrigley, S., Hayes, M., Thomas, R., & Chrystal, E., (1997). *Phytochemical Diversity: A Source of New Industrial Products*. Royal Society of Chemistry.

Wu, D., & Sun, D. W., (2013). Advanced applications of hyper spectral imaging technology for food quality and safety analysis and assessment: A review-Part I: Fundamentals. *Innovative Food Science and Emerging Technologies, 19*, 1–14.

Yantarasri, T., Sornsrivichai, J., & Chen, P., (1996). X-ray and NMR for nondestructive internal quality evaluation of durian and mangosteen fruits. In: *International Postharvest Science Conference Postharvest* (Vol. 96, No. 464, pp. 97–102).

Zhang, L., & McCarthy, M. J., (2013). Assessment of pomegranate postharvest quality using nuclear magnetic resonance. *Postharvest Biology and Technology, 77*, 59–66.

Zhou, B. N., Mattern, M. P., Johnson, R. K., & Kingston, D. G., (2001). Structure and stereochemistry of a novel bioactive sphingolipid from a Calyx sp. *Tetrahedron, 57*(47), 9549–9554.

Zhou, R., Mo, Y., Li, Y., Zhao, Y., Zhang, G., & Hu, Y., (2008). Quality and internal characteristics of Huanghua pears (PyruspyrifoliaNakai, cv. Huanghua) treated with different kinds of coatings during storage. *Postharvest Biology and Technology, 49*(1), 171–179.

Zion, B., Chen, P., & McCarthy, M. J., (1995). Nondestructive quality evaluation of fresh prunes by NMR spectroscopy. *Journal of the Science of Food and Agriculture, 67*(4), 423–429.

CHAPTER 9

X-RAY IMAGING FOR QUALITY DETECTION IN FRUITS AND VEGETABLES

R. R. SHARMA,[1] S. VIJAY RAKESH REDDY,[2] and G. GAJANAN[1]

[1]ICAR-Indian Agricultural Research Institute, New Delhi – 110 012, India

[2]ICAR-Central Institute for Arid Horticulture, Bikaner, Rajasthan, India

ABSTRACT

Digital image processing with associated sensing elements operating in the broad range of the electromagnetic spectrum has emerged as a prominent technique in almost every field for making crucial decisions. Although the imaging technique originated more than a century ago, tremendous growth has been seen in the last two decades, especially with the growth in the hardware and software industries providing a significant boost. Imaging was originally developed for applications in the medical field and remote sensing. As technology grew, the technique has occupied an irreplaceable part in every walk of life. X-ray imaging has been explored for inspecting the interior of horticultural commodities. The intensity of energy exiting the product is dependent upon the incident energy, absorption coefficient, density of the product, and sample thickness. Some of the internal disorders that could be detected non-destructively include cork spot, bitter pit, watercore, and brown core for apple; blossom end decline, membranous stain, black rot, seed germination and freeze damage for citrus; and hollow heart, bruises, and perhaps black heart for potato.

9.1 INTRODUCTION

Quality is a complex perception of many attributes that are simultaneously evaluated by the consumer either objectively or subjectively. For horticulture produce, the quality evaluation is mostly carried out widely through manual inspection, which is time-consuming, laborious as well as costly (Van et al., 2017). The manual inspection may easily be influenced by physiological factors including subjective and inconsistent evaluation results (Du and Sun, 2006). Horticultural products often develop internal defects, cracks, or cavities that reduce their commercial value (Van et al., 2017). Currently, the most widespread method to detect these defects is a destructive evaluation where a number of samples is randomly selected from each batch and cut open for inspection (Shewfelt, Brueckner, and Florkowski, 2012). This has some inherent disadvantages like financial losses due to the destruction of a large number of samples and the fact that only a small subset of total products is being checked.

Conventional quality assessment approaches employed are normally destructive and off-line in nature, efforts have thus been made to develop diverse non-contact, rapid, environmental-friendly, and accurate methods for non-invasive examination of various food products especially fruits and vegetables (Van et al., 2017; Caballero et al., 2017). The development of imaging techniques has been established since the early 1980s. Nowadays, applications of imaging techniques have been extended to quality and safety assessment and monitoring of agricultural cum food products (Ma et al., 2016; Sun, 2016). The various imaging technologies include fluorescence imaging, Raman imaging, magnetic resonance imaging (MRI), and x-ray imaging, etc. A typical x-ray inspection system involves an x-ray source, image acquisition, image segmentation, and classification. To improve image quality, the selection of the x-ray source and its operating parameters largely depends on the product being inspected (Mathanker, Weckler, and Bowser, 2013).

The current applications in the field of detecting and curbing internal disorders in fruits and vegetables include spongy tissue in mango, soft nose in mango, cork spot, bitter pit, watercore, and brown core in apple, blossom end decline, membranous stain, black rot, seed germination, and freeze damage for citrus, and hollow heart, bruises, and perhaps the black heart of potato, bruise damage in apple and avocado, etc. Commercial routine is the detection of a hollow heart in potato tubers (Butz, Hofmann, and Tauscher, 2005). The discovery of residual pits in processed cherries or olives is important for the processing industry and identification of insect infestation in pistachio nuts has been the subject of investigation (Butz, Hofmann, and Tauscher, 2005). Furthermore, low-energy x-rays, often referred to as soft x-rays, are employed for food inspection, and they are less harmful to operators than the high-energy x-rays used in medical diagnostics. In this book chapter, we tried to explain the basic principle, components, applications, advantages, and disadvantages of the X-ray imaging system.

9.2   PRINCIPLE OF X-RAY IMAGING

Different objects absorb X-rays in different ways depending on their composition, density, and thickness. For example, when a human body is exposed to X-rays, the absorption by bones is relatively higher compared to the soft tissues as well as cartilages. As a result, the bones appear white in an X-ray image while the soft tissues appear grey, and air spaces appear black. Greater is the amount of X-ray absorption by the target object, brighter it would appear in the X-ray image. When any object is exposed to X-rays, they interact with the material to lose their energy exponentially through a technical process known as attenuation. When soft X-ray beams containing photons are incident over an object, they get transmitted; absorbed, or scattered resulting in an exponential reduction of the photon intensity (Curry, Dowdey, and Murry, 1990). The attenuation coefficient describes the ease of penetration of any material by a beam of sound, light, or other energy forms. Larger values of the attenuation coefficient indicate greater loss of intensity when passed through any medium and *vice versa*.

The intensity of X-rays depends on the number of photons as well as their energy. The number of photons can be increased by increasing the tube current, while their energy could be augmented by escalating the peak voltage between the two electrodes (Kotwaliwale et al., 2007). There are various methods of attenuating X-rays, while it depends on how the photons interact with matter. Generally, X-ray attenuation occurs basically by three methods: *viz.* (i) photoelectric absorption, (ii) Compton scattering, and (iii) pair production. When the X-ray photons collide with an electron bonded to an atom, photoelectric absorption occurs resulting in the knocking of electrons from their orbit. Compton scattering occurs when X-ray photons collide with loosely held electrons resulting in the energizing of electrons by photon energy and travel of less energy photons in a different direction. Pair production does not usually occur at the energy levels typically used in the food industry (Peariso, 2008).

9.3   COMPONENTS OF X-RAY IMAGING SYSTEM

The major components of X-ray imaging system include: (1) a source, that generates X-rays; (2) a detector and a camera that captures the real-time image of the target object; (3) an image digitizer that converts the captured video image into a digital image; and (4) an image processing system, including analog-to-digital (A/D) converter, frame grabber, and a computer to store and process the digital images. Basic components are shown schematically in Figure 9.1.

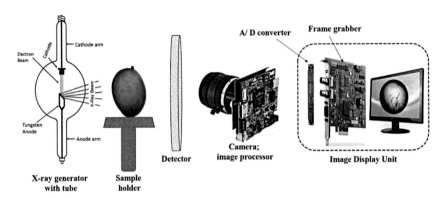

FIGURE 9.1   Illustration of the typical x-ray imaging unit.

9.3.1   X-RAY SOURCE/GENERATOR

The essential parts of an X-ray generator include X-ray tube, a high voltage generator, and the control console. The X-ray tube basically consists of cathode and anode, rotor, and stator, tube casing, and tube cooling system. The design of the X-ray tube dictates the field uniformity, focal spot size, and energy spectrum of the X-rays which are important for contrast and spatial resolution of the images (Zink, 1997).

The X-ray generator supplies the necessary power required for generating X-rays, and they produce X-rays by hitting high energy electrons with tungsten or molybdenum. The important characteristics of X-rays that determine the operating conditions are the voltage (or energy), current, and exposure period. The high voltage provided by the generator is selected according to the application and it regulates the energy of the outcoming X-rays. The peak voltage is called kilovoltage peak (kVp), and the current is based on the number of X-ray photons produced, which is measured in milliamperes (mA). Exposure time is the time during which high voltage is applied to produce X-rays. The amount of X-ray energy produced is controlled by adjusting the voltage potential, X-ray tube current, and exposure time. X-ray tubes and radioactive substances are the two principal sources of X-rays. Radioactive sources produce monochromatic X-rays, while X-ray tubes produce polychromatic beams. X-ray tubes are most widely used for applications of X-ray radiography in the horticultural and food industry (Kotwaliwale et al., 2014).

9.3.2 IMAGE INTENSIFIER

An image intensifier is one of the components that convert the low-intensity X-rays into visible light where it amplifies the low-level X-ray photons to a visible image. The working principle of an image intensifier is that X-ray photons get converted into optical photons inside a vacuum case and later further gets converted to photoelectrons. Due to acceleration by the electric field, the photoelectrons are collected at the output phosphor and many optical photons are produced by each accelerated electron. Image intensifiers are indicated by conversion factors that represent their efficiency of converting X-rays to visible light (Wang and Blackburn, 2000).

9.3.3 DETECTORS AND CAMERA

Various types of detectors used for the detection of X-rays include gas detectors, solid-state detectors, and charge-coupled device (CCD) detectors. Traditionally, X-ray images are acquired either on photographic plates or films. X-ray films are most commonly used in medical applications because it provides a permanent record of the internal fractures, etc. with a typical spatial resolution of 10 to 100 micron (Xradia, 2010). Though the photographic plates have similar resolution and sensitivity as that of films, they are reusable. However, with the advancement of technologies, X-ray images are acquired through digital methods nowadays. The major benefit of using digitized X-ray images is that they allow online quality monitoring of objects, *viz.* food or agricultural produce, or biological materials, along the production/processing line coupled with the ease of storage and transmission and relatively lower time requirement for digital imaging. Haff and Slaughter (Haff and Slaughter, 2004) reported that the use of digital imaging saves time by a factor of four as compared to film radiographs. The various cameras used in X-ray imaging are CCD, line scan cameras, and flat-panel sensors, because these cameras could be used in real-time imaging applications.

9.3.4 IMAGE PROCESSING SYSTEM (A/D CONVERTER, FRAME GRABBER, AND COMPUTER)

The A/D converter is basically a device that converts the continuously varying analog signal to digital signal in binary form. The frame grabber captures video signals and converts them to digital form, which are displayed, transmitted, or stored. A computer attached to the imaging system serves as the software and hardware component that stores and analyzes the images using image processing algorithms.

9.4 TYPES OF X-RAY MACHINES

Based on the purpose and kind of the produce, the x-ray machines used for various industrial applications are broadly classified into three types; viz. vertical beam, horizontal beam, and a combination of the two (Mettler, 2009).

a. **Vertical Beam System:** This system is one of the most common types used wherein the X-ray generator is fixed on the top of the cabinet and the beam passes downwards through the target objects till they strike the detector placed at the bottom of the cabinet. These are most commonly used for products with relatively shallow depth compared to their length and widths. For an appropriate inspection of the complete volume of the target object, it should fit within the conical beam of the X-ray system (Figure 9.2).

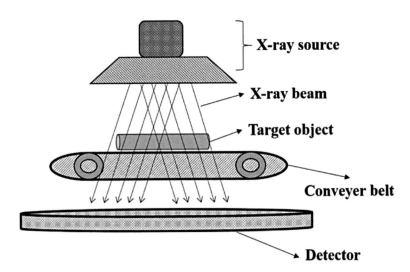

FIGURE 9.2   Illustration of vertical beam x-ray imaging system.

b. **Horizontal Beam System:** This system has the X-ray generators placed on the sides of the conveyor belts and scans the product from sideways. These type systems are more suited for objects with greater depth compared to their length and width. The number of X-ray sources/beams used depends on the density of the object passing over the conveyor system. For example, a single horizontal X-ray beam system is sufficient for low-density packings such as cartons, plastic bottles, and plastic jars. For medium dense packaging such as metal cans, the beam from a single source generator could be split into two by funneling through a dual diverging collimator creating two separate angled beams. This split dual arrangement helps in the detection of small contaminants at the bottom/sidewalls of the container. For high-density packaging containing glass jars/ bottles, dual X-ray generators are used which produce two separate angled beams. This system gives two digital images obtained in two different angles that help in the increased efficiency of contaminant detection (Figure 9.3).

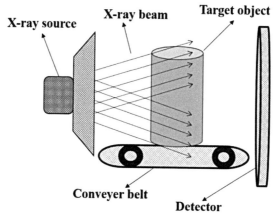

FIGURE 9.3   Illustration of horizontal beam x-ray imaging system.

c. **Combined Beam System:** This is a combined system of the above mentioned two systems which is composed of both vertical as well as horizontal X-ray beam systems. These are used when either of the two couldn't accomplish the given monitoring task when used alone. Also, these are the most advanced X-ray systems available in the food industry for the best possible detection and monitoring of food products (Figure 9.4).

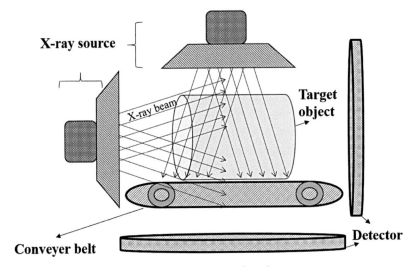

FIGURE 9.4   Illustration of a combined beam x-ray imaging system.

The soft x-ray systems are employed in horticulture industry which has wavelengths in the range of 0.01 to 10 nm, corresponding to frequencies in the range 30 to 30,000 Petahertz ($3 \times 10E16$ Hz to $3 \times 10E19$ Hz) and energies in the range 120 eV to 120 keV (Kotwaliwale et al., 2014). These x-rays are travel in straight lines, and are not deflected either by electric field or magnetic field, due to their high penetrating power. They could blacken a photographic film, cause glowing in exposed fluorescent materials, could produce photoelectric emission and/or ionization of a gas. Typically, in most of the x-ray studies, a product was imaged at one energy level. However, in some of the studies, a product was imaged at two different energy levels; which are referred to as dual-energy x-ray imaging. However, in certain studies, a product was imaged from many angles to construct a three-dimensional (3D) image referred to as computed tomography (CT). X-ray computed tomography (X-ray CT) is a proven method for evaluating a cross-section of an object using a movable X-ray source and detector assembly to accumulate data from a thin projected slice of a sample. The basic principle behind the CT is that the internal structure of an object can be reconstructed from multiple projections of the object (Curry, Dowdey, and Murry, 1990). The basic parts of X-ray CT are X-ray tubes, collimators, turntable, and multichannel detectors installed in a shielded chamber. A collimated X-ray beam was flashed over the produce and the radiation amount used was measured through the computer. Most of the x-ray studies in the food and agricultural sector use imaging systems up to 50 keV energy level, commonly referred to as soft x-ray systems (Mathanker, Weckler, and Bowser, 2013; Kotwaliwale et al., 2014). Neethirajan (Neethirajan, Jayas, and White, 2007) reported that the soft X-ray method was rapid and took only 3–5 s to produce an X-ray image.

9.5 APPLICATION OF X-RAY IMAGING IN THE HORTICULTURE INDUSTRY

The agri-horticultural as well as food industries have started installing X-ray imaging systems for monitoring of product quality in production/processing lines for maintenance of consumer goodwill over their brand names. Due to the impressive performance of the X-ray imaging systems compared to the traditional metal detectors, they are being replaced by most of the food industries. In the food and food processing industry, implementation of an X-ray imaging system as a part of an effective overall inspection system improves the product quality and to a great extent reduces the possibility of product recall, which would result in better brand name and greater business opportunity. Some of the major applications of X-ray imaging system in quality detection of horticultural produce include as given in following subsections (Table 9.1).

9.5.1 DETECTION OF INTERNAL DISORDERS

Fruit disorders occur due to abnormal growth patterns or abnormal external or internal conditions created by adverse environmental conditions such as deviation from the normal state of temperature, light, moisture, nutrient, harmful gases, or inadequate supply of growth regulators. The physiological disorder is a non-pathological malady occurring in fruits and vegetables resulting in huge economic losses. Though most of these disorders could be seen physically with naked eyes, there are certain disorders which are manifested inside the fruit with no physical symptoms on the surface *viz.* spongy tissue and jelly seed disorder in mango, Wooly breakdown of cold-stored nectarines, etc. Sonego et al. (1995) investigated the incidence of wooliness in nectarines by using X-ray CT and revealed that the areas exhibiting wooliness appeared darker indicating the presence of gas inclusions, which is confirmed through X-ray CT. Similarly, Herremans et al. (2013) examined the use of high-resolution X-ray micro CT to detect the Braeburn' browning disorder in apple. He revealed that this technology has the ability to classify fruit tissue as healthy and disordered with a success rate of 97%. Some other works proving the use of X-ray imaging technology for detection of internal disorders include detection of bruised apples (Diener, Mitchell, and Rhoten, 1970), potatoes exhibiting hollow heart (Finney and Norris, 1978) and detection of split-pit peaches (Han, Bowers, and Dodd, 1992). X-rays were also used to detect structural discontinuities caused by voids and cracks and density variation. An X-ray line scan was used to determine bruising, density, and water content of apples, select mature lettuce heads, and X-ray CT was used to detect density changes in tomatoes, seeds in oranges and pomelos, defects in watermelons and cantaloupes, internal changes of peaches and stones in apricots (Barcelon, Tojo, and Watanabe, 1999b). Postharvest assessment of internal decay (chestnuts), microstructural changes induced by roasting (coffee beans) and insect pest behavior (pecan nuts), detection of watercore disorder (apples) and characterization of 'Braeburn' browning disorder (apples), quantification, and characterization of internal structure (pomegranate) and determining maturity (tomatoes) are some other additional applications. These applications are mainly based on the variations of X-ray transmission according to the mass density and the mass absorption coefficient of the tissues under study.

9.5.2 NUTRITIONAL QUALITY

Fruits and vegetables are considered as protective foods since they are relatively low in calories and fat (avocado and olives being the exceptions), but rich in carbohydrates, fiber, vitamins, and minerals. Several postharvest handling techniques were deployed for maintaining their quality and shelf life which fail in one or the other aspect for curbing the desired problems. Traditionally, the quality of fruits and vegetables are measured through destructive means of sampling resulting in the creation of manual errors, inaccuracy, and more time-consuming. Thus, there is an urgent need for finding alternate solutions for solving this problem. X-ray imaging or X-ray CT could be used as an alternative for evaluating the quality parameters of fresh fruits and vegetables. From a consumer point of view, the fruit quality is measured mostly in terms of their size, shape, color, flavor, firmness, texture, taste, and freedom from defects/foreign materials. Numerous advanced techniques are commercially available for evaluating the above said external quality factors. However, the internal quality factors, such as maturity, sugar, acidity, oil content, internal defect, and tissue breakdown, are

TABLE 9.1 Application of X-Ray Imaging Technology for Non-Destructive Evaluation of Horticultural Commodities

Particular	Crop	Target Problem	Specification	Results	References
Physiological Disorder	Nectarines	Wooliness	X-ray micro-computed tomography	Detected darker areas which exhibiting wooliness	Sonego et al., 1995
	Apple	Braeburn Browning Disorder	X-ray micro-computed tomography	Detects the healthy fruit tissue and disordered with a success rate of 97%.	Herremans et al., 2013
		Bruised apples	X-rays	Used to detect bruised apples	Diener, Mitchell, and Rhoten, 1970
	Pears Cv. Conference	Core breakdown	X-ray CT	Detection of browning and healthy fruits	Lammertyn et al., 2003
	Potato	Hollow heart	X-rays	Detection of hallow heart in potato tubers	Finney and Norris, 1978
	Mango	Jelly Seed	X-rays	Detection of jelly seed and pulp	Sharma and Krishna, 2017
	Apple	Bruised apples	X-ray system containing X-ray-sensitive camera tube and a video amplifier chain	Sorting out the bruised apple fruit	Diener, Mitchell, and Rhoten, 1970
Nutritional Quality	Pomegranate	Processing quality	MicrofocusX-ray micro-computed tomography	Measurement of volume of juice and total aril volume	Arendse et al., 2017

TABLE 9.1 *(Continued)*

Particular	Crop	Target Problem	Specification	Results	References
Nutritional Quality	Mangoes	Physic-chemical characteristics	X-ray (150 keV and 3 mA)	Detects the moisture content, soluble solids, titrable acidity and pH of mangoes	Barcelon, Tojo, and Watanabe, 1999a
	Apple cv. Red delicious	Moisture content	X-ray computed tomography	Used for measuring the moisture content in apple fruit	Tollner et al., 2005
	Tomato	Maturity Judging	X-ray CT)	Detecting the maturity stage accurately with 99% precision	Brecht et al., 1991
	Mango	Nutritional composition	X-ray computed tomography	Non-destructive analysis of moisture content, soluble solids, titratable acidity, and pH in mangoes	Barcelon, Tojo, and Watanabe, 1999a
	Mango	Ripening	X-ray computed tomography	Detecting the ripening process in mango	Barcelon, Tojo, and Watanabe, 2000
	Apple	Internal quality	X-ray computed tomography	Studied the moisture content, soluble solids, titratable acidity, and pH of apple	Barcelon, Tojo, and Watanabe, 1999b
	Sweet Potato	Sweet potato weevil	X-ray CT	Detection of weevil larvae in sweet potato tuber	Thai et al., 1997
Nutritional Quality	Nuts	Pecan Weevil	X-ray CT	Used for monitoring and understanding the behavior of weevil	Kotwaliwale et al., 2007

TABLE 9.1 (Continued)

Particular	Crop	Target Problem	Specification	Results	References
Morphological Characterization	Cucumber	Internal structure	X-ray CT	It emerged as non-destructive technique for characterization and quantification of the internal structure of cucumber fruit during storage	Tanaka et al., 2018
	Apple	Bruise injury	X-ray CT	X-ray CT successfully employed for detection of internal bruise injury in apple fruits	Diels et al., 2017
	Persimmon	Internal quality	X-ray CT	The porosity, thermal conductivity and moisture distributions were visualized	Tanaka et al., 2018
	Pomegranate	Peel Thickness	X-ray CT	The size and distribution of pores in the peel of pomegranate fruit has an important influence on the moisture and gas transport phenomena that occur in the fruit tissue during storage	Ambaw et al., 2017
	Pear	Internal structure	Micro- and Nano-computed tomography (CT)	X-ray CT is useful for characterizing and quantifying microstructural differences related to mealiness in seeded and parthenocarpic 'Forelle' pears	Schoeman et al., 2018

more difficult to evaluate without destroying the fruit. Thus, there is a great need for developing methods that could better predict the internal quality of fruits and vegetables. Barcelon, Tojo, and Watanabe (1999a) studied the use of X-ray CT for internal quality evaluation of mango fruits and suggested its use for quantifying the moisture content, soluble solids, titratable acidity, and pH in mangoes. Similarly, he investigated the application of x-ray imaging for understanding the ripening process in mangoes.

Arendse et al. (2017) investigated the physicochemical properties of pomegranate fruit using microfocus X-ray micro CT and found a non-destructive (ND) method for estimating the total aril volume and juice content from intact pomegranate fruit cv. 'Wonderful.' Kim and Schatzki (2000) developed an algorithm for detecting watercore in apples using X-ray imaging. It is very difficult to detect watercore in apples at the early stages based on visual inspection, because watercore does not affect the external appearance of the apple until it is very severe. Ogawa, Kondo, and Shibusawa (2003) studied the internal quality evaluation of apples, pears, and peaches using X-ray imaging. The soft X-ray generator used for the study was XT60–60 and the selected voltage and current was 60 keV and 3 mA, respectively. X-ray images were captured and analyzed and it was possible to detect split pits of apples and peaches. Pineapple has a common internal physiological disorder called translucency results in a water-soaked appearance of the flesh of the pineapple. The size and skin toughness of pineapple poses a challenge for the use of near-infrared (NIR) spectroscopy, light transmission, or acoustic methods. Hence, X-ray imaging was explored by Haff et al. (2006) to determine translucency in pineapple and found to give promising results with 86–95% accuracy. Deshpande, Pillai, and Cheeran (2010) developed an X-ray imaging technique to determine the spongy tissue disorder in Alphonso mangoes. Tollner et al. (2005) conducted a series of tests at the University of Georgia and Vidalia Onion Research Center to detect the quality of onion using X-ray machines. In 2001, two 100-onion batches were inspected using X-ray machines and visually inspected for internal quality. The accuracy of the test was greater than 93% and false positives were less than 6%.

9.5.3 MORPHOLOGICAL AND STRUCTURAL CHARACTERIZATION OF FRUITS AND VEGETABLES

The evaluation of internal quality of fresh produce is extremely important for the agricultural industry and consumers, especially for those meant for the export market. There are numerous ND imaging methods developed in the recent years for the morphological and structural characterization of horticultural produce. X-ray CT is one such useful technique meant for ND mapping of radiodensity distribution, as it provides high-quality images of the internal structure of horticultural produce. This is possible because of the differences in attenuation of X-ray on impacting different materials creating a contrast to differentiate low- and high-density materials (Karathanasis and Hajek, 1996). Arendse et al. (2017) measured the geometrical characters (volume and size) of pomegranate fruit non-destructively. In addition, Kelkar, Boushey, and Okos (2015) proposed a method to determine the density of foods using X-ray linear attenuation. Moreover, Jha et al. (2010) reported the potential utility of X-ray CT to characterize the internal and external properties of mango. DonisGonzález et al. (2014) investigated the internal decay of chestnuts, internal defects in pickling cucumbers, translucency disorder in pineapple, pit presence in tart cherries, and plum curculio infestation of tart cherries. They suggested that there is a great potential for ND inline sorting of various several horticultural commodities based on their internal quality. Donis-González et al. (2014) developed a classification method based on CT images for fresh chestnuts. Donis-González et al. (2014) also investigated the presence of undesirable fibrous tissue in carrots through CT images. Using high resolution CT, the pore space within apple (Mendoza et al., 2007), core breakdown in pears (Lammertyn et al., 2003), microstructure in kiwifruit (Cantre et al., 2014) and bread (Demirkesen et al., 2014) were also investigated. Tanaka et al. (2018) investigated the internal changes occurring in the cucumber fruit at 15°C and 25°C with 90% RH for 7 Days. The radiodensity of cucumber fruit changed in the endocarp and mesocarp tissues, but it did not change in the placenta tissue during 7 days of storage at 25°C. It is concluded that X-ray CT can be used to estimate the physical properties of cucumber fruit without destructive sampling. Similarly, Diels et al. (2017) also investigated the impact of artificial pendulum injury on measuring its bruise damage injury by using x-ray CT.

9.5.4   INSECT PESTS DETECTION

The presence of arthropod pests in or on horticultural commodities has caused major disruptions in the storage, processing, and shipment of these products. Management of these pests has posed a problem to humans for thousands of years. The fresh produce industry must begin to place increased emphasis on other methods of pest management. Integrated postharvest management (IPM) is a systematic approach towards pest management where good management practices are combined with one or more contributory treatments to result in an acceptable level of disease and insect control. There are several important postharvest insect pests such as mango fruit fly, stone weevil, anar butterfly, codling moth of apple, sweet potato weevil, etc. that cause disruption of export market demand. However, several recent advances have emerged out for the detection and destruction of these postharvest insect pests such as chemical fumigation, vapor heat treatment, irradiation, cold plasma technique, etc. Recently, Thomas et al. (1995) investigated the application of x-ray imaging technique in detection of mango stone weevil where they collected mango fruits (healthy and infested) and scanned them with digital x-ray machine (Voltage = 54 kV; Current = 100 mA; Exposure Time = 4 ms) for getting a clear visible picture of presence/absence of weevil in them.

Keagy, Parvin, and Schatzki (1996) studied the possibility of X-ray image analysis as a ND method for the determination of Navel orange worm. Pistachio nuts were X-rayed (25 keV) with a Faxitron series X-ray system. They concluded that it is possible to determine insect-infested nuts but 100% accuracy is not possible. Pecan nuts are infested by pecan weevil (*Curculio caryae* Horn) larvae. The nuts damaged by larvae could be easily detected by low specific gravity if the larvae have consumed the nutmeat and exited the nut. But, it is difficult to determine if the larvae are present inside the nut. Kotwaliwale et al. (2007) explored the potential of X-ray imaging to determine the pecan nut quality. By applying contrast stretching, features of pecan nuts such as shell, air gap, nutmeat, defects, and presence of insects were clearly seen in X-ray images. Their study showed that defects and insects could be determined in pecan nuts by X-ray imaging.

Chestnut has a hard reddish-brown shell that makes it very difficult to inspect using conventional techniques. Lü, Cai, and Wang (2010) studied the potential of X-ray imaging for ND monitoring of chestnuts that are insect-damaged, air-dried, or decayed. The irregular shape and size of chestnuts and the presence of many furrows on the surface of nutmeat made it difficult for segmentation to be performed. Hence to accurately evaluate the quality, a dynamic threshold segmentation algorithm was developed. The results showed that for healthy nuts, the classification accuracy was 87% with 13% false positives. For diseased nuts, classification accuracy was 93.5% with 6.5% false positives. The authors concluded that X-ray imaging has great potential to monitor the quality of chestnuts, and it took only 27 ms to inspect one nut, which is fast enough to be implemented on a real-time inspection unit.

9.6 ADVANTAGES OF X-RAY IMAGING SYSTEM

- X-ray is a ND method to determine the internal quality characteristics of any product.
- Increased productivity can be obtained because X-ray units can be installed along the production lines to inspect different products, thereby increasing the efficiency and productivity of the entire process line.
- X-ray monitoring is ideal for high-speed monitoring and can handle 1500 products per minute or up to 10 tonnes per hour for bulk applications depending on the product.
- X-ray inspection of food products helps manufacturers and food processors to comply with global food safety standards such as hazard analysis critical control points (HACCP), the Global Food Safety Initiative (GFSI), and good manufacturing practices (GMP).
- X-ray is more convenient and relatively cheap compared to X-ray CT or MRI imaging equipment.
- X-ray imaging does not require any sample preparation or very minimal preparation is needed.
- X-ray imaging has become very affordable, accurate, and easier to install and use in the industrial setup due to the technological advancement in the areas of high-voltage power supplies, solid-state detectors, and computer speed and processing.

9.7 LIMITATIONS OF X-RAY IMAGING SYSTEM

- One disadvantage of X-ray imaging is that it requires a high-voltage power supply to produce X-rays.
- X-ray imaging applications require protective shielding and employees need to keep track of the radiation levels they are exposed to by use of a dosimeter.
- The detection of split pits and normal pits in peaches using X-ray imaging was evaluated by Han, Bowers, and Dodd (1992) with 98% accuracy, but in order to achieve this, fruits need to be placed in a specific orientation such that X-rays could penetrate from top to bottom or front to back during imaging. Hence, in a real-time situation, it becomes complicated to maintain a specific orientation of fruit to monitor the quality characteristics.

- X-ray imaging to detect watercore in apples explored by Kim and Schatzki (2000) showed that system could categorize sound and watercore-affected apples less than 5% to 8% false positive and negative ratio, independent of apple orientation. But the limitation was that the algorithm would work only when the stem-calyx axis makes a fixed angle with the X-ray beam.
- It was possible to detect glass contaminants in peat but they could not distinguish between stone and glass (Ayalew et al., 2004). X-ray imaging has certain limitations that hinder many applications to move forward from the research phase to the commercial phase. When solutions for these limitations are developed, X-ray imaging would be available for many commercial applications in the agriculture and food industry.

9.8　FUTURE APPLICATIONS

With the advancement of the technologies and globalization, the applications of X-ray imaging in the medical, engineering, and biological industries are increasing exponentially. Many food processing industries have resented to install inline/online X-ray imaging systems as a quality monitoring tool for attaining maximum product safety and to avoid costly product recalls and issues related to health claims. The design and implementation of X-ray installation in any food industry should be a strategic decision to be able to effectively implement the process at the critical control points (CCP) in order to achieve maximum benefits. The implementation of X-ray imaging should be proactive rather than reactive, that is, the technique needs to be used to prevent the occurrence of contamination rather than just detecting and eliminating the defects. Thus, the X-ray imaging system was found to play a crucial role in the HACCP of food industries. High-speed industrial X-ray units and efficient image processing algorithms have resulted in X-ray imaging available for various applications in a wide range of industries. In the future, food, and agriculture industries will continue to invest in X-ray imaging techniques to meet the demands of the food regulatory authorities and consumers on food safety initiatives. The performance of the X-ray imaging system needs to be assessed based on real-time conditions to make it a feasible technology in the near future.

KEYWORDS

- **charge-coupled device**
- **computed tomography**
- **critical control points**
- **Global Food Safety Initiative**
- **internal disorders**
- **x-ray imaging**

REFERENCES

Ambaw, A., Arendse, E., Du Plessis, A., & Opara, U. L., (2017). Analysis of the 3D microstructure of pomegranate peel tissue using x-ray micro-CT. In: *VII International Conference on Managing Quality in Chains (MQUIC2017) and II International Symposium on Ornamentals in 1201*, pp. 197–204.

Arendse, E., Fawole, O. A., Magwaza, L. S., & Opara, U. L., (2017). Non-destructive estimation of pomegranate juice content of intact fruit using x-ray computed tomography. In: *VII International Conference on Managing Quality in Chains (MQUIC2017) and II International Symposium on Ornamentals* (pp. 297–302).

Ayalew, G., Holden, N. M., Grace, P. M., & Ward, S. M., (2004). Detection of glass contamination in horticultural peat with dual-energy x-ray absorptiometry. *Computers and Electronics in Agriculture, 42*, 1–17.

Barcelon, E. G., Tojo, S., & Watanabe, K., (1999a). Relating x-ray absorption and some quality characteristics of mango fruit (*Mangifera indica* L.). *Journal of Agricultural and Food Chemistry, 47*, 3822–3825.

Barcelon, E. G., Tojo, S., & Watanabe, K., (1999b). X-ray computed tomography for internal quality evaluation of peaches. *Journal of Agricultural Engineering Research, 73*, 323–330.

Barcelon, E. G., Tojo, S., & Watanabe, K., (2000). Nondestructive ripening assessment of mango using an x-ray computed tomography. *International Agricultural Engineering Journal, 9*(2), 73–80.

Brecht, J. K., Shewfelt, R. L., Garner, J. C., & Tollner, E. W., (1991). Using x-ray-computed tomography to nondestructively determine maturity of green tomatoes. *Hort. Science, 26*, 45–47.

Butz, P., Hofmann, C., & Tauscher, B., (2005). Recent developments in noninvasive techniques for fresh fruit and vegetable internal quality analysis. *Journal of Food Science, 70*, 131–141.

Caballero, D., Pérez-Palacios, T., Caro, A., Amigo, J. M., Dahl, A. B., Ersboll, B. K., & Antequera, T., (2017). Prediction of pork quality parameters by applying fractals and data mining on MRI. *Food Research International, 99*, 739–747.

Cantre, D., East, A., Verboven, P., Araya, X. T., Herremans, E., Nicolaï, B. M., & Heyes, J., (2014). Micro structural characterization of commercial kiwifruit cultivars using x-ray micro-computed tomography. *Postharvest Biology and Technology, 92*, 79–86.

Curry, T. S., Dowdey, J. E., & Murry, R. C., (1990). *Christensen's Physics of Diagnostic Radiology* (p. 4). Baltimore: Williams and Wilkin.

Demirkesen, I., Kelkar, S., Campanella, O. H., Sumnu, G., Sahin, S., & Okos, M., (2014). Characterization of structure of gluten-free breads by using x-ray microtomography. *Food Hydrocolloids, 36,* 37–44.

Deshpande, A. S., Pillai, M. P., & Cheeran, A. N., (2010). X-ray imaging for non-destructive detection of spongy tissue in *Alphonso* mango. In: Kale, K. V., Mehrotra, S. C., & Manza, R. R., (eds.), *Computer Vision and Information Technology: Advances and Applications* (pp. 252–256). New Delhi, India: I.K. International Publishing House Pvt. Ltd.

Diels, E., Van, D. M., Keresztes, J., Vanmaercke, S., Verboven, P., Nicolai, B., & Smeets, B., (2017). Assessment of bruise volumes in apples using x-ray computed tomography. *Postharvest Biology and Technology, 128,* 24–32.

Diener, R. G., Mitchell, J. P., & Rhoten, M., (1970). Using an x-ray image scan to sort bruised apples. *Agricultural Engineering, 51,* 356–361.

Donis-González, I. R., Guyer, D. E., Fulbright, D. W., & Pease, A., (2014). Postharvest noninvasive assessment of fresh chestnut (*Castanea* spp.) internal decay using computer tomography images. *Postharvest Biology and Technology, 94,* 14–25.

Du, C. J., & Sun, D. W., (2006). Learning techniques used in computer vision for food quality evaluation: A review. *Journal of Food Engineering, 72,* 39–55.

Finney, E. E., & Norris, K. H., (1978). X-rays scans for detecting hollow heart in potatoes. *American Potato Journal, 55,* 95–105.

Haff, R. P., & Slaughter, D. C., (2004). Areal time X-ray assessment of wheat for infestation by the granary weevil, *Sitophilus granarius* (L.). *Transactions of ASAE, 47,* 531–537.

Haff, R. P., Slaughter, D. C., Sarig, Y., & Kader, A., (2006). X-ray assessment of translucency in pineapple. *Journal of Food Processing and Preservation, 30,* 527–533.

Han, Y. J., Bowers, S. V., & Dodd, R. B., (1992). Non-destructive detection of split-pit peaches. *Transactions of ASAE, 35,* 2063–2067.

Herremans, E., Verboven, P., Bongaers, E., Estrade, P., Verlinden, B. E., Wevers, M., & Nicolai, B. M., (2013). Characterization of '*Braeburn*' browning disorder by means of x-ray micro-CT. *Postharvest Biology and Technology, 75,* 114–124.

Jha, S. N., Narsaiah, K., Sharma, A. D., Singh, M., Bansal, S., & Kumar, R., (2010). Quality parameters of mango and potential of non-destructive techniques for their measurement: A review. *Journal of Food Science and Technology, 47,* 1–14.

Karathanasis, A. D., & Hajek, B. F., (1996). Chapter 7. In: Bigham, J. M., (ed.), *Methods of Soil Analysis, Part-3-Chemical Methods, Soil Sci. Soc. Amer. Book Ser. 5. The Soil Sci. Soc. and Amer. Soc. Agronomy* (pp. 163–263). Madison, WI.

Keagy, P. M., Parvin, B., & Schatzki, T. F., (1996). Machine recognition of navel orange worm damage in x-ray images of pistachio nuts. *LWT-Food Science and Technology, 29,* 140–145.

Kelkar, S., Boushey, C. J., & Okos, M., (2015). A method to determine the density of foods using x-ray imaging. *Journal of Food Engineering, 159,* 36–41.

Kim, S., & Schatzki, T. F., (2000). Apple watercore sorting system using x-ray imagery: I. Algorithm development. *Transactions of ASAE, 43,* 1695–1702.

Kotwaliwale, N., Singh, K., Kalne, A., Jha, S. N., Seth, N., & Kar, A., (2014). X-ray imaging methods for internal quality evaluation of agricultural produce. *Journal of Food Science and Technology, 51,* 1–15.

Kotwaliwale, N., Weckler, P. R., Brusewitz, G. H., Kranzler, G. A., & Maness, N. O., (2007). Non-destructive quality determination of pecans using soft x-rays. *Postharvest Biology and Technology, 45*, 372–380.

Lammertyn, J., Dresselaers, T., Van, H. P., Jancsók, P., Wevers, M., & Nicolaı̈, B. M., (2003). Analysis of the time course of core breakdown in 'conference' pears by means of MRI and x-ray CT. *Postharvest Biology and Technology, 29*, 19–28.

Lü, Q., & Cai, J. Y., & Wang, F., (2010). Real time non-destructive inspection of chestnuts using X-ray imaging and dynamic threshold. In: *World Automation Congress (WAC)* (pp. 365–368).

Ma, J., Sun, D. W., Qu, J. H., Liu, D., Pu, H., Gao, W. H., & Zeng, X. A., (2016). Applications of computer vision for assessing quality of agri-food products: A review of recent research advances. *Critical Reviews in Food Science and Nutrition, 56*, 113–127.

Mathanker, S. K., Weckler, P. R., & Bowser, T. J., (2013). X-ray applications in food and agriculture: A review. *Transactions of the ASABE, 56*, 1227–1239.

Mendoza, F., Verboven, P., Mebatsion, H. K., Kerckhofs, G., Wevers, M., & Nicolaï, B., (2007). Three dimensional pore space quantification of apple tissue suing x-ray computed microtomography. *Planta, 226*, 559–570.

Mettler, T., (2009). Safe line x-ray inspection guide. *Chapter 3: Choosing the Right System*, 17–18.

Neethirajan, S., Jayas, D. S., & White, N. D. G., (2007). Detection of sprouted wheat kernels using soft x-ray image analysis. *Journal of Food Engineering, 81*, 509–513.

Ogawa, Y., Kondo, N., & Shibusawa, S., (2003). Inside quality evaluation of fruit by x-ray image. In: *Proceedings of the IEEE/ASME International Conference on Advanced Intelligence Mechatronics* (pp. 20–24) Kobe, Japan.

Peariso, D., (2008). X-ray examination of foods for foreign materials. In: *Preventing Foreign Material Contamination of Foods*. Hoboken, NJ: Wiley-Blackwell.

Schoeman, L., Cronje, R. J., Fourie, W. J., Theron, K. I., Steyn, W. J., Human, T., & Crouch, E. M., (2018). Influence of pollination on seed content and mealiness development in 'Forelle' pear as detected with x-ray micro- and nano-computed tomography (CT). *Acta Horticulture, 1201*, 371–378.

Sharma, R. R., & Krishna, K. R., (2017). *Non-Destructive Evaluation of Jelly seed Disorder in Mango*. https://www.biotecharticles.com/Applications-Article/Non-destructive-Evaluation-of-Jelly-Seed-Disorder-in-Mango-PDF-4061.html (accessed on 23 May 2020).

Shewfelt, R., Brueckner, B., & Florkowski, S. E. P. J., (2012). *Postharvest Handling: A Systems Approach*. Academic Press,

Sonego, L., Ben-Arie, R., Raynal, J., & Pech, J. C., (1995). Biochemical and physical evaluation of textural characteristics of nectarines exhibiting woolly breakdown: NMR imaging, x-ray computed tomography and pectin composition. *Postharvest Biology and Technology, 5*, 187–198.

Sun, D. W., (2016). *Computer Vision Technology for Food Quality Evaluation* (2nd edn.). San Diego, California, USA: Academic Press/Elsevier.

Tanaka, F., Imamura, K., Tanaka, F., & Uchino, T., (2018). Determination of thermal diffusivity of persimmon flesh tissue using three-dimensional structure model based on x-ray computed tomography. *Journal of Food Engineering, 221*, 151–157.

Tanaka, F., Nashiro, K., Obatake, W., Tanaka, F., & Uchino, T., (2018). Observation and analysis of internal structure of cucumber fruit during storage using x-ray computed tomography. *Engineering in Agriculture, Environment, and Food, 11*, 51–56.

Thai, C. N., Tollner, E. W., Morita, K., & Kays, S. J., (1997). X-ray characterization of sweet potato weevil larvae development and subsequent damage in infested roots. *Proceedings Sensors for Nondestructive Testing: Measuring Quality of Fresh Fruits and Vegetables-Florida*, pp. 361–368.

Thomas, A. D. H., Rodd, M. G., Holt, J. D., & Neill, C. J., (1995). Real-time industrial visual inspection: A review. *Real-Time Imaging, 1*, 139–158.

Tollner, E. W., Gitaitis, R. D., Seebold, K. W., & Maw, B. W., (2005). Experiences with a food product x-ray inspection system for classifying onions. *Applied Engineering in Agriculture, 21*, 907–912.

Van, D. M., Verboven, P., Dhaene, J., Van, H. L., Sijbers, J., & Nicolai, B., (2017). Multisensor X-ray inspection of internal defects in horticultural products. *Postharvest Biology and Technology, 128*, 33–43.

Wang, J., & Blackburn, T. J., (2000). X-ray image intensifiers for fluoroscopy. *Radio Graphics, 20*, 1471–1477.

Xradia (2010). X-ray detectors. https://www.zeiss.com/microscopy/int/products/x-ray-microscopy.html (accessed 30 May 2020).

Zink, F. E., (1997). X-ray tubes. *Radio Graphics, 17*, 1259–1268.

CHAPTER 10

RADIO FREQUENCY IDENTIFICATION (RFID)

H. A. MAKROO, S. BASHIR, A. JABEEN, and D. MAJID

Department of Food Technology, IUST Kashmir – 192122, India

ABSTRACT

This chapter includes a brief introduction of radio frequency identification (RFID) systems and their working principle followed by their types and various applications. The RFID system is a process of transforming a unique and predefined protocol from a devise to a reader through radio frequency (RF) waves. It is one of the promising monitoring features for quality assessment of fruits during storage and transportation. The RFID systems are composed of three different parts viz RFID tag, reader and data processing sub-system. It can provide thorough information of the produce only by placing its tags on the containers, boxes, and embedded into any object such packet of the product. Based on the power source of the Tag the RFID system can be either active tag or passive tag RFID system. Whereas based on the frequency of the radio waves the system can be categories as near-field or far-field RFID. The RFID system can be used in many ways, supply chain management and product traceability are the major areas where RFID is being used.

10.1 INTRODUCTION

With the advent of globalization, the world has changed into a global village and therefore, transportation of commodities has become exponentially important affecting the people's lives day by day. Amongst all other commodities, the transportation of fresh produce (fruits and vegetables) is a very important branch of logistics. Around lakhs of refrigerators and containers for carrying out fresh fruits are registered throughout the world (Jedermann et al., 2006). However, the actual status of the fresh fruits and vegetable while transporting is not known and the loss of the value occurs regularly while transporting the produce. China, the largest producer of fruits in the world faces a serious economic and quality degradation during the transportation course. A loss of up to 35% has been reported just during the transportation time, i.e., the time required to take the produce from the producer to the consumer or retailer.

Moreover, there is an apprehension of foodborne infections as the fresh produce is more prone to the microbial growth and metabolic activities leading to foodborne infections. In addition to safety issues, earlier studies have shown that fruits stored in ambient temperature suffer from physical, chemical, and decrease in their sensory characteristics. For the effective fruit supply chain management (FSCM), it is necessary to adopt efficient cold chain with proper monitoring features from farm to the fork in order to keep a check on wastage (Parfitt et al., 2010; Kummu et al., 2012).

The RFID system is a process of transforming a unique and predefined protocol from a devise to a reader through radio frequency (RF) waves (Borriello, 2005). RFID is one of the promising monitoring features for quality assessment of fruits during storage and transportation. It is the more convenient method of product identification as it does not requires visual contact. It can provide thorough information of the produce only by placing RFID tags on the containers, boxes, and embedded into any object such as passport (Finkenzeller, 2010). Although the cost of RFID system is, high but still they are popularly used for different purposes in various areas (Kumari et al., 2015; Realini and Marcos, 2014). Their domains of applications are expanding rapidly ranging from logistics, identification to pallet tracking. The progress of RFID system is presented in Table 10.1.

10.2   WORKING PRINCIPLE

Radio waves (RF) are that the portion of the electromagnetic spectrum having a frequency in between 3 to 3×10^5 kHz. The electromagnetic spectrum is divided into different sections or types of radiations having different frequencies and wavelengths. The two extremes of the spectrum, on one side, are electric waves having an extremely low frequency and very long wavelength whereas on the other extreme are gamma rays. The gamma rays have extremely high frequency and very short wavelengths.

TABLE 10.1   Decades of Radio Frequency Identification

1940–1950	1950–1960	1960–1970
• Radar refined and used major World War II development effort. • RFID invented in 1948.	• Early explorations of RFID technology, laboratory experiments.	• Development of the theory of RFID. • Start of application field trials.

1970–1980	1980–1990	1990–2000
• Explosion of RFID development. • Tests of RFID accelerate. • Very early adopter implementations of RFID.	• Commercial applications of RFID enter mainstream.	• Emergence of standards. • RFID widely deployed. • RFID become s a part of everyday life

Source: Roberts (2006).

The huge markets for food safety testing and pathogen detecting sensors offer great opportunities for making novel sensing techniques applicable in food quality regulations. The automatic identification system is one of the most important and heavily demanded application areas of the sensor technology. With the passage of time, the improvisation has brought many novelties in the automatic identification system. Such as Barcodes, Card technologies, (magnetic, smart, or optical) and radio frequency identification (RFID) systems (Harvey Lehpamer, 2008; Jones et al., 2005). The method of labeling food packages that can indicate about the freshness and or other information of the product inside the package is among the important possible applications of the RFID sensing systems (Potyrailo et al., 2012).

In recent times, the RFID has been increasingly utilized in supply chain management (SCM) and logistics. As RFID permits non contact reading of information on products, places, time, and transactions, thus it provides the manufacturers and retailer appropriate and precise information about the stock of the product. SCM is one of the major areas where RFID is being used (Wang, Kwok, and Ip, 2010). The RFID systems have different parts with their specific role in the overall functioning; the major parts are listed in Table 10.2.

TABLE 10.2   Three Major Parts of RFID System

Part of RFID System	Specific Role
RFID tag or transponder	It is located on the object to be identified and is the data kept in the RFID system
RFID reader or transceiver	It may be able to both read data from and write data to a transponder
Data processing sub-system	It utilizes the data obtained from the transceiver in some useful manner

RFID tag or transponder as well as the reader both have an antenna each. As depicted in Figure 10.1, the reader sends out RF signals that form a magnetic field as these waves get coupled with the tag antenna. In reply the tags share the stored information, in particular, the unique identification or serial number, however, some additional information available in the memory of the tag may be also transmitted (Kumar et al., 2009; Weis et al., 2004).

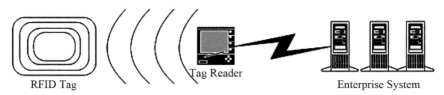

FIGURE 10.1   Schematic communication in RFID system. (Reprinted with permission from Roberts, 2006. © Elsevier.)

Earlier the tags used active transmission however, due to less cost and without requiring any separate power supply (battery) nowadays passive transmission (tags) are very common in practice. The power required for the functioning of passive tags is received from the energy in the reader field, hence they communicate by backscattering modulation with the reader (Zhen-Hua, Jin-Tao, and Bo, 2007). A passive tag can be powered by two ways, either near-field (NF) or far-field coupling. The near and far field (FF) coupling works on the base of magnetic field induction or EM waves capture respectively, by both ways the passive tag receives sufficient power to get operational (Kumar et al., 2009). Although in most RFID application systems the transfer of information is unidirectional, i.e., the information flows from the tag and received by the reader; however, all the RFID system protocols involve both ways communication in between the tag and the reader (Zhen-Hua et al., 2007).

RFID system is similar to Wi-Fi and Bluetooth wireless technologies. However, all these three technologies are different in functionality thus, are used for different purposes. The three have shared ground in between with some hybrid starting to appear. Wi-Fi and Bluetooth both can be used in RFID and are not competitors to each other (Harvey, 2008).

When compared with optical bar-code systems the RFID tags have various major advantages, such as automatic data reading, via non-conducting materials like paper or cardboard, without line of sight, very high rate of tag reading (few hundred every seconds) and range of coverage up to several meters stability against dirt or moisture (Wang et al., 2010).

As the tags are silicon-based microchip, thus its functionality can be enhanced beyond just storing the identification details or serial number. It can be used as an integrated sensor, to read or write storage details, for facilitating encryption and control of access (Weis et al., 2004). Therefore, depending up on the functionality the tags are of various designs, few examples are depicted in Figure 10.2.

FIGURE 10.2 A passive RFID tag, an RFID tag with a printed barcode, and dust-sized RFID microchips. (Reprinted with permission from Weis et al., 2004. © Springer Nature.)

The microchip of the tag receives the power for operation from the magnetic field. The microchip of the tag then modulates the waves received and sends back to the reader; consequently, the reader converts them into the digital data (Jones et al., 2004). Due to the wide application of RFID systems, the tags may be of various types of sizes and shapes sometimes bare or embedded in other material such as epoxy resin or glass, etc., or protective covering. Building access cards are the examples of tags which are encapsulated in credit card-sized packages (Ron, 2005). Some tags are injectable having approximately 1 cm long and 0.1 cm in diameter and generally used in animal tracking (injected beneath the animal skin). Whereas some are designed to be used in harsh environmental conditions, such as tags used for container tracking and can measure 12×10×5 cm. The smallest devices commercially available are having dimensions less than half of a millimeter or even the thickness less than that of a paper (Roberts, 2006). Some examples are shown in Figure 10.3 and mentioned in Table 10.3.

10.3 TYPES OF RFID SYSTEM

As RFID systems have wide applications and uses, thus they are categorized in many categories as shown in the flowsheet (Figure 10.4) (Sarma, Brock, and Engels, 2001).

10.3.1 ELECTRICAL POWER SOURCE OF TAGS

1. Active tag RFID system; and
2. Passive tag RFID system.

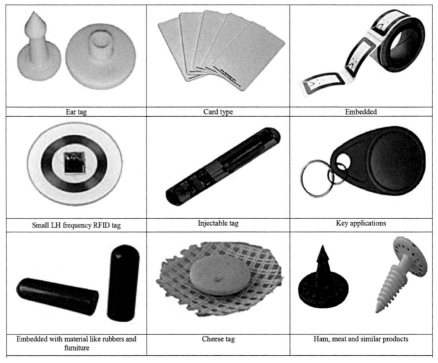

FIGURE 10.3 Examples of different kinds of RFID tags and the area of application. (Reprinted with permission from Costa et al., 2013. © Springer Nature.)

TABLE 10.3 Various Types of RFID Tags and Their Suitable Application

Tag Type	Specific Application
Ear tag	• Suitable for management and scientific testing applications in live stock such as cattle, sheep, etc.
Card	• Applicable in controlled access with data memory
Embedded	• Embedded in usual barcode for extra security • Reading in absence • Sorting operations
Small LH frequency clear RFID tag	• For identification and tracking of animals, used in the collar of the animals.
Injectable tag	• Can be injected
Key applications	• Used for the key protection and for the easy finding of keys if lost
Embedded with material like rubbers and furniture	• Used in furniture etc and it is capable of bear extreme environmental conditions
Cheese tag	• Used in Cheese during storage and ageing for proper monitoring

The active and passive tags are major types of tags in RFID system. Even though both have their own specific benefits, thus it is important to differentiate between them. Passive tags are the most commonly used and they are also known as ultra-high frequency (UHF) tags or as RAIN RFID. As discussed before the passive tags require a powered reader for signal transmission, they are best for tracking during the continuous flow of huge volumes of low-cost items. Such as tracking of pallets, bins, or returnable transport items by the food manufacturers.

On the other hand, the active or Wi-Fi-based tags unlike the passive tags are internally powered; hence utilize their own power for the signal transmission to the access points (wireless) (Ron, 2005). This gives real-time location information or tracking high value-value, high impact, and mobile assets. While active tags are more costly than passive tags, they have a much greater read range up to 300 feet (Tom, 2016). Table 10.4 summarizes some main differences between the active and passive tags.

Depending on the type of tag of the system, a reader may be either active or passive. There are also designs of semi-passive tags where the chip's circuitry is run by the battery runs however, for the communication the power is drawn from the reader (Wang et al., 2010). Based on the frequency RFID systems are categorized in two categories, one using magnetic induction where frequencies range from 0.1 to 30 MHz, while the other one operates using electromagnetic waves or microwaves having a frequency between 2.45 to 5.8 GHz (Kumar et al., 2009).

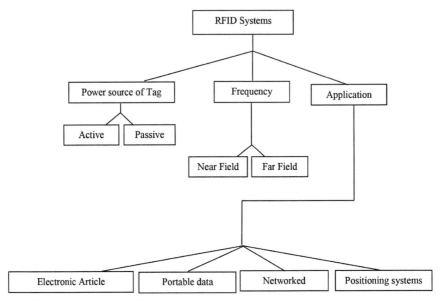

FIGURE 10.4   Types of RFID systems.

TABLE 10.4   Distinguish between Active and Passive Tag

Active Tags	Passive Tags
• Self-powered mostly by a battery on board. • The signal transmitted is generally stronger thus the reader has a higher range to trace the tag. • The size and cost are relatively higher due to the onboard battery, thus used mostly for bigger items tracked over large distances. • Due to the onboard power source (Battery) it has a limited life, however, using modern batteries it may be a long as 10 years. • Can constantly broadcast a signal or can remain dormant until they come in a range of a receiver. • Readers can trace active RFID tags across 20 to 100 meters. • Can read as well as write the data • Works through propagation coupling.	• Unlike Active tags they don't possess their own power source therefore get power from the signal the reader, thus require a higher-power reader. • Low in cost; and the cost is still reducing by the use of modern technologies to make them easily applicable for common materials and products. • Not only low in cost but passive tags can also be quite small to about the size of a quarter. • Too small memory (2 Kbits) to grasp any extra information other than identification and history. • The performance is decreased in electromagnetically "noisy" environments. • Can only read the data, and have a shorter read range • Works through inductive coupling

10.3.2 RFID SYSTEM APPLICATIONS

10.3.2.1 ELECTRONIC ARTICLE SURVEILLANCE (EAS)

Generally, EAS are 1-bit systems used to identify if the object is present or absent. EAS are mostly used as an antitheft tool in retail stores. Tags are fixed with item and unauthorized leave (without prior deactivation of tag) of the item from the store triggers an alarm to inform the staff in the store for the same. EAS has been used for the past many years in stores for a variety of items including clothing, electrical appliances and goods, books, etc. (Roberts, 2006).

10.3.2.2 PORTABLE DATA CAPTURE

These RFID systems are used for capturing data followed by downloading and processing captured data. They are used for portable devices and applicable when the data required from the tagged object varies. Some devices incorporate sensors such as temperature, movement (seismic) and radiation to record information.

10.3.2.3 NETWORKED SYSTEMS

This type of system is used to sense the movement of the tagged item or object while the reader is fixed at a position. This is a recent novelty of the identification technology, which is directly linked with an enterprise system.

10.3.2.4 POSITIONING SYSTEMS

The automatic location and navigation information can be collected by using this type of RFID system. These systems are used when the details about the navigation and location are collected for such as animals, automobiles, or even people.

10.3.3 FREQUENCY

10.3.3.1 NEAR-FIELD (NF) RFID

The NF coupling in between tag and the reader works on the Faraday's law of magnetic induction. A magnetic field is generated as a result of alternating current passed by a reader throughout a reading coil. If a small coil of tag gets incorporated in it, an alternating voltage will appear across. The tag chip can be powered with charge accumulated in a capacitor by rectifying the voltage generated. The NF coupling is the simplest method for the working of a passive RFID system. However, NF communication has few physical disadvantages, for example, the operational distance of NF coupling decreases as the operational frequency increases (Kaur et al., 2011).

10.3.3.2 FAR-FIELD (FF) RFID

The FF emission RFID tags capture electromagnetic waves transmitting from a dipole antenna fixed to the reader. This energy (electromagnetic waves) is captured as an alternating potential difference across the arms of the dipole by the tag antenna (small dipole). This potential is rectified by the dipole and links it to a capacitor, ultimately results in charge accumulation for the power of its electronics. The designers use the technology of backscattering for the commercial FF RFID tags. Generally, tags that work on FF principles operate at a frequency >100 MHz typically in the band of UH frequency (such as 2.45 GHz).

The amount of the energy received by the tag from the reader and how sensitive the reader's receiver is to the reflected signal controls the range of a FF RFID system. The attenuation of signals is based on the inverse square law, and the attenuation in RFID occurs twice, once when the waves travel for the reader to the tag and next on the return from the tag to the reader. Therefore, the actual signal returning is very small. As the technology has shrunk the size of semiconductors, therefore the energy needed to power a tag at a specific frequency is less (few microwatts). So, with modern semiconductors, it is possible to make tags that can be read at very larger distances as compared to those used a decade back. A typical FF reader can easily examine tags in the range of 3 m, however, few some RFID companies claim their products does have a read range of 6 m.

10.4   RFID APPLICATIONS

1. Supply chain management;
2. Traceability.

FSCM is the major area in which RFID is used successfully. The RFID systems have a larger read range as compared to that of the barcode system.

The RFID also have higher capacity and have the characteristics of many readings at the same time, additionally, the RFID systems are highly stable in moist conditions (Dolgui and Proth, 2008). The recent important development in the RFID systems is the combination TTI with RFID tags. The combination of the duo yields relatively better monitoring and overall better SCM. The RFID tags in combination with TTI's are capable to record temperature details at a constant span of time period hence assist in independent data recoding. This results in the development of more precise and convenient temperature tracking system than a traditional data logger. Although TTI RFID system contributes a lot to ensure a proper control of produce during transportation and storage but it is necessary to make sure that the RFID tags inside the packages, pellets or containers are easily readable.

Fruits and vegetable freshness is mainly influenced by the production of ethylene, therefore chemical sensors integrated with RFID tags are used for the monitoring within the package. A chemical sensor (such as Pt, Ag, Tin-oxide, Au-doped) included with the RFID reader can be used for onboard sensing of fresh fruits (Vergara et al., 2007). However, with the advancement of sensor development, a gas and vapor sensor has been developed without integration by Potyrailo et al. (2013). This sensor has a characteristic of self-correction, whenever uncontrolled temperature fluctuations occur. Researchers have been successful in developing small and cheap chemical sensors with modified optical potentiometric properties and are directly applied on RFID tags (Kassal et al., 2014). Bhadra et al. (2015) developed a wireless passive sensor for monitoring acidic and basic volatile concentration based on the pH electrode. Further smart tags with chemical sensors have been developed by various researchers based on potentiometric responses (Steinberg et al., 2014), photometric responses (Kassal et al., 2014).

1. **Humidity Sensors:** Their use has been found beneficial in the agri-food sector as relative humidity is an important factor having a major influence on the shelf life of fruits and vegetables. The humidity sensors integrated and interfaced with RFID system have been developed. Their main objective is to monitor the relative humidity and transmit the record to the user. Opera et al., (2007) made a capacitive humidity sensor using polyethylene-naphthalate and polyamide on flexible RFID labels having RH range of between 20 and 90%. Salmeron et al. (2014) developed a humidity sensor based on polyamide substrate to be coupled with ultra-high frequency (UHF) RFID tags, the twisted electrodes on polyamide substrate were used to give a sensitivity of 100% RH. Abad et al. (2009) linked humidity sensor

with the chip of a smart tag for cold chain monitoring of foods. The use of RFID smart tags in SCM facilitates the measurement or monitor humidity and temperature without tempering the package hence, preserving the integrity of the package.

2. **Gas Sensors:** They furnish the information about the change in the concentration of gasses without destructing the package. Martinez-Olmos et al. (2013) used an oxygen coupled sensor coupled to RFID tag for monitoring the color transformation of the luminophore. Jang et al., (2014) developed an oxygen indicator with a long life span and evaluated that how well oxygen scavengers/absorbers did remove all oxygen within a package. Espinosa et al. (2010) developed RFID tag coated with a metal oxide-based gas-sensitive coating for monitoring climacteric fruits during the transport of vending. Hong and Park (2000) developed a system for monitoring the degree of fermentation during storage and transportation. The system was based on the principle of the color change of a polymeric film on the absorption of carbon dioxide produced during the fermentation.

KEYWORDS

- electronic article surveillance
- fruit supply chain management
- radio frequency
- radio frequency identification
- supply chain management
- ultra high frequency

REFERENCES

Abad, E., Palacio, F., Nuin, M., De Zarate, A. G., Juarros, A., Gómez, J. M., & Marco, S., (2009). RFID smart tag for traceability and cold chain monitoring of foods: Demonstration in an intercontinental fresh fish logistic chain. *Journal of Food Engineering, 93*(4), 394–399.

Bhadra, S., Thomson, D. J., & Bridges, G. E., (2015). Monitoring acidic and basic volatile concentration using a pH-electrode based wireless passive sensor. *Sensors and Actuators B: Chemical, 209*, 803–810.

Borriello, G., (2005). RFID: Tagging the world. *Communications of the ACM, 48*(9), 34–37.

Costa, C., Antonucci, F., Pallottino, F., Aguzzi, J., Sarriá, D., & Menesatti, P., (2013). A review on agri-food supply chain traceability by means of RFID technology. *Food and Bioprocess Technology, 6*(2), 353–366.

Dolgui, A., & Proth, J. M., (2008). RFID technology in supply chain management: State of the art and perspectives. *IFAC Proceedings Volumes, 41*(2), 4464–4475.

Espinosa, E., Ionescu, R., Zampolli, S., Elmi, I., Cardinali, G. C., Abad, E., & Llobet, E., (2010). Drop-coated sensing layers on ultra low power hotplates for an RFID flexible tag micro lab. *Sensors and Actuators B: Chemical, 144*(2), 462–466.

Finkenzeller, K., (2010). *RFID handbook: Fundamentals and Applications in Contactless Smart Cards, Radio Frequency Identification, and Near-Field Communication.* John Wiley & Sons.

Hong, S. I., & Park, W. S., (2000). Use of color indicators as an active packaging system for evaluating kimchi fermentation. *Journal of Food Engineering, 46*(1), 67–72.

Jang, N. Y., & Won, K., (2014). New pressure-activated compartmented oxygen indicator for intelligent food packaging. *International Journal of Food Science and Technology, 49*(2), 650–654.

Jedermann, R., & Lang, W., (2007). Semi-passive RFID and beyond: Steps towards automated quality tracing in the food chain. *International Journal of Radio Frequency Identification Technology and Applications, 1*(3), 247–259.

Jones, P., Clarke-Hill, C., Hillier, D., & Comfort, D., (2005). The benefits, challenges, and impacts of radio frequency identification technology (RFID) for retailers in the UK. *Marketing Intelligence and Planning, 23*(4), 395–402.

Jones, P., Clarke-Hill, C., Shears, P., Comfort, D., & Hillier, D., (2004). Radio frequency identification in the UK: Opportunities and challenges. *International Journal of Retail and Distribution Management, 32*(3), 164–171.

Kassal, P., Steinberg, I. M., & Steinberg, M. D., (2013). Wireless smart tag with potentiometric input for ultra low-power chemical sensing. *Sensors and Actuators B: Chemical, 184*, 254–259.

Kaur, M., Sandhu, M., Mohan, N., & Sandhu, P. S., (2011). RFID technology principles, advantages, limitations and its applications. *International Journal of Computer and Electrical Engineering, 3*(1), 151.

Kumar, P., Reinitz, H. W., Simunovic, J., Sandeep, K. P., & Franzon, P. D., (2009). Overview of RFID technology and its applications in the food industry. *Journal of Food Science, 74*(8).

Kumari, L., Narsaiah, K., Grewal, M. K., & Anurag, R. K., (2015). Application of RFID in agri-food sector. *Trends in Food Science and Technology, 43*(2), 144–161.

Kummu, M., De Moel, H., Porkka, M., Siebert, S., Varis, O., & Ward, P. J., (2012). Lost food, wasted resources: Global food supply chain losses and their impacts on freshwater, cropland, and fertilizer use. *Science of the Total Environment, 438*, 477–489.

Lehpamer, H. (2012). Component of the RFID System. *RFID Design Principles,* 133–201.

Martínez-Olmos, A. J. P. A., Fernández-Salmerón, J., Lopez-Ruiz, N., Rivadeneyra, T. A., Capitan-Vallvey, L. F., & Palma, A. J., (2013). Screen printed flexible radiofrequency identification tag for oxygen monitoring. *Analytical Chemistry, 85*(22), 11098–11105.

Oprea, A., Bârsan, N., Weimar, U., Bauersfeld, M. L., Ebling, D., & Wöllenstein, J., (2008). Capacitive humidity sensors on flexible RFID labels. *Sensors and Actuators B: Chemical, 132*(2), 404–410.

Parfitt, J., Barthel, M., & Macnaughton, S., (2010). Food waste within food supply chains: Quantification and potential for change to 2050. *Philosophical Transactions of the Royal Society B: Biological Sciences, 365*(1554), 3065–3081.

Potyrailo, R. A., & Surman, C., (2013). A passive radio-frequency identification (RFID) gas sensor with self-correction against fluctuations of ambient temperature. *Sensors and Actuators B: Chemical, 185*, 587–593.

Potyrailo, R. A., Nagraj, N., Tang, Z., Mondello, F. J., Surman, C., & Morris, W., (2012). Battery-free radio frequency identification (RFID) sensors for food quality and safety. *Journal of Agricultural and Food Chemistry, 60*(35), 8535–8543.

Realini, C. E., & Marcos, B., (2014). Active and intelligent packaging systems for a modern society. *Meat Science, 98*(3), 404–419.

Roberts, C. M., (2006). Radio frequency identification (RFID). *Computers and Security, 25*(1), 18–26.

Salmerón, J. F., Rivadeneyra, A., Agudo-Acemel, M., Capitán-Vallvey, L. F., Banqueri, J., Carvajal, M. A., & Palma, A. J., (2014). Printed single-chip UHF passive radio frequency identification tags with sensing capability. *Sensors and Actuators A: Physical, 220*, 281–289.

Sarma, S., Brock, D., & Engels, D., (2001). Radio frequency identification and the electronic product code. *IEEE Micro, 21*(6), 50–54.

Steinberg, M. D., Žura, I., & Steinberg, I. M., (2014). Wireless smart tag with on-board conductometric chemical sensor. *Sensors and Actuators B: Chemical, 196*, 208–214.

Tom, O., (2016). *RFID: A Taste of Traceability* (pp. 1–7). Retrieved from: http://www.foodqualityandsafety.com/article/rfid-taste-traceability/?sing (accessed on 23 May 2020).

Vergara, A., Llobet, E., Ramírez, J. L., Ivanov, P., Fonseca, L., Zampolli, S., & Wöllenstein, J., (2007). An RFID reader with onboard sensing capability for monitoring fruit quality. *Sensors and Actuators B: Chemical, 127*(1), 143–149.

Wang, L., Kwok, S. K., & Ip, W. H., (2010). A radio frequency identification and sensor-based system for the transportation of food. *Journal of Food Engineering, 101*(1), 120–129.

Weinstein, R., (2005). RFID: A technical overview and its application to the enterprise. *IT Professional, 7*(3), 27–33.

Weis, S. A., Sarma, S. E., Rivest, R. L., & Engels, D. W., (2004). Security and privacy aspects of low-cost radio frequency identification systems. In: *Security in Pervasive Computing* (pp. 201–212). Springer, Berlin, Heidelberg.

Zhen-Hua, D., Jin-Tao, L., & Bo, F., (2007). Radio frequency identification in food supervision. In: *Advanced Communication Technology, the 9th International Conference on* (Vol. 1, pp. 542–545). IEEE.

CHAPTER 11

APPLICATION OF COMPUTER VISION SYSTEM IN FRUIT QUALITY MONITORING

S. BASHIR, A. JABEEN, H. A. MAKROO, and F. MEHRAJ

Department of Food Technology, IUST Kashmir – 192122, India

ABSTRACT

Quality of a fruit is defined as a combination of features that distinguish an individual from its similar ones based on its standard. It determines how safe is it to consume a product postharvest, and hence its extent of acceptance and grading. Quality is defined on the basis of certain external features such as shape, weight, color, firmness, temperature, smoothness, taste, apparent spots and a few internal features like ripeness, sugar content, biochemical composition, pigments, dry matter and pectic ingredients. Scrutiny of fruits and vegetables for determining the quality is done either by human vision, i.e., manual observation or by machine vision, i.e., using sensors. Human monitoring requires a panel of 10–15 trained experts to check the shape, dimensions, age, ripeness, and other features of fruits and vegetables. Machine based scrutiny involves sensors and instruments and involvement of mathematical algorithms to measure quality of fruit. The conventional machine-based practice of examining the quality may damage the fruit itself, hence known as destructive method, however the advanced technique used for quality monitoring and control works without destroying the product, hence termed as non-destructive or non-invasive methods of monitoring. Non-destructive monitoring methods used in commercial applications include mechanical methods like impact, vibration analysis and electronic nose; optical methods like visible/near-infrared spectroscopy, time and space resolved spectroscopy and electronic eye; magnetic methods like magnetic resonance imaging and electrical conductivity method; acoustic methods like ultrasonics and impulse response method; and other dynamic methods like X-ray and CT scan.

11.1 INTRODUCTION

The term 'Quality' is defined as the degree of excellence; it defines the standard or value of an item or commodity (Chauhan et al., 2017). The foodstuff quality is defined as the combination of the features of a food product that distinguish the individual food product from other similar products; it determines the extent of acceptance of a particular unit to the consumer. Quality of fruit or vegetable is determined by its safety to consume, which is in turn resolved into its sensory properties, its nutritional value, and its apparent defects. The superior sensory and nutritional quality of a food product implies high economical value, as the economic value is proportional to the quality of the product and it is equally true for fruits and vegetables (Abbott, 1999). Juran (1951) defined Quality as "fitness for use," where fitness is defined by the customer himself. This concept is based on the ISO 9000 standard, which defines the word 'quality' as the "degree to which a set of inherent characteristics fulfill the requirements of a customer."

Practically Quality is defined as a mathematical function of a series of quality attributes; the consumer assesses these quality attributes and consciously or unconsciously assigns a score (Bart et al., 2014). On the basis of this individual score, an overall quality score is calculated for future use and purchase decisions. The assessment of quality attributes is thus an essential component of quality appreciation.

Quality is defined on the basis of the evaluation of various parameters, which include external as well as internal features. External quality attributes relate to the appearance of the product which mainly includes shape, dimensions, color, luster, firmness, smoothness, perception of touch and taste and the surface should lack defects such as visible spots, imperfections, dryness, etc. (Chauhan et al., 2017). The dimensions of the fruits/vegetables are assessed by using various parameters like size, diameter, and volume (Moreda et al., 2009). External Quality factors determine the purchase behavior of consumers by and large, as these properties may be mostly inspected by the eye; therefore, most of the commercial quality measurement systems are based on external quality attributes only. The texture is other external sensory parameters, which are attributed by the perception of touch and the texture of a product can be defined by various properties such as firmness, stickiness, springiness, adhesiveness, etc.

Firmness is an indirect quality measuring parameter defined in terms of maturity and is measured in terms of ripeness of fruits. The parameter firmness is helpful to establish optimal transportation conditions and transportation time (Mirzaee et al., 2010; Khalifa, Komarizadeh, and Tousi, 2011).

Internal quality factors include ripeness, sugar content, acid, and fat content, pigments, and color, dry matter, pectic ingredients; Good quality implies the absence of intrinsic defects like bumps, pits, cracks, voids, fissures, etc. (Chauhan et al., 2017). These factors are used to measure internal attributes of fruits and vegetables and thus determine the quality; these attributes also determine the organoleptic satisfaction of consumers. Besides the above-listed attribute, there exist certain more attributes that define quality, but those parameters are often difficult to measure and even to define. These elusive quality attributes include freshness of fruit, its safety to consume, the nutritional value and health maintaining properties, chemical residues, production system, authenticity, convenience, and ethical aspects (Zou, Huang, and Malcolm, 2016). The factors may affect the quality perception of consumers on a more abstract and cerebral level.

Inspection of fruits and vegetables for quality monitoring can be performed either manually, by human vision and other sensory aspects; or by instruments and sensors, popularly known as machine vision. In a human vision-based monitoring system, superficial visual scrutiny is used to check the shape, dimensions, age, ripeness, and other features of fruits and vegetables (Jha and Matsuoka, 2000). To assess those features of fruit and vegetables, one needs to maintain expert panels (Brezmes and Eduard, 2016; Meilgard, Carr, and Vance, 2006), which typically consist of 10–15 experts, who are trained to score a list of quality descriptors that they have either derived themselves or obtained from the literature. Thus, the sensors used are human sensors, who quantitatively determine the features of the product. A series of standards is available to select experts, train them, then evaluate the performance and statistically analyze the data. Besides Experts, general Consumer panels typically consisting of hundreds of consumers that represent the group of consumers are considered relevant for the test. They only provide preference scores that can be used to subdivide the consumers into preference segments and associate them, for example, with certain culti-vars of a fruit or vegetable species. The preference scores, however, only become meaningful when compared to quality attributes measured by other trained expert panels. The advantage of having an expert consumer panel is that the characteristics of fruit and vegetables measured by them are similar to those of the ultimate consumers. However, even when the quality scores are acquired from the well-trained team of professional experts, the scores are likely to possess a huge inconsistency and are expected to change with time. Besides, this type of monitoring procedure is time-consuming, and the expensive by cost.

Instrumental/Computerized techniques are the alternate methods for measuring features of fruit and vegetables. These techniques do not require human panels for monitoring. However, their success depends critically on how well the measurement principle mimics human perceptions of a particular property. Examples of such techniques include Laboratory colorimeters, which are based on the three-component theory of human color perception, elaborated by the International Commission of Illumination in 1931 (Smith and Guild, 1931). Mealiness, a texture attribute describing the sensation of dryness and granularity due to cell debonding rather than fracturing during mastication, is another such example (Barreiro et al., 1998).

Quality attributes determine fruit and vegetable quality, corresponding to their human analogs, and are measured using both destructive as well as non-destructive (ND) practices. Destructive practices include the conventional procedure of examining the quality which may damage the fruit itself while examining whereas ND methods are machine-based examination methods, which are advantageous as they do not tear-away the fruit tissue. ND practice is a procedure that does not affect the features of the fruit and are more efficient as machines are more consistent than humans. Besides, ND techniques are often fast, and reduce waste. Because of the wide variation in the characteristics of fruits and vegetables, the grading of every individual unit of product is needed to meet the expectations of a consumer. The important advantage is that such practices are that it can be used for grading of every individual unit of a fruits/vegetables with respect to quality prior to selling the same to consumers. Color grading and dimensional analysis by superficial visual inspection is an age-old process used by humans to eradicate products that would not meet the minimal necessities for quality, and hence at the same time improve standardization. Over the times, manual practices have been replaced by computerized ones, to get high-speed and better grading. These computer-based automated sensors, for measuring the external attributes such as color, dimensions, and appearance, are nowadays used widely by fruit growers and packaging houses across the globe. Besides, ND techniques are helpful for developing prototypes of sensors that determine the quality attributes for postharvest packing, to optimize the methods of storage. As the same fruit can be monitored over time, the inter fruit variability can be separated easily from the time effect (De Ketelaere et al., 2006; Hertog et al., 2007). The introduction of ND quality monitoring methods to measure internal features thus launched the exciting and innovative marketing opportunities for horticultural products. The introduction of machine-based vision has, in addition, amplified the automated streamlining and grading in food industries, and reduced the labor.

ND practices are advantageous over outdated destructive practices, as ND practices are mostly based on physical properties, which associate well with the excellence of fruits and vegetables in terms of quality (Khalifa, Komarizadeh, and Tousi, 2011). The methods widely accepted in fruit and vegetable industries to determine the quality and in packaging houses include mechanical, optical, electromagnetic, dynamic, and acoustic practices (Roohinejad et al., 2009). This chapter is a review of ND practices used currently for computing internal and external quality attributes of fruits and vegetables.

11.2 MECHANICAL METHODS

Mechanical practices have been the oldest ways used to gain information about the internal quality of fruit and vegetables in a ND way. Mechanical methods are used to assess the fruit texture, firmness, and maturity, by measuring the toughness and elastic properties such as turgor pressure, water content, the mechanical strength of cell wall, middle lamella, etc. (Hertog et al., 2007; Sheeja and Ajay, 2016). Typically, mechanical techniques are divided into two broad categories:

1. Local force-deformation characteristics of a fruits/vegetables after application of an external force; and
2. Response of the fruit as a whole after it has been excited.

For both classes, several means of inserting energy into the fruit as well as for measuring the resulting response are used. Mechanical methods of quality assessment include low mass impact analysis (e.g., mechanical thumb), vibrational sensors (e.g., microphone test), and gas sensors (e.g., electronic nose (e-nose)).

11.2.1 IMPACT ANALYSIS

Substances deform locally after being compressed by an external load and, if the load is below a certain threshold, will eventually converge back to the original state, once this load has been removed. The above principle is used to measure the textural quality of fruit such as stiffness, firmness, etc. (Vursavus and Zehan, 2016).

As per Hertz's contact theory, the maximal deformation, maximal force, and contact time between impactor and sample provide information about the mechanical properties of the sample. The external load is typically provided using a low-mass impact device with a spherical tip made from a material with known constant physical properties. An integrated force sensor measures the maximal force and the contact time of the impact, and is processed further to derive the firmness values (Mizrach, Nahir, and Ronen, 1985). In order to study the properties properly, fruits/vegetables must be impacted with a constant energy. Besides an essential requirement is the precise control of the impact device, so that it can handle different sizes and shapes of fruits and vegetables. A number of methods based on the above principles have been investigated under laboratory conditions and a number of commercial sensors, for quality control of fruit and vegetables have been designed so far (Delwiche, Tang, and Mehlschau, 1989).

Impact analysis is applicable to determine the quality of a broad range of fruits and vegetables such as peaches, tomatoes, apples, oranges, mangoes, and kiwi. The main advantage of impact method over others (such as conventional vibration method) is that this method is not sensitive to overall fruit shape and can determine the quality of fruits with complex shapes such as Mangoes (De Ketelaere et al., 2006). On the contrary, for fruits with a high firmness, the deformation due to low-mass impact will be very small and the contact time very short, so that the signal-to-noise ratio is low, that implies the biggest disadvantage to this method.

A substitute to low-mass impact device is the process where the impact duration can be measured by dropping a substance from a small height onto a flat surface, which is connected to a force cell (Delwiche, McDonald, and Bowers, 1987; De Baerdemaeker, Lemaitre, and Meire, 1982). An additional such method is to use the compressed air to develop the load on the substance surface. The leading benefit of the latter method is that it is a contactless method; however, it requires a substitute to a force sensor for deriving the response. Precise laser distance sensors have been used for this method, where the working principle is based on laser triangulation. This method is used to measure the deformations of the order of magnitude of micrometers (Hung, Prussia, and Ezeike, 1999; Prussia et al., 1994). On the basis of Hooke's law, the force-deformation relation is used to estimate local stiffness.

The principle of this method when combined to the destructive Magness-Taylor tester method (Moreda et al., 2009; Magness and Taylor, 1925) results in a mechanical thumb (shown in Figure 11.2) is used to measure fruit firmness. This method was first developed by Schomer and Olsen (1962). The job is accomplished by penetrating a cylindrical head into the skin of a fruit or vegetable, and the maximum penetration force is measured. The mechanical thumb has a contact head whose penetration depth is limited to 1.27 mm. This head is a replacement of the standard plunger of the Magness-Taylor tester. This Mechanical thumb measures difference in fruit firmness by comparing the color of fruits and vegetables, and thus determines the maturity. Scientists developed a tiny flat-head pin, around 3 mm in diameter, and pierced the same into the skin of oranges and tomatoes. The firmness of oranges and tomatoes was estimated by measuring the force applied on pin and deformation of the skin. Using this technique, red tomatoes, and green tomatoes (ripe and unripe) can be easily categorized (Mizrach, Nahir, and Ronen, 1992).

11.2.2 VIBRATION ANALYSIS

The shape, size, composition, and rigidity affect the vibrational behavior of a sample; therefore, the vibration of a sample has been utilized to analyze the superiority of fruits and vegetables. The in-depth study conducted by different researchers in order to understand the frequency response of substances to various input parameters is utilized to estimate quality metrics in fruits/vegetables. As per the research (Abbott et al., 1968; Cooke, 1972; Cooke and Rand, 1973; Finney, 1970, 1971) the fruit will vibrate according to a well-defined pattern of mode shapes, each associated with a certain resonance frequency. The research was mainly focused on spherical fruit and vegetables, and a parameter (acoustic) stiffness of the fruit, is defined as:

$$S = f_R^2\, m^{2/3}$$

where f_R is the resonant frequency of the vibration in hertz and m is the mass of fruit in grams. The stiffness is basically a scaled version of the elastic modulus of the fruit; it is a measure of the force required to cause a certain deformation of the fruit while squeezing it. This parameter of Stiffness is used as a firmness indicator.

Another estimate of the ripeness of fruits is by analyzing the sound that results when the fruit is tapped gently. If tapping results in higher tones, then the fruit is harder and is hence associated to less mature fruit. This phenomenon for studying fruit quality was first described by Clark and Mikelson (1942). Typically, a small impactor with high stiffness is used to excite the fruit so that the impact duration is short and a broad range of frequencies is excited. The broadly used way to record, the output vibration of the fruit is with a microphone placed optimally with respect to the impactor, so that the vibration mode, the shape of interest (usually the spherical mode) is identified. This technique used for quality measurement and grading is fast, as up to 10 fruit per second can be evaluated for quality.

The above-explained fruits/vegetables monitoring principle is based on mechanical resonance where the fruits/vegetables vibrates at its natural frequency after it is excited. Here the transient behavior of vibrations is observed up to the stage where the fruit reaches to equilibrium, during this stage a surface wave travels along the surface of the fruit. This transient behavior is characterized by the speed of the surface wave and is used to derive the fruit firmness.

Another approach to measuring frequency is contact methods that utilize either accelerometers or piezoelectric transducers (Peleg, 1993; Shmulevich, Galili, and Rosenfeld, 1996). Here a range of sinusoidal input frequencies is applied at input of the fruit and the detection of the response signal on the other side is performed with a pick-up rod or an accelerometer. However, this method is time-consuming and requires the attachment of sensors to the fruit itself, which is less desirable, and not appropriate for high-speed grading. Laser Doppler Vibrometry is an alternate contactless method to measure vibrations of excited fruit (Landahl and Terry, 2012). However, because of its high cost, this method is restricted for laboratory use only.

The advantage of vibration analysis is that it is based on the integrated response of the whole fruit, whereas impact analysis is a local method. The applications of vibration analysis on fruit and vegetables mainly include apple, melon, and tomato. The specimens used are typically spherical fruits since spherical shapes facilitate the interpretation of the mechanical resonance pattern observed. This method is rarely used for the analysis of irregular shaped and soft fruits and vegetables since a large portion of the impact energy will be absorbed by deformations and a minor fraction will be transformed into vibration energy.

11.2.3 GAS SENSORS

The sensors that aim to perceive an aroma in a similar way as the human olfactory system are termed as Gas Sensors. An e-nose is one such biomimetic instrument. Electronic-nose comprises of electrochemical sensors that attempt to replicate the working of olfactory system. It involves an array of electro-chemical sensors with partial specificity, data processing system, a tool with an appropriate pattern-recognition algorithms and library for referencing, as shown in Figure 11.1 (Gardner and Bartlett, 1994; Gardner, 1991; Freund and Lewis, 1995; Abe et al., 1988; Kowaiski and Bender, 1972). The volatile components in the air interact with the gas sensors and produce a physical response that is transduced into an electrical signal for further processing. The reaction of the sensors to the tester is called as 'electronic fingerprint' (Zerbini, 2006). The sensors distinguish the smell on the basis of the reaction pattern of a tester to sensors in comparison to the already recognized fingerprint. The results are calculated by multi-variable statistical analysis methods which include Linear Discriminant Analysis, Discriminant Function Analysis, Hierarchical Cluster Analysis, soft independent modeling of class analogy (SIMCA) and partial least squares (PLS) method, etc.

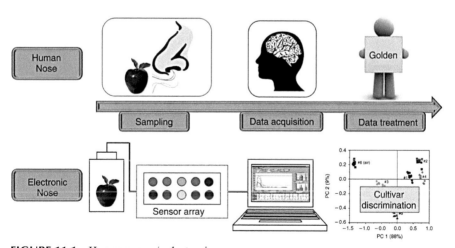

FIGURE 11.1 Human nose v/s electronic nose.

Similar to human olfactory receptors, the sensors are semi-selective. This feature is essential, as it involves combining several sensors to discriminate more aromas than the number of sensors used. The following desirable features are supposed to be fulfilled for integration in an e-nose as ideal sensors (Nicolai et al., 2008):

1. It should be highly sensitive to chemicals.
2. It should be least sensitive to moisture and heat.
3. It should have an average selectivity.
4. It should be highly stability.
5. It should have high reproducibility and repeatability.
6. Its response and recovery time should be minimum.
7. It should be easy to calibrate.
8. It should have small dimensions.
9. It must react to every substance in the sample space.
10. It should be robust and durable.
11. It should have easy output data processing.

Different sensor principles have been used in e-noses, including MOSFETs, conducting organic polymers, quartz microbalances, surface acoustic wave (SAW) sensors, gas chromatography (GC), and colored dyes.

e-Nose is suitable for defining the age of a fruit and superiority estimate. e-nose is useful in monitoring the aroma of fruits including melons, pears, nectarines, apples, tomatoes, mangos, and other citrus fruit. Other applications aim to differentiate cultivars or ripeness stages, and to detect fungal infection. A typical latest example of e-nose is 'RipeSense,' a disposable ripeness sensor developed by a New Zealand company. The sensor originally indicates the red color, progresses towards orange, and finally settles at yellow color in response to aroma produced while the fruit is ripening. This concept can be used during fruits/vegetables packaging and can help the customer in estimating the ripeness and hence the age of the fruits/vegetables.

11.3 OPTICAL METHODS

Optical practices are based on reflection, transmission, absorption, scattering of light, and fluorescence in various media. The presence of soluble solid content, amount of acids and starches, and age of fruits can be estimated by utilizing those ND optical techniques. These procedures include image analysis, analysis of mass, contour, dimensions, shade of fruit, and external defects using the following methods: Visible infrared (VIR) spectroscopy, laser spectroscopy, reflectance, transmittance, and absorption spectroscopy, etc.

11.3.1 VISIBLE/NEAR-INFRARED SPECTROSCOPY (NIRS)

Fruit and vegetables are opaque to radiation in the visible and near-infrared regions (VIR) of the electromagnetic spectrum. In these media, a complex interaction between absorption and scattering of the white light directs the light-matter communication. Absorption and scattering are based on the spectral and spatial specifications at the microscopic level of the complex refractive index (Bohren and Huffman, 1983). The tissue structures, made up of the cells and intracellular and extracellular environments, especially the cell nuclei, mitochondria, vesicles, membranes, and cell walls are responsible for the VIR light scattering by fruit and vegetable tissue. The photons are mostly scattered by structures inside the tissues with the same dimensions as that of wavelength of photon. On the other hand, absorption is caused mainly by the C-H, O-H, and N-H bonds present in compounds like water, sugars, chlorophylls, carotenoids, etc. The photon is absorbed when it has the sufficient energy to excite the vibrational states of the compound. The fundamental vibrations of the bonds occur in the infrared region. The absorption in the VIR region is caused by overtones and combinations of these fundamental vibrations. The absorption peaks in the VIR region are thus broad and overlapping. Moreover, the peaks can shift a few nanometers due to hydrogen bonding. The combination of these effects means that the VIR spectra of complex mixtures such as fruit and vegetables are hard to interpret. Advanced chemometric procedures are used to infer knowledge about the concentrations of the major components from these spectra. For water molecules, the absorption rate is relatively low in the VIR region compared to the UV and mid-infrared (MIR) regions, thus the electromagnetic radiation can penetrate quite deep into the biological tissue. Conversely, the effect of light scattering is remarkably bigger, and this enables electromagnetic radiation to diffuse in the sample volume and to be reemitted at the tissue boundaries.

Practically, VIR spectroscopy is the most suitable option for sorting and grading of fruits and vegetables. This technique is helpful to examine the age of fruits/vegetables by evaluating the skin color. As the skin color of fruits/vegetables is proportional to the amount of chlorophyll content of the skin, the variation in shades of fruit skin are used to estimate the ripeness.

The theory behind this fact is that the chlorophyll content decreases with the fruit ripening (De Jager and Roelofs, 1996). In addition, VIR spectroscopy is used in determining the fruit firmness measurement and in recognition of perceptible stains. The quantity of soluble solids, dry matter, hygroscopicity, firmness, sugar content, acidity, etc., of fruits and vegetables can be determined using this method. Fruit sorting lines equipped with VIR spectroscopy sensors are already available in the market and have been used successfully to non-destructively measure the soluble solids contents of various fruits, including apple, cherry, kiwifruit, mandarin, melon, pineapple, banana, tomato, mango, and peach.

The drawback of VIR spectroscopy is that a new calibration model for each fruit species and cultivar is required and these calibration models are made on the basis of a large collection of data, integrating the data of various orchards, different seasons, with variety of cultivation systems. Another drawback of this method is that acidity, texture, and other fruit properties are much more difficult to measure.

11.3.2   TIME AND SPACE RESOLVED SPECTROSCOPY

An improvement to the classical approach of VIR includes advanced techniques, such as space-resolved spectroscopy (SRS) and time-resolved spectroscopy (TRS). These techniques meet the requirement for the optical characterization of highly diffusive media (Patterson, Chance, and Wilson, 1989; Kienle et al., 1996; Yodh and Chance, 1995). These methods involve the penetration of light into a diffusive medium which in turn depends on the optical properties of the medium and on the distance between the source and detector. The core characteristic of SRS and TRS is their ability to retrieve information on the path length traveled by a photon in a diffusive medium, which is generally much larger than the geometrical distance between the source and the detector. The typical values of the optical properties of fruit and vegetables in the VIR region correspond to average photon path lengths. SRS collects photons at multiple source-detector distances using an optical fiber arrangement (Nguyen et al., 2011; Peng and Lu, 2008). TRS analyzes the variation in time-of-flight (ToF) of photons in relation to the speed of light in the medium and at some fixed distance between source and detector. This is done with pulsed laser sources and fast detection techniques (Torricelli et al., 2008).

It is postulated that the absorption properties are associated with the chemical composition and are determined by the concentration of water, sugars, and pigments like chlorophyll and anthocyanins in a fruit. The scattering properties are associated with the microstructural parameters such as the topology of the intercellular space and the size and shape of the cells. This could enable a means for non-destructively assessing texture, as these features affect the overall mechanical properties. TRS is said to be a potential practice for the sorting of fruits and vegetables in a ND manner. In the case of SRS, the results are based on the outputs of a multispectral or a hyperspectral camera, which is used to obtain diffuse reflectance spectra resolved in space for the estimation of quality parameters in fruit and vegetables (Qin and Lu, 2008; Qin, Lu, and Peng, 2009).

The use of SRS or TRS, in combination with proper models of photon migration, enables the complete optical characterization and concurrent ND measurement of the optical properties of a diffusive medium such as absorption and scattering. These techniques are important for most fruit and vegetables because the information derived by TRS and SRS refer to the intrinsic properties and are not so much concerned with surface features as is traditional spectroscopy.

TRS and SRS are used for sorting, grading, age calculation, and detection of intrinsic faults. In addition, it is useful in predicting shelf-life. The main advantage is that it gives a valuable knowledge about the intrinsic features of fruits/vegetables as the analyzes is based on the bulk properties and not on the surface properties alone. Besides ripeness and age of fruits can be estimated even when the fruit skin is misleading. TRS and SRS are thus used for the ND estimation of fruit quality such as apples, pears, nectarines, etc.

11.3.3   MACHINE VISION

Visual inspection is the basic way of sorting and grading fruit and vegetables, however in order to save time and effort, human vision is replaced by the machine vision. This is done to systematize the visual examination and make it a new objective. A machine vision system comprises a digital camera that is linked to an image processing software tool. Monochrome digital cameras produce a 2D array of intensity values corresponding to the different positions in the image. Similarly, a color image consists of three 2D arrays of intensity values: one for the red band, one for the green band, and one for the blue band. To detect objects in the image, the pixels should be clustered and then categorized based on some features. The typical way to categorize pixels is by assigning a threshold to the intensity values of one of the wave-band images. The image analysis is used in the food industry as it enables the accurate discrimination of colors. One of the proposed methods for grading mangoes (Nandi, Tadu, and Koley, 2014) is shown in Figure 11.2. Another such application of discrimination of color red from green is used for the prediction of the lycopene content in tomatoes (Arias et al., 2000).

A major disadvantage of this approach is that a change in the illumination intensity will change the classification result if the threshold is not adapted accordingly. To avoid this problem, it is better to separate the information on the perceived color from the information on the luminosity.

Applications of machine vision to fruit and vegetables are vast. Sorting lines with machine vision systems for detecting color, size, shape, and surface defects are made commercially available by all major sorting line manufacturers.

Alternative segmentation approach exploits the properties of the object's contours, such as concaveness or an intensity gradient. When the pixels have not all been correctly classified, morphological operations such as erosion and dilatation can be used to remove small groups of pixels that have been misclassified as an object or to fill small holes in detected objects. For these detected objects, morphological parameters such as area, largest length, or shortest width can be calculated easily. Alternatively, geometrical models such as spheres, rectangles, or ellipsoids can be fitted to the detected object, thus identifying the shape parameters.

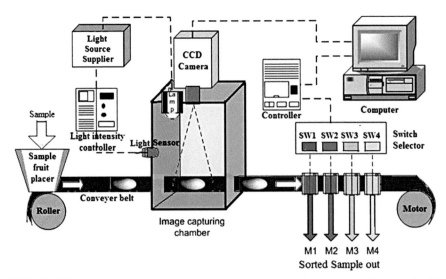

FIGURE 11.2 A machine vision based automatic fruit grading system. (Reprinted with permission from Nandi, Tadu, and Koley, 2014. © Springer Nature.)

11.4   ELECTROMAGNETIC METHODS

11.4.1   MAGNETIC RESONANCE IMAGING (MRI)

Magnetic resonance imaging (MRI) is a ND, non-intrusive spectroscopic technique based on the interaction of electromagnetic radiation in the radio frequency (RF) range with matter. The magnetic resonance based practices include nuclear magnetic resonance (NMR) spectroscopy, NMR relaxometry and MRI. In these techniques protons, such as those present in the water of fruit or vegetables are targeted. The spins of the protons within the material are aligned by applying a strong magnetic field. By monitoring the proton dynamics afterward, the information on the spatial distribution of proton density, relaxation parameters, and self-diffusion parameters inside the sample are obtained. The relaxation and self-diffusional properties often provide complementary information and enhanced contrast and are linked to proton mobility. As per the Principle of Magnetic Resonance, the magnetic energy is absorbed by the nuclei when placed in an alternating magnetic field. The number of protons present in a particular sample is used to calculate the amount of energy absorbed by the nuclei (Farrar and Becker, 1971). MRI is successfully used for study of biochemical materials like water, but nowadays can be used for fat, oil, or salt, and hence allows us to differentiate the components. Furthermore, MRI is sensitive to many quality parameters which affect the produce, particularly those that affect the water concentration or mobility. Its ND character makes it particularly attractive for scanning intact fruit and vegetables but also for monitoring their quality over time, for example, during storage. The sensitivity to water is high in MRI, but it is used predominantly to investigate relative differences or changes in water content instead of quantification, given that the latter is not straightforward due to temperature changes and other components containing protons.

MRI has been used efficiently over ages to estimate, the characteristics of fruit and vegetables. It is used to detect the presence of internal defects, such as voids and cavities, worm damage, pits, bruises, water core disorder, etc. in apples, internal browning in apples, mealiness of apples or peaches, core breakdown in pears, and chilling injury in citrus fruit and zucchini squash. It is also used to measure physical characteristics like mass, contour, and dimensions, etc. and hence determine the firmness. Besides, it can detect the amount of soluble solids and acids present in an individual unit. A distinction between ripe and immature fruit and vegetables could be made, for example, by measuring the free water content in tomato or pineapple and the oil content in avocado. Pathogen infection has also been successfully detected (Chen, McCarthy, and Kauten, 1989; Hall, Evans, and Nott, 1998). The Figure 11.3 shows images of a healthy apple (A) and an apple with having water core symptom (B). The water core symptom is a disorder characterized by water-soaked areas in the fruit. The disorder affected areas are visible, as these areas lighten up, because the mobility of water is altered from that of healthy tissue. Thus, MRI possesses a vast potential for grading, sorting, and parametric estimation of fruits/vegetables (Ruiz-Altisent et al., 2010). However, due to the high volume and low value of fruit and vegetables, the high cost of MRI equipment and its relatively low acquisition speed, industrial implementations of MRI were non-existent.

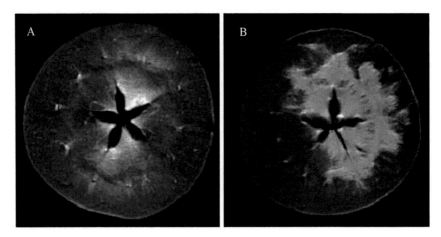

FIGURE 11.3   MRI scan (A) healthy fruit (B) fruit with disorder.

11.4.2   ELECTRICAL METHOD

Electrical practices for sorting and grading of fruits and vegetables comprise of two techniques, viz. electrochemical impedance spectroscopy (EIS) and di-electric analysis (DEA). In this method the physical state of a fruit, flesh, defined in terms of its conducting ability, is measured as a function of frequency. Measuring the electrical conductivity is robust and easy practice for estimation of the physical and biological state of tissues (Ando, Mizu-tani, and Wakatsuki, 2014; Damez et al., 2007; Greenham, 1966; Yamamoto and Yamamoto, 1976). The changes in the electrical equivalent circuits of constituents are used to estimate the frequency response and the impedance, and thus the physical properties of corresponding constituent can be quantified. The measure of electrical properties and the high frequency areas are thus used in estimation of moisture and determination of bulk density. This practice is thus used to determine the composition of fruits and vegetables and hence determine the diseased or damaged fruit tissues and can also detect levels of ripeness and help in estimating fruit and vegetable quality.

11.5 ACOUSTIC METHODS

Acoustic wave transmission is an alternate practice to examine the fruit and vegetables, their age, density, moisture content and firmness. These waves compare the firmness of fruits/vegetables. An efficient practice of analyzing fruits and vegetables using acoustic waves is through Surface wave transmission analysis, i.e., using ultrasonic techniques. The principle involves the transmission of fixed wave energy in fruits and vegetables and calculation of the energy in response. Acoustic practices include ultrasonics/ultrasound and impulse response methods.

11.5.1 ULTRASONICS/ULTRASOUND

Ultrasound/ultrasonics is defined as sound waves of frequency 20 kHz and above, that are undetectable to the human ear. An ultrasonic wave moves through a medium as a series of alternating compressions and rarefactions with its velocity being determined by the density and elasticity of the medium (McClements, 1997). The structure and composition of the waves change the nature of the medium, hence knowing the interaction impact of sound-medium, the waves can be analyzed. The presence of cavities and pores in fruits/vegetables results in scattering and absorption of the sound when traveling through the plant tissue. This complicates the interpretation and understanding of the data. Absorption and scattering by the medium result in variations of this velocity and hence attenuate the sound waves. Measurement of this attenuation is useful in deciding the quality of fruits/ vegetables. Ultrasound technique is relatively simple and cheap in its instrumentation and has been applied to many different food processing operations since 1970's (Povey and McClements, 1988). The measurements have been applied to a range of physiochemical quantities, such as firmness, maturity, and chilling injury.

11.5.2 IMPULSE RESPONSE METHOD

Acoustic impulse response is a practice used to evaluate the consistency in the texture of fruits and vegetables (Abbott et al., 1997; Muramatsu et al., 1997). The acoustic impulse resonance frequency (AIF) is one such practice that measures the fundamental frequency of a fruits/vegetables. This method involves estimation of resonance frequency obtained from fruits/vegetables, where resonance frequency of fruits/vegetables changes with the ripening. The job is furnished by hitting the fruits/vegetables and then recording the sound it produces and carrying out the Fourier transformation on the recorded signal. The impact response is thus created by the fruit on applying pressure, and this response is used to judge the intrinsic properties and age factor in fruits/vegetables. The impact response is defined in terms of a parameter, known as the Stiffness factor, which is calculated for a specific fruit from this resonance frequency and the given mass of the unit. The impact response is a highly sensitive parameter and is altered with the variation of many characteristics of fruit such as the dimensions of the fruit, the impact angle of the force, location of fruit growth, etc. Besides the factor that affects impact response, the most are the water content of the fruits/vegetables (Zerbini, 2006). The technique is an efficient practice for determining the shelf life of fruits and vegetables.

11.6 DYNAMIC METHOD

11.6.1 X-RAY METHOD

X-rays were discovered by Roentgen in 1895. In the electromagnetic spectrum, the X-rays fall between gamma rays and ultraviolet rays and cover wavelengths from 10 nm down to 0.01 nm, corresponding to a frequency of 3×10^{16} Hz to 3×10^{19} Hz and energies in the range 120 eV to 120 keV. Although X-rays interact with a material similar to other types of electromagnetic radiation, most imaging applications are based on the absorption of the X-ray photons by the material, which depends on the local density, the atomic number, and the energy of the X-rays. To produce X-rays, electrons are accelerated in a vacuum tube over a potential gradient and are directed onto a specific target, usually made of metal. Those electrons that hit the target, release X-rays as they slow down. The X-ray photons produced in this way have a continuous energy spectrum from around zero up to the energy of the electrons. Additional photons at specific energy levels are emitted through X-ray fluorescence when orbital electrons are knocked out of the inner electron shell of the metal atom and electrons from higher energy levels fill up the vacant positions. After passing through the object, the X-rays enter crystal scintillators that convert them to flashes of light that are detected and processed electronically to produce an image In X-ray radiography, a single image of transmitted X-rays through an object is acquired. The resulting image is thus superimposed information or a projection of the 3D object volume in a 2D plane (Salvo et al., 2003).

Beer Lambert's law states, "The ratio of transmitted to incident photons is proportional to the integral of the absorption coefficient of the object along the path that the photons follow through the sample." When the object contains a feature that is sufficiently large and has sufficiently different absorption properties than those of the surrounding material in the object, the feature can be distinguished on radiographic images. Radiography equipment is available commercially for industrial use and is used mainly for the detection of foreign objects with high contrast in foods.

Radiography has the advantage that it is fast and can be implemented inline on sorting lines (Jiang et al., 2008; Hansen et al., 2005; Kim and Schatzki, 2001). X-ray radiography has been used to investigate internal disorders in fruit and vegetables (Haff and Toyofuku, 2008), especially nuts, onions, apples, pear, peach, cherry, tomato, and orange. Using radiography methods, affected water core, larval feeding damage caused by codling moths, infestation damage due to pests inside fruit, rotting, biological faults in the tissues, mutilation due to freezing, internal color changes, detection of pits, existence of unwanted elements, etc. has been successfully detected.

The drawback is that the success of the method depends on sufficient contrast between the features to be detected in the scanned product and their environment. The detection resolution of conventional radiography equipment is limited. Recent advances in X-ray radiography methods are dedicated to improving image contrast, more accurate and fast image segmentation methods, image texture analysis, and machine learning classification (Nielsen et al., 2013; Mathanker et al., 2010; Toyofuku and Schatzki, 2007). Although most applications rely on the absorption contrast between defects and the product, in transmission radiography, newer developments such as phase contrast and dark-field imaging using grating interferometry have been proposed.

11.6.2   COMPUTED TOMOGRAPHY (CT) SCAN

Computed tomography (CT) Scan came into existence in the late 1970s. Both X-ray radiography and x-ray computed tomography (X-ray CT) facilitates the ND analysis of the intrinsic structure of an object. Whereas, X-ray radiography yields 2-dimensional images, CT is used to create an image in three dimensions. CT uses a mathematical algorithm to compute a 3D image, known as tomogram, from multiple radiographs of the object taken from different angles (Salvo et al., 2003). Earlier CT scanners would work at a lower pixel resolution, i.e., around 1 mm, however with advances in technology microfocus X-ray sources and high-resolution detection systems were developed, that could achieve a pixel resolution in the micrometer range.

CT scan sources are generally X-ray sources that produce a polychromatic, divergent beam that puts constraints on the achievable image resolution, field of view, and image quality. In the present day, CT uses synchrotron radiation sources, which can deliver a very intense X-ray beam with high flux, resulting in high contrast and resolution. The synchrotron radiation sources produce a parallel beam having good 3D coherence, which makes a quantitative reconstruction free of artifacts. The source-detector assembly revolves at very high speeds covering the substances and thus 3D images are recorded in very less time. However, the cost is high, as the system needs to precisely control the movements of mechanical and electronic parts.

Early CT research aspired to determine maturity or ripeness-related parameters of fruit (Brecht et al., 1991; Tollner et al., 1992), but a good correlation was found between X-ray absorption and water content, or density as a derivation thereof. More successful applications of CT addressed the detection of internal defects such as core breakdown disorder, tissue browning, and cavity formation, detection of codling moth feeding tunnels. CT is thus used for non-invasive visualization of individual cells in the 3D microstructure of fruit and vegetables such as pears, apples, cherry, and chestnut.

KEYWORDS

- di-electric analysis
- electrochemical impedance spectroscopy
- magnetic resonance imaging
- nuclear magnetic resonance
- space-resolved spectroscopy
- time-resolved spectroscopy

REFERENCES

Abbott, J. A., (1999). Quality measurement of fruits and vegetables. *J. Postharvest Biol. Technol., 15*, 207–225,

Abbott, J. A., Bachman, G. S., Childers, R. F., Fitzgerald, J. V., & Matusik, F. J., (1968). Sonic technique for measuring texture of fruits and vegetables. *Food Technology, 22*, 635–646.

Abbott, J. A., Lu, R., Upchurch, B. L., & Stroshine, R. L., (1997). Technologies for non-destructive quality evaluation of fruit and vegetables. *Hort. Rev., 20*, 1–120.

Abe, H., Kanaya, S., Takahashi, Y., & Sasaki, S. I., (1988). Extended studies of the automated odor sensing system based on plural semiconductor gas sensor with computerized pattern recognition techniques. *Anal. Chim. Acta., 215*, 155–168.

Ando, Y., Mizutani, K., & Wakatsuki, N., (2014). Electrical impedance analysis of potato tissues during drying process. *J. Food Eng., 121*, 24–31.

Arias, R., Lee, T. C., Logendra, L., & Janes, H., (2000). Correlation of lycopene measured by HPLC with the L*, a*, b* color readings of a hydroponic tomato and the relationship of maturity with color and lycopene content, *J. Agric. Food Chem., 48*, 1697–1702.

Barreiro, P., Ortiz, C., Ruiz-Altisent, M., De Smedt, V., Schotte, S., et al., (1998). Comparison between sensory and instrumental measurements for mealiness assessment in apples: A collaborative test. *J. Texture Stud, 29*, 509–525.

Bart, M., Nicolai, Thjis, D., Bart, D. K., et al., (2014). Non-destructive measurement of fruit and vegetable quality. *Annual Review: Food Science and Technology* (pp. 285–312).

Bohren, C. F., & Huffman, D. R., (1983). *Absorption and Scattering of Light by Small Particles.* Wiley New-York.

Brecht, J. K., Shewfelt, R. L., Garner, J. C., & Tollner, E. W., (1991). Using x-ray computer tomography (x-ray CT) to non-destructively determine maturity of green tomatoes. *Hortic. Sci., 26*, 45–47.

Brezmes, J., & Eduard, L., (2016). *Electronic Noses for Monitoring the Quality of Fruit.* Electronic noses and tongues in food science.

Chauhan, O. P., Lakshmi, S., Pandey, A. K., et al., (2017). Non-destructive quality monitoring of fresh fruits and vegetables. *Defense Life Science Journal, 2*(2), 103–110.

Chen, P., McCarthy, M. J., & Kauten, R., (1989). NMR for internal quality evaluation of fruits and vegetables. *Trans. ASABE, 32*, 1747–1753.

Clark, H. L., & Mikelson, W., (1942). *Fruit Ripeness Tester.* US Patent No. 2277–37.

Cooke, J. R., & Rand, R. H., (1973). A mathematical study of resonance in intact fruits and vegetables using a 3-media elastic sphere model. *J. Agric. Eng. Res., 18*, 141–157.

Cooke, J. R., (1972). An interpretation of the resonant behavior of intact fruits and vegetables. *Trans. ASABE, 15*(6), 1075–1080.

Damez, J. L., Clerjon, S., Abouelkaram, S., & Lepetit, J., (2007). Dielectric behavior of beef meat in the 1–1500 kHz range: Simulation with the Fricker/Cole-Cole model. *Meat Science, 77,* 512–519.

De Baerdemaeker, J., Lemaitre, L., & Meire, R., (1982). Quality detection by frequency spectrum analysis of the fruit impact force. *Trans. ASABE, 25,* 175–178.

De Jager, A., & Roelofs, F. M. M., (1996). Prediction of optimum harvest date of jonagold, in the postharvest treatment of fruit and vegetables-Current status and future prospects. *Proc. of the Sixth International Symposium of the European Concerted Action Program (COST)* (Vol. 94, pp. 21–31). Luxembourg.

De Ketelaere, B., Stulens, J., Lammertyn, J., & De Baerdemaeker, J., (2006). Identification and quantification of sources of biological variance: A methodological approach. *Postharvest Biol. Technol.*, pp. 1–9.

Delwiche, M. J., McDonald, T., & Bowers, S. V., (1987). Determination of peach firmness by analysis of impact forces. *Trans. ASABE, 30*, 249–254.

Delwiche, M., Tang, S., & Mehlschau, J. J., (1989). An impact force response fruit firmness sorter. *Trans ASAE, 32*(1), 0321–0326.

Farrar, T. C., & Becker, E. D., (1971). *Pulse and Fourier Transform NMR: Introduction to Theory and Methods.* Academic Press, New York.

Finney, E. E., (1970). Mechanical resonance within red delicious apples and its relation to fruit texture. *Trans. ASABE, 13*, 177–180.

Finney, E. E., (1971). Random vibration techniques for non-destructive evaluation of peach firmness. *J. Agric. Eng. Res., 16*, 81–87.

Freund, M. S., & Lewis, N. S. (1995). A chemically diverse conducting polymer-based "electronic nose." *Proc. of Natl. Acad. Sci., 92*(7), 2652–2656.

Gardner, J. W., & Bartlett, P. N., (1994). A brief history of electronic noses. *Sens. Actuators B, 18*, 211–220.

Gardner, J. W., (1991). Detection of vapors and odors from a multisensory array using pattern recognition: Principle component and cluster analysis, *Sens. Actuators, 4*, 109–115.

Greenham, C. G., (1966). Bruise and pressure injury in apple fruits. *J. Experimental Botany, 17*, 404–409.

Haff, R. P., & Toyofuku, N., (2008). X-ray detection of defects and contaminants in the food industry. *Sens. Instrum. Food Qual., 2*, 162–173.

Hall, L. D., Evans, S. D., & Nott, K. P., (1998). Measurement of textural changes of food by MRI relaxometry. *Magn. Reson. Imaging, 16*, 485–492.

Hansen, J. D., Schlaman, D. W., Haff, R. P., & Yee, W. L., (2005). Potential postharvest use of radiography to detect internal pests in deciduous tree fruits. *J. Entomol. Sci., 40*, 255–262.

Hertog, M. L. A. T. M., Lammertyn, J., De Ketelaere, B., Scheerlinck, N., & Nicolai, B. M., (2007). Managing quality variance in the postharvest food chain. *Trends Food Sci. Tech.*, pp. 320–332.

Hung, Y., Prussia, S. E., & Ezeike, (1999). Nondestructive firmness sensing using a laser air-puff detector. *Postharvest Biol. Technol., 16*, 15–25.

Jha, S. N., & Matsuoka, T., (2000). Review: Non-destructive techniques for quality evaluation of intact fruits and vegetables. *Food Sci. Technol. Res.*, pp. 248–251.

Jiang, J. A., Chang, H. Y., Wu, K. H., Ouyang, C. S., Yang, M. M., et al., (2008). An adaptive image segmentation algorithm for x-ray quarantine inspection of selected fruits. *Comput. Electron. Agric, 60*, 190–200.

Juran, J. M., (1951). *Quality Control Handbook.* McGraw-Hill, New York,

Khalifa, S., Komarizadeh, M. H., & Tousi, B., (2011). Usage of fruit response to both force and forced vibration applied to assess fruit firmness: A review. *AJCS*, pp. 516–522.

Kienle, A., Lilge, L., Patterson, M. S., Hibst, R., Steiner, R., & Wilson, B. C., (1996). Spatially resolved absolute diffuse reflectance measurements for non-invasive determination of the optical scattering and absorption coefficients of biological tissue. *Appl. Opt., 35*, 2304–2314.

Kim, S., & Schatzki, T. F., (2001). Detection of pinholes in almonds through x-ray imaging. *Trans. ASABE, 44*, 997–1003.

Kowaiski, B. R., & Bender, C. F., (1972). Pattern recognition: A powerful approach to interpreting chemical data. *J. Amer. Chem. Soc., 94*, 5632–5639.

Landahl, S., & Terry, L. A., (2012). Avocado firmness monitoring with values obtained by means of laser Doppler vibrometry. *4th Int. Conf. Postharvest Unlimited 2011* (Vol. 945, pp. 239–245).

Magness, J. R., & Taylor, G. F., (1925). *An Improved Type of Pressure Tester for Determination of Fruit Maturity*. USDA Agric. Circular.

Mathanker, S. K., Weckler, P. R., Wang, N., & Bowser, T., (2010). Maness, local adaptive thresholding of pecan x-ray images: Reverse water flow method. *Trans. ASABE, 53,* 961–969.

McClements, D. J., (1997). Ultrasonic characterization of foods and drinks: Principles, methods, and applications. *Crit. Rev. Food Sci., 37,* 1–46.

Meilgard, M. C., Carr, B. T., & Vance, C. G., (2006). *Sensory Evaluation Techniques* (4th edn., p. 464). CRC Press, Boca Raton, FL.

Mirzaee, E., Rafiee, S., Keyhani, A., & Emam Djom-eh, Z., (2010). Physical properties of apricot to characterize best post-harvesting options. *Aust. J. Crop. Sci.,* pp. 95–100,

Mizrach, A., Nahir, D., & Ronen, B., (1985). Fruit and vegetable sorting by their rigidity. *Trans. ASABE No. 85.*

Mizrach, A., Nahir, D., & Ronen, B., (1992). Mechanical thumb sensor for fruit and vegetable sorting. *Trans. ASABE, 35,* 247–250.

Moreda, G. P., Ortiz, C. J., Garcia, R. F. J., & Ruiz-Altisent, M., (2009). Non-destructive technologies for fruit and vegetable size determination: A review. *J. Food Eng.,* pp. 119–136.

Muramatsu, N., Sakurai, N., et al., (1997). Comparison of a non-destructive acoustic method for firmness measurement of kiwi fruit. *Postharv. Biol. Technol., 12,* 221–228.

Nandi, C. S., Tadu, B., & Koley, C., (2014). Machine Vision based technique for automatic mango fruit sorting and grading based on maturity level and size. S*mart Sensors: Measurement and Instrumentation.* Springer.

Nguyen, D. T. N., Watte, R., Aernouts, B., Herremans, E., et al., (2011). Spatially resolved spectroscopy for non-destructive quality inspection of foods. *Meet. Pap. Annu. Int. ASABE Meet.* Louisville.

Nicolai, B. M., Berna, A., Beullens, K., Vermeir, S., Saevels, S., & Lammertyn, J., (2008). High throughput flavor profiling of fruit. *Fruit and Vegetable Flavor: Recent Advances and Future Prospects.* CRC Press-Woodland Publ. Ltd.

Nielsen, M. S., Lauridsen, T., Christensen, K. B., & Feidenhans, L. R., (2013). X-ray dark field imaging for detection of foreign bodies in food. *Food Control, 30,* 531–535.

Patterson, M. S., Chance, B., & Wilson, B. C., (1989). Time resolved reflectance and transmittance for the non-invasive measurement of tissue optical properties. *Appl. Opt., 28,* 2331–2336.

Peleg, K., (1993). Comparison of non-destructive measurement of apple firmness. *J. Agric. Eng. Res., 55,* 227–238,

Peng, Y., & Lu, R., (2008). Analysis of spatially resolved hyperspectral scattering images for assessing apple fruit firmness and soluble solids content. *Postharvest Biol. Technol., 48,* 52–62.

Povey, M. J. W., & McClements, D. J., (1988). Ultrasonics in food engineering: Part I: Introduction and experimental methods. *J. Food Eng., 8,* 217–245.

Prussia, S. E., Astleford, J. J., Hewlett, B., & Hung, Y. C., (1994). *Non-Destructive Firmness Measuring Device.* US Patent No. 5372030.

Qin, J. W., & Lu, R. F., (2008). Measurement of the optical properties of fruits and vegetables using spatially resolved hyper spectral diffuse reflectance imaging technique. *Postharvest Biol. Technol., 49,* 355–365,

Qin, J., Lu, R., & Peng, Y., (2009). Prediction of apple internal quality using spectral absorption and scattering properties. *Trans. ASABE, 52*, 499–507.

Roohinejad, S. H., Mirhosseini, H., Saari, N., et al., (2009). Evaluation of GABA, crude protein and amino acid composition from different varieties of Malaysian brown rice. *Aust. J. Crop. Sci., 3*, 184–190.

Ruiz-Altisent, M., Ruiz-Garcia, L., et al., (2010). Sensors for product characterization and quality of specialty crops: A review. *Comput. Electron. Agric., 74*, 176–194.

Salvo, L., Cloetens, P., Maire, E., Zabler, S., Blandin, J. J., et al., (2003). X-ray micro-tomography an attractive characterization technique in materials science. *Nucl. Instrum. Methods Phys. Res., 200*, 273–286.

Schomer, H. A., & Olsen, K. L., (1962). A mechanical thumb for determining firmness of apples. *J. Am. Soc. Hortic. Sci., 81*, 61–66,

Sheeja, P. S., & Ajay, G. A. J., (2016). Nondestructive quality evaluation for fruits and vegetables. *Int. J. Modern Trends Eng. Res., 3*, 2349–9745.

Shmulevich, I., Galili, N., & Rosenfeld, D., (1996). Detection of fruit firmness by frequency analysis, *Trans. ASABE, 39*, 1047–1055.

Smith, T., & Guild, J., (1931). The C.I.E. colorimetric standards and their use. *Trans. Opt. Soc., 33*, 73–134.

Tollner, E. W., Hung, Y. C., Upchurch, B. L., & Prussia, S. E., (1992). Relating x-ray absorption to density and water content in apples. *Trans. ASABE, 35*, 1921–1928.

Torricelli, A., Spinelli, L., et al., (2008). Time-resolved reflectance spectroscopy for non-destructive assessment of food quality. *Sens. Instrum. Food Qual., 2*, 82–89.

Toyofuku, N., & Schatzki, T. F., (2007). Image feature-based detection of agricultural quarantine materials in x-ray images. *J. Air. Transp. Manag., 13*, 348–354.

Vursavus, K. K., & Zehan, K., (2016). Modeling of impact parameters for non-destructive evaluation of firmness of greenhouse tomatoes. *Agronomy Research, 14*, 1458–1508.

Yamamoto, T., & Yamamoto, Y., (1976). Electrical properties of the epidermal stratum. *Medical and Biological Engineering, 14*, 151–158.

Yodh, A., & Chance, B., (1995). Spectroscopy and imaging with diffusing light. *Phys. Today, 48*, 34–40.

Zerbini, P. E., (2006). Emerging technologies for non-destructive quality evaluation of fruit. *J. Food and Ornamental Plant Research, 14*, 13–23.

Zou, X., Huang, X., & Malcolm, P., (2016). Non-invasive sensing for food reassurance. *The Analyst.*

INDEX